Springer Tracts in Modern Physics 98

Springer Tracts in Modern Physics

82 **Electronic Structure of Noble Metals, and Polariton-Mediated Light Scattering**
With contributions by B. Bendow, B. Lengeler

83 **Electroproduction at Low Energy and Hadron Form Factors**
By E. Amaldi, S. P. Fubini, G. Furlan

84 **Collective Ion Acceleration** With contributions by C. L. Olson, U. Schumacher

85 **Solid Surface Physics** With contributions by J. Hölzl, F. K. Schulte, H. Wagner

86 **Electron-Positron Interactions** By B. H. Wiik, G. Wolf

87 **Point Defects in Metals II:** Dynamical Properties and Diffusion Controlled Reactions
With contributions by P. H. Dederichs, K. Schroeder, R. Zeller

88 **Excitation of Plasmons and Interband Transitions by Electrons** By H. Raether

89 Giant Resonance Phenomena in **Intermediate-Energy Nuclear Reactions**
By F. Cannata, H. Überall

90* **Jets of Hadrons** By W. Hofmann

91 **Structural Studies of Surfaces**
With contributions by K. Heinz, K. Müller, T. Engel, and K. H. Rieder

92 **Single-Particle Rotations in Molecular Crystals** By W. Press

93 **Coherent Inelastic Neutron Scattering in Lattice Dynamics** By B. Dorner

94 **Exciton Dynamics in Molecular Crystals and Aggregates** With contributions by V. M. Kenkre and P. Reineker

95 **Projection Operator Techniques** in Nonequilibrium Statistical Mechanics
By H. Grabert

96 **Hyperfine Structure in 4d- and 5d-Shell Atoms** By S. Büttgenbach

97 **Elements of Flow and Diffusion Processes in Separation Nozzles** By W. Ehrfeld

98 **Narrow-Gap Semiconductors** With contributions by R. Dornhaus, G. Nimtz, and B. Schlicht

99 **Dynamical Properties of IV–VI Compounds** With contributions by H. Bilz, A. Bussmann-Holder, W. Jantsch, and P. Vogl

100* **Quarks and Nuclear Forces** Edited by D. C. Fries and B. Zeitnitz

101 **Neutron Scattering and Muon Spin Rotation** With contributions by R. E. Lechner, D. Richter, and C. Riekel

102 **Theory of Jets in Electron-Positron Annihilation** By G. Kramer

103 **Rare Gas Solids** With contributions by H. Coufal, E. Lüscher, H. Micklitz, and R. E. Norberg

104 **Surface Enhanced Raman Vibrational Studies at Solid/Gas Interfaces** By I. Pockrand

105 **Two-Photon Physics at e^+e^- Storage Rings** By H. Kolanoski

106 **Polarized Electrons at Surfaces** By J. Kirschner

107 **Electronic Excitations in Condensed Rare Gases**
By N. Schwentner, E.-E. Koch, and J. Jortner

---

* denotes a volume which contains a Classified Index starting from Volume 36

Narrow-Gap Semiconductors

Contributions by
R. Dornhaus G. Nimtz B. Schlicht

With 244 Figures

Springer-Verlag
Berlin Heidelberg New York Tokyo

Dr. Ralf Dornhaus
I. Physikalisches Institut der Rheinisch-Westfälischen Technischen Hochschule Aachen
D-5100 Aachen 1, Fed. Rep. of Germany

Professor Dr. Günter Nimtz
Dipl.-Phys. Burghard Schlicht
II. Physikalisches Institut der Universität Köln, Zülpicherstrasse 77
D-5000 Köln 41, Fed. Rep. of Germany

Manuscripts for publication should be addressed to:

Gerhard Höhler
Institut für Theoretische Kernphysik der Universität Karlsruhe
Postfach 6380, D-7500 Karlsruhe 1, Fed. Rep. of Germany

Proofs and all correspondence concerning papers in the process of publication should be addressed to:

Ernst A. Niekisch
Haubourdinstrasse 6, D-5170 Jülich 1, Fed. Rep. of Germany

Second Printing 1985

ISBN 3-540-12091-2 Springer-Verlag Berlin Heidelberg New York Tokyo
ISBN 0-387-12091-2 Springer-Verlag New York Heidelberg Berlin Tokyo

Library of Congress Cataloging in Publication Data. Nimtz, G. (Günter), 1936– . Narrow-gap semiconductors. (Springer tracts in modern physics; 98) Includes bibliographies and index. 1. Narrow gap semiconductors. I. Schlicht, B. (Burghard), 1951– . II. Dornhaus, R. (Ralf), 1944– . III. Title. IV. Series. QCI.S797 vol. 98 [QC611.8N35] 539s 83-530 [537.622]

Offset printing and bookbinding: Brühlsche Universitätsdruckerei, Giessen
2153/3130 – 5 4 3 2 1 0

Preface

This volume is devoted to various narrow gap semiconductors which are very important for both basic solid state physics and infrared devices. The first contribution reviews the properties of IV-VI compound lead salts, the second one those of mercury cadmium telluride. The latter contribution was originally published first 1976 in Volume 78 of the Springer Tracts in Modern Physics. Since that time the research on this II-VI alloy has not ceased, however, the main interest has now switched to topics as for instance the epitaxial growth of crystals, the transport properties and the recombination behaviour of p-type crystals, and last but not least to the analysis of surface and interfaces. As that volume 78 is sold out, this contribution is reprinted with supplements to various sections that contain some of the data published quite recently. We added also some new sections, which consider the changed spectrum of research on this alloy. (These supplements are indicated by vertical stripes on the page margin.) We would like to mention three other review articles published quite recently on $Hg_{1-x}Cd_xTe$ dealing with "Magnetooptical Properties" by M.H. Weiler, "Nonlinear Optical Effects" by T.W. Kruse and J.F. Ready, and "Growth, Properties and Applications" by H. Maier and J. Hesse.*

We would like to apologize for using in the lead salt article a citation system different from that in the $Hg_{1-x}Cd_xTe$ one. The reason for doing so against the publisher's esthetic arguments is, that the system used in the lead salt article seems to us more comfortable for both the reader and the author.

Finally we would like to acknowledge gratefully the patience and the support in preparing this volume by the Editors Prof. Dr. G. Höhler and Prof. Dr. E.A. Niekisch. To Miss Therese Becker, Mrs. Gisela Hahn, Mrs. Hilde Latta, Mrs. Inge Kirschke, Miss Ruth Küpper, and Mrs. Inge Simons we wish to express our appreciation and thanks for their assistance in the preparation of this volume.

Cologne, April 1983

R. Dornhaus
G. Nimtz
B. Schlicht

---

* WEILER /367/, KRUSE and READY /368/, MAIER and HESSE /369/

Contents

Narrow-Gap Lead Salts

By G. Nimtz and B. Schlicht (With 111 Figures)

The Properties and Applications of the $Hg_{1-x}Cd_xTe$ Alloy System

By R. Dornhaus and G. Nimtz (With 133 Figures)

Narrow-Gap Lead Salts

By G. Nimtz and B. Schlicht

1. Introduction

The narrow-gap semiconductors among the family of IV-VI compounds and their pseudobinary alloys show some extraordinary aspects of solid-state physics. The polymorphism of IV-VI compounds indicate the inherent structural instability which is based on the average five valence electrons. The most interesting point, however, is that the bonds are based on the itinerant electrons in the valence band which is separated from the conduction band by a narrow energy gap. In such a situation one expects the electronic properties to be strongly correlated with the lattice instability. Another interesting feature follows from the many-valley band structure in combination with a strongly non-isotropic effective mass in lead telluride: the Auger mechanism is expected to play a dominant role in free-carrier recombination.

During the last twenty years many experimental and theoretical studies were performed to elucidate the basic properties of the IV-VI compounds. In addition many activities in the field of infrared devices have been successful in producing laser diodes and photodetectors for the wavelength range between 4 and 30 μm.

It is our intention to present in this article the basic data of the lead salt narrow gap semiconductors currently available. We should like to encourage further efforts towards a better understanding of this fascinating semiconductor family, and supply data for the development of efficient devices. We are aware of the fact that this article is not a complete presentation of all investigations published so far.

2. The Crystal

This chapter is devoted to the basic physical properties of the four lead salt narrow-gap alloys $Pb_{1-x}Sn_xTe$, $Pb_{1-x}SnSe$, $PbS_{1-x}Se_x$, and $PbTe_{1-x}S_x$ and their binary compounds. In the first section the lattice, elastic and related data are presented. The lattice instability representing one of the outstanding physical properties of these IV-VI binary and ternary compounds is also discussed in this section. In the second section phase diagrams are introduced, and the third section deals with lattice imperfections and their properties.

2.1 Basic Properties

The narrow-gap lead salt alloys are more or less ionic crystals. The ionicity decreases in going from PbS to PbSe to PbTe. The ionic nature of the bonding of the lead chalcogenides is impressively shown by the spatial valence charge distributions.

Fig. 2.1. The total valence electron charge density for PbSe derived from pseudopotential band-structure calculations. The values are given in units of electrons per unit-cell volume. The charge around the cation sites is mainly of s character whereas around the anion sites it clearly exhibits contributions from both s and p states /75S1/

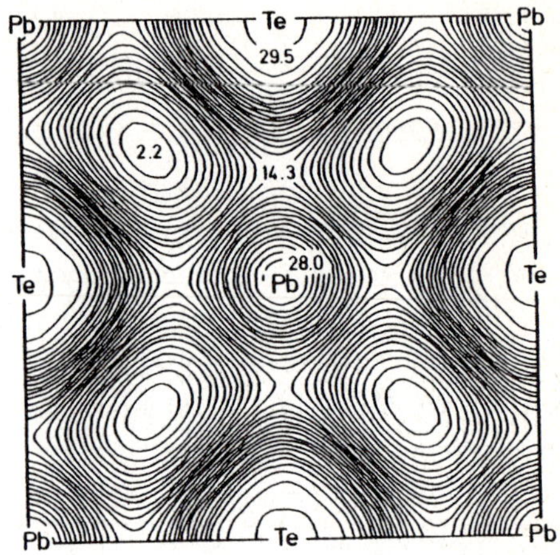

Fig. 2.2. The total valence electron charge density for PbTe derived from pseudopotential band-structure calculations. The values are given in units of electrons per unit-cell volume. The charge around the cation sites is mainly of s character whereas around the anion sites it clearly exhibits contributions from both s and p states /75S1/

SCHLÜTER et al. /75S1/ have calculated the contours of constant charge density in a (100) plane for various bands. The results for the sum of the five valence bands are shown in Fig. 2.1 for PbSe and in Fig. 2.2 for PbTe. In Table 2.1 the calculated approximate fractional charges inside touching spheres around anions and cations decomposed into l = 0 , 1 , 2 angular components, are given for the valence bands /75S1/. From these an approximate charge transfer of 1.9 (PbSe) and 1.5 (PbTe) electrons from cation to anion can be estimated (in the "totally ionic" model two electrons of the cation are transferred to the anion).

Most of the binary compounds, e.g. PbTe, PbS, PbSe, SnTe are found to be isomorphous with rocksalt, its lattice is shown in Fig. 2.3. The low molecular weight material SnSe, on the other hand, is found in an orthorhombic structure (Fig. 2.4). The nature of chemical binding in these crystals is mixed ionic-covalent. The phy-

Table 2.1. Calculated fractional charges (in percent) inside touching spheres around anions and cations. The sphere radii were chosen as r_{cation} = 1.29 Å, r_{anion} = 1.94 Å for PbTe and r_{cation} = 1.22 Å, r_{anion} = 1.83 Å for PbSe (from /75S1/)

	PbSe						PbTe					
	Pb			Se			Pb			Te		
Band	s	p	d	s	p	d	s	p	d	s	p	d
1	1	1	0	97	1	0	3	1	0	95	1	0
2	55	0	0	18	24	3	70	0	0	13	13	4
3 - 5	9	7	1	1	78	4	5	9	2	0	80	4
1 - 5	16	4	1	23	52	4	18	6	1	22	50	3
	21			79			25			75		
Simple ionic	20	0	0	20	60	0	20	0	0	20	60	0

sical background of the crystal structure of IV-VI compounds was recently discussed extensively in papers by LITTLEWOOD /80L1/. The covalent binding is assumed to be predominant which is concluded for instance from analysis of carrier scattering /70R/. Investigations of the carrier scattering in various lead chalcogenides have shown that both optical and acoustical phonons are important, the latter being an indication of predominant covalent binding.

All the compounds with NaCl structure have a structural phase transition due to their mixed ionic-covalent chemical binding. With decreasing ionicity the NaCl structure becomes more and more unstable, accompanied by both an increasing electronic

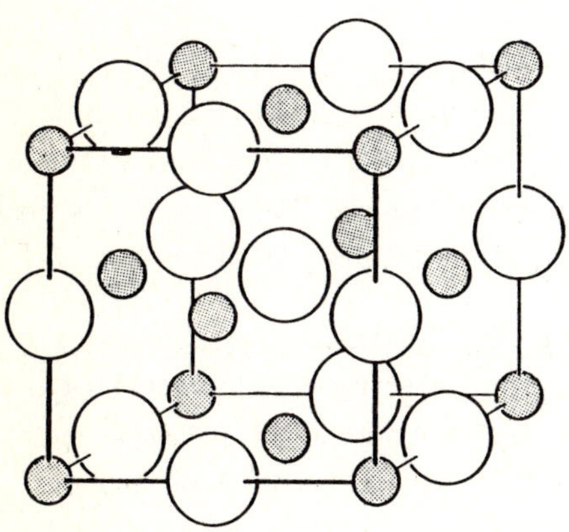

Fig. 2.3. The rocksalt structure

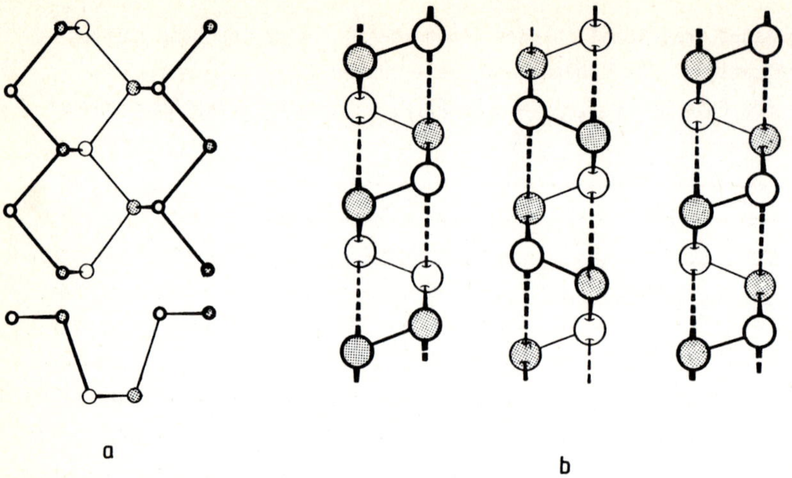

Fig. 2.4a,b. Sketch of an orthorhombic lattice (a) ground plan and front elevation, thick and thin circles indicate the atom is above or below the plane of the paper, respectively, (b) projection on the (100) plane

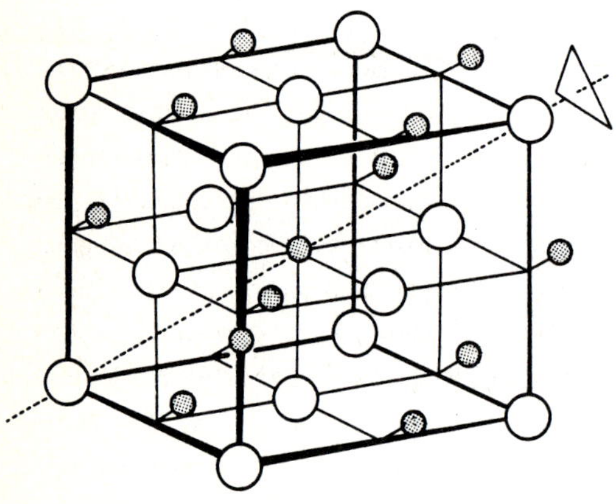

Fig. 2.5. The As-type rhombohedral structure

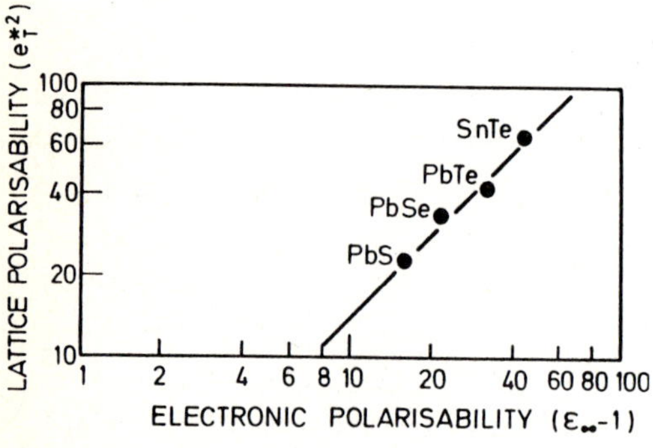

Fig. 2.6. Lattice polarisability $(e_T^*)^2$ vs electronic polarisability $(\varepsilon_\infty - 1)$ for some IV-VI compounds with rocksalt structure /73L2/

and lattice polarisability /77L5/. This behaviour is shown in Fig. 2.6 and will be discussed in more detail in Sect. 5.4.

The cubic to rhombohedral phase transition can be explained as a result of a large electron-TO phonon coupling due to the resonant nature of the half-filled p states. Below the transition temperature T_C (for most of the compounds $T_C < 0$ holds as will be shown below) the two neighbouring ions along the <111> axis make displacements in opposite directions forming a rhombohedral, As-type crystal structure as shown in Fig. 2.5. The cubic rocksalt structure of the lead chalcogenides is stabilised by perpendicular p bonds, with the s states fully occupied and making only a small contribution to the bonding. The lack of hybridisation between the s and p bands in the lead salts arises from the large energy splitting between s and p states. The rhombohedral structure is an intrinsic instability of the unsaturated p bonds in the rocksalt structure. Because there are only six p electrons per atom pair, and yet there are six neighbours for each atom, the p bonds are unsaturated. The rocksalt structure is thus a resonantly bonded system /68K1 , 73L2/ in which the bonding electrons are relatively free to move from the bond on one side of the atom to the bond on the opposite side in response to an external field. In the rhombohedral phase the six cubic nearest neighbours are displaced along the <111> direction with respect to the central atom so that each atom now has three nearest neighbours and three second-nearest neighbours in the distorted phase. The nearest neighbours bond strongly together in a layer perpendicular to the <111> direction. The bonds in the rhombohedral structure are now more nearly saturated than in the cubic system, because the bonding electrons are localised in the nearest-neighbour bonds.

The cubic-rhombohedral phase transition can be understood as the result of the electron-TO phonon coupling for modes propagating in the <111> direction. Coupled to this is a shear which distorts the cube into a rhombohedron, and lengthens the <111> axis.

The orthorhombic structure found in SnSe arises because of the strong sp hybridisation in IV-VI compounds with low molecular weights /80L1/. The orthorhombic structure also seems to present the high pressure phase of IV-VI compounds and their alloys, which were found in the rocksalt structure at atmospheric pressure /e.g.: 40B , 48B , 54B , 67M , 68K3 , 68W , 80S , 81S1/. Data of the lattice constants of the pressure induced orthorhombic structure of several IV-VI compounds are given in Table 2.2. From a study of this high pressure phase transition in $Pb_{1-x}Sn_xTe$ SUSKI et al. /81S1/ conclude that a metallic-covalent transition takes place. They assume a promotion of an s electron to p states due to the decrease in interatomic distance.

Consequences for the transport properties as well as the accompanied softening of the transverse optical phonon due to the structural phase transition are discussed in Sects. 4.2 and 5.4 respectively. Experimental data of the transition temperature are shown for $Pb_{1-x}Sn_xTe$ in Fig. 2.7. These data published by MURASE

Table 2.2. Lattice constants of the pressure-induced orthorhombic (Pnma) crystal structure /67M/

Material	T (K)	a (Å)	b (Å)	c (Å)	P (kbar)
PbS	300	11.28	3.98	4.21	25
PbSe	300	11.61	4.00	4.39	43
PbTe	300	11.71	4.36	4.42	42
SnTe	300	11.59	4.37	4.48	20

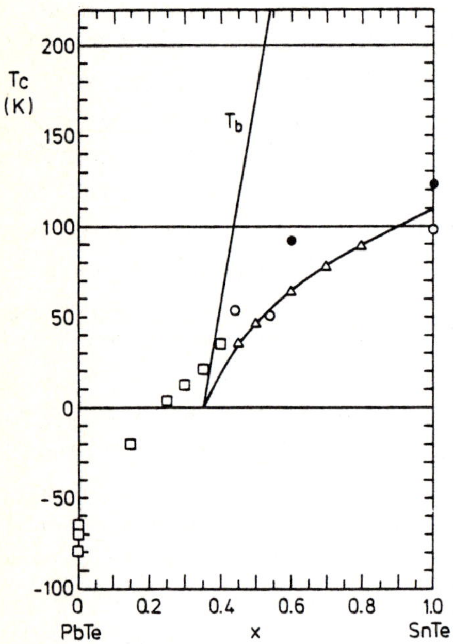

Fig. 2.7

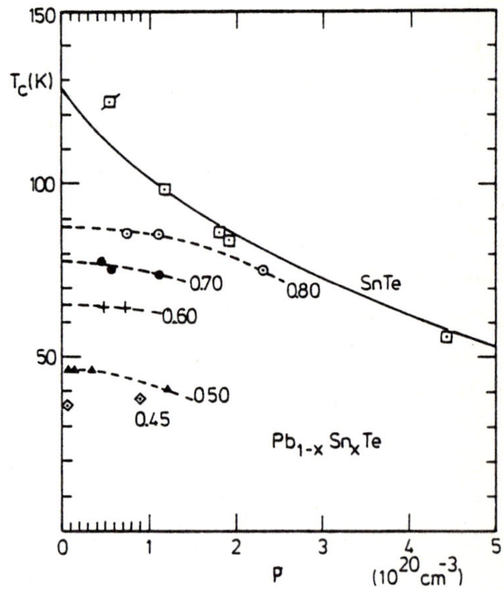

Fig. 2.8

Fig. 2.7. Composition dependence of the transition temperature for sample series grown by different methods /82M1/. Band inversion occurs at the T_b line /68D/. The data (□) represent the Curie temperature deduced from dielectric constant. Transition temperature data of bulk material are from (△) /82M1/, (○) /79K2/. The values (●) are obtained from epitaxial layers on BaF_2 substrates /82M1/

Fig. 2.8. $Pb_{1-x}Sn_xTe$. Transition temperature as a function of carrier concentration for various tin compositions x /76K2, 82M1/: x = 1 (⊡), x = 0.8 (⊙), x = 0.7 (●), x = 0.6 (+), x = 0.5 (▲), x = 0.45 (◈)

and NISHI /82M1/ present values of the Curie temperature (see Sect. 5.4) and the transition temperature obtained from bulk material. The Curie temperature was determined by a tangent line to $1/\varepsilon_0 = f(T)$ at $T \approx 90$ K, whereas the transition temperature was obtained from the resistivity anomaly due to an increment of electron-phonon scattering near the transition temperature. The anomaly is assumed to originate from an increasing relaxation rate due to soft TO phonons and screened LO phonons /76K2, 76K3, 80K2, 82M1/ (see also Sect. 4.2). Data of the temperature of band inversion as a function of composition are also presented in Fig. 2.7.

The transition temperature also appears to be a function of the free-carrier concentration /69C2, 75I1, 76K2, 78K2/. This property is shown in Fig. 2.8 for various mixtures of $Pb_{1-x}Sn_xTe$. As will be discussed in Sect. 5.4 this phenomenon is caused by electron-TO phonon coupling due to the optical deformation potential. DAUGHTON and DE FACIO /78D2/ published calculated values for T_C assuming a model for the structural phase transition, which was developed by KRISTOFFEL and KONSIN /68K3, 73K3/ and by NATORI /76N2, 76N3/, (see also Sect. 5.4). For the time being a quantitative agreement is not obtained either between experimental results or between various theoretical data. One reason for this situation may be that the lattice instability does depend not only on carrier density and on band gap but also on the crystal quality, e.g., the number of lattice defects, or on strain which may be induced in thin films (see also Sect. 5.4).

The lattice constants as well as density, melting temperature, and thermal expansion coefficient of the binary compounds are given in Table 2.3. It was found that both lattice constant and energy gap vary approximately linearly with composition in the ternary alloys as shown in Fig. 2.9. The large difference in the lattice constants between PbTe and PbS causes the ternary compound $PbTe_{1-x}S_x$ to be brittle. On the other hand, the crystalline properties of $Pb_{1-x}Sn_xTe$, $PbS_{1-x}Se_x$ and $Pb_{1-x}Sn_xSe$ are comparable to those of the binary compounds. The mixed crystal $Pb_{1-x}Sn_xSe$ crystallizes in the cubic rocksalt structure on the lead-rich side up to $x \leqslant 0.43$. A two-phase region is observed for $0.43 \lesssim x \lesssim 0.75$ and eventually for $x \gtrsim 0.75$ an orthorhombic structure is established /61K/. The lattice constant as a function of the composition is plotted in Fig. 2.10. The lattice constants show a linear dependence on composition only in the cubic crystal phase of the $Pb_xSn_{1-x}Se$ alloy.

Elastic data of the binary compounds are given in Table 2.4. There is not much data published on the IV-VI alloys. For $Pb_{1-x}Sn_xTe$ ($x = 0.13$) DAUGHTON et al. /76D/ have calculated elastic constants from phonon dispersion data.

The pressure dependence of the elastic constants of PbTe and SnTe were studied by MILLER et al. /81M/. Some third-order elastic constants as obtained in this study are presented in Table 2.5.

Table 2.3. Lattice constants, density, melting point, and linear thermal expansion coefficient

Compound	a (10^{-10}m)	T (K)	Ref.	density (kg/m^3)	T (K)	Ref.	T_M (°C)	Ref.	α (10^{-5}/K)	T (K)	Ref.
PbS	5.936	299	/69D1/	7500	299	/67W/	1111	/74H1/	2.027	300	/64N/
(cubic)	5.929	300	/67M/						1.755	100	/64N/
									0.754	30	/64N/
PbSe	6.124	299	/69D1/	8100	288	/67W/	1082	/70S ,	1.940	300	/64N/
(cubic)	6.117	300	/67M/					74H1/	1.737	100	/64N/
									0.765	30	/64N/
PbTe	6.462	299	/69D1/	8160	293	/67W/	924	/70S ,	1.980	300	/64N/
(cubic)	6.443	300	/67M/	8242	300	/80M/		74H1/	1.770	100	/64N/
				8219	300	/81M/			0.902	30	/64N/
									2.04	300	/68H2/
									1.97	77-300	/68H2/
SnSe	a = 4.46	300	/80M/	6179	300	/80M/	874	/73H1 ,			
(ortho-	b = 4.14	300	/80M/				857	74L2/			
rhombic)	c = 11.47	300	/80M/								
SnTe	6.327	300	/74H1/	6445	300	/74H1/	805	/74H1/			
(cubic)	6.303	300	/67M/	6383	300	/81M/	807	/70S/			
				6410	300	/76S1/					

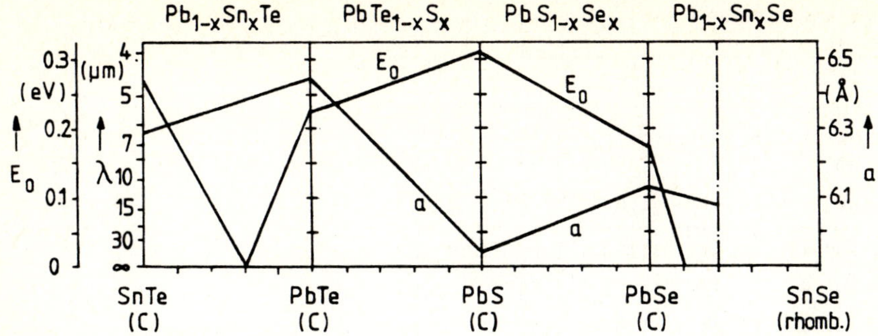

Fig. 2.9. Schematic representation of energy gap E_0, corresponding electromagnetic wavelength λ, and lattice constant a versus composition for various ternary alloys /79P/

The crystal binding energies of binary compounds were estimated from the heat of crystal formation by TANAKA and MORITA /79T1/. The values are: PbS: 32.6 eV, PbSe: 32.5 eV, PbTe: 31.4 eV, SnTe: 31.9 eV. Data of the Debye temperature of the binary compounds as well as some of their ternary alloys are listed in /83L/.

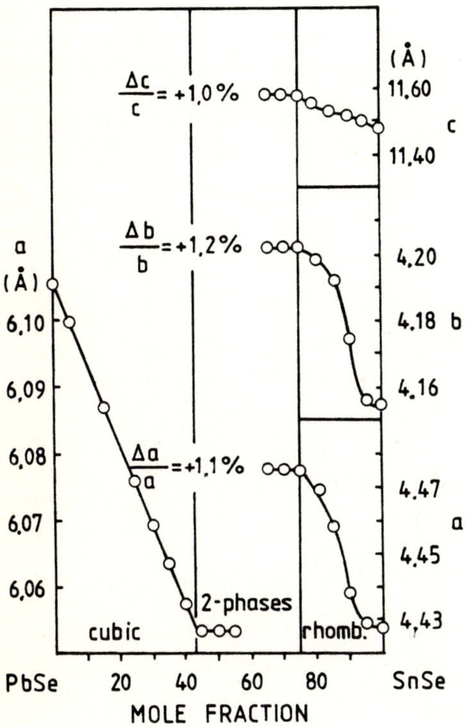

Fig. 2.10. Lattice constants vs mole fraction for the $Pb_{1-x}Sn_xSe$ alloy /61K/

Table 2.4a. Elastic moduli (10^{10} N/m^2)

Substance	T (K)	c_{11}	c_{12}	c_{44}	Ref.
PbS		13.9	3.47	1.69	/63C1/
(cubic)	300	12.7	2.98	2.48	/51B/
	300	12.61	1.624	1.709	/76P/
PbSe	298	12.37	1.93	1.59	/71L2/
(cubic)	195.8	13.37	1.98	1.642	/71L2/
	77.4	13.98	1.97	1.695	/71L2/
	4.2	14.18	1.94	1.749	/71L2/
PbTe	300	10.53	0.70	1.322	/81M/
(cubic)	300	10.80	0.77	1.343	/68H2/
	300	10.40	0.437	1.30	/62C/
SnTe	300	10.43	0.178	1.133	/81M/
(cubic)	293	10.93	0.21	0.97	/76S1/

Table 2.4b. Sound velocity (m/s), T = 300 K (wave vector , polarisation)

PbS	$k_{[100]}$, transv. 1500	$k_{[100]}$, long. 4080	$k_{[011]}$, transv. 1820	/63C1/
PbSe	$k_{[100]}$, transv. 1415	$k_{[100]}$, long. 3860	$k_{[011]}$, transv. 1690	/63C2/
PbTe	$k_{[100]}$, transv. 1260	$k_{[100]}$, long. 3590	$k_{[011]}$, transv. 1610	/62C/
	$k_{[100]}$, transv. 1277	$k_{[100]}$, long. 3620	$k_{[111]}$, long. 2676	/68H2/
SnTe	$k_{[110]}$, transv. [001] 1388	$k_{[110]}$, long. 3012	$k_{[110]}$, transv. [110] 2900	/76S1/

Table 2.5. Third-order elastic constants /81M/

(10^{11} N/m^2)	PbTe	SnTe
C_{111}	- 18.5	
C_{112}	0.35	
C_{123}	- 0.97	
C_{144}	0.44	
C_{166}	- 0.98	
C_{456}	0.12	
$C_{111} + 2C_{112}$	- 17.8	- 17.2
$(C_{111} - C_{123})/2$	- 8.77	- 8.35
$C_{144} + 2C_{166}$	- 1.52	- 1.34

2.2 Phase Diagrams and Crystal Growth

In view of some extensive review articles on phase diagrams and crystal growth published recently /70M1 , 70R , 72N , 74H1 , 78L1 , 79P , 80M/ this section is constrained to present the basic data and the main methods. A general view on the solubility of pseudobinary lead and tin chalcogenides gives Fig. 2.11.

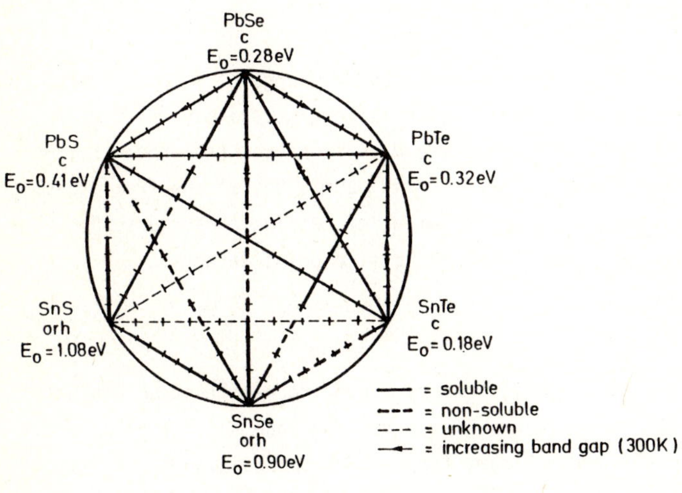

Fig. 2.11. Solubility of pseudobinary lead and tin chalcogenides /80M/

2.2.1 $Pb_{1-x}Sn_xTe$

The pseudobinary temperature-composition phase diagram for $Pb_{1-x}Sn_xTe$ is shown in Fig. 2.12 /68C1/. The solidus compositions were determined by electron microscope analysis of the first-to-freeze ends of Bridgman-grown crystals. The separation of the liquidus and solidus lines is small. The temperature-composition phase diagram for the Pb-Sn-Te ternary system was investigated by various authors /74L1, 74H1/.

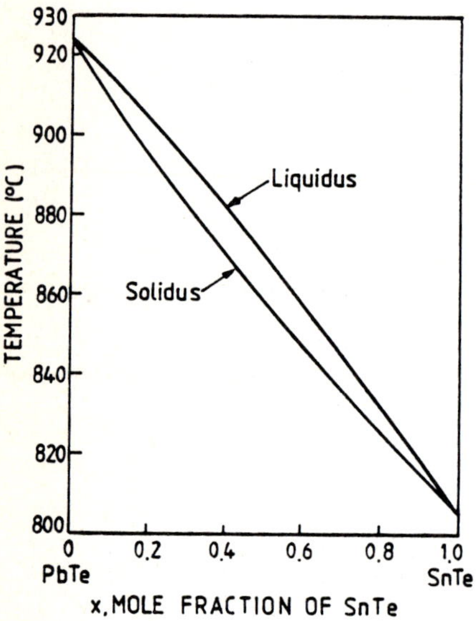

Fig. 2.12. Temperature-composition pseudobinary phase diagram for $Pb_{1-x}Sn_xTe$ /74H1/

Due to lattice point defects alloy crystals exist over a range of compositions. Accordingly the formula for the alloy composition is more accurately written as $(Pb_{1-x}Sn_x)_{1/2+\delta}Te_{1/2-\delta}$, where δ represents the deviation from stoichiometry. More details of the ternary system are discussed in /74H, 80M/. Deviations from stoichiometry contribute electrons or holes to the number of extrinsic carriers. Details of this phenomena are presented in Sect. 2.3 on lattice point defects.

2.2.2 $Pb_{1-x}Sn_xSe$

As discussed in Sect. 2.1 the PbSnSe alloy exists in the rock-salt-structure only over a limited range of alloy compositions with a maximum SnSe solubility of about 0.43 mole fraction of SnSe. The pseudobinary temperature-composition phase diagram is shown in Fig. 2.13. The phase diagram of this alloy is more complicated than the analogous telluride one, since PbSe and SnSe possess different crystal structures. A gap of miscibility exists for about $0.4 \lesssim x \lesssim 0.8$ /61K/. The peritectic point of

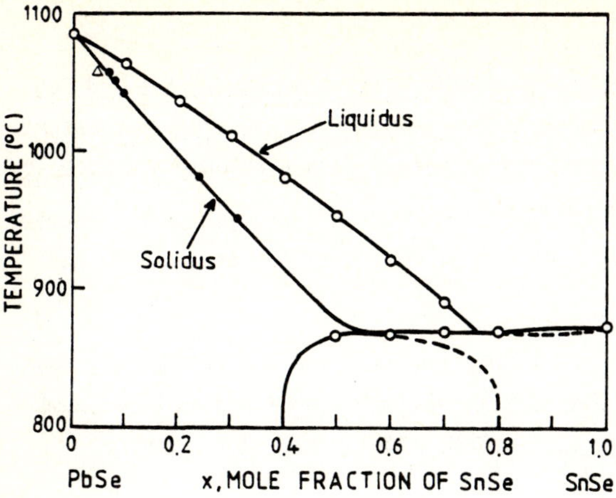

Fig. 2.13. The pseudobinary temperature-composition phase diagram for PbSe - SnSe /74H1/

the liquidus occurs at 870°C and x = 0.75 and that of the solidus at x = 0.5. The temperature-composition phase diagram for the ternary Pb-Sn-Se system was studied by /58H , 66S , 74L2/.

2.2.3 $PbS_{1-x}Se_x$

The pseudobinary phase diagram as determined by various authors is shown in Fig. 2.14. Both binary compounds PbS and PbSe have the rock-salt structure. The diagram shows a small liquidus-solidus separation.

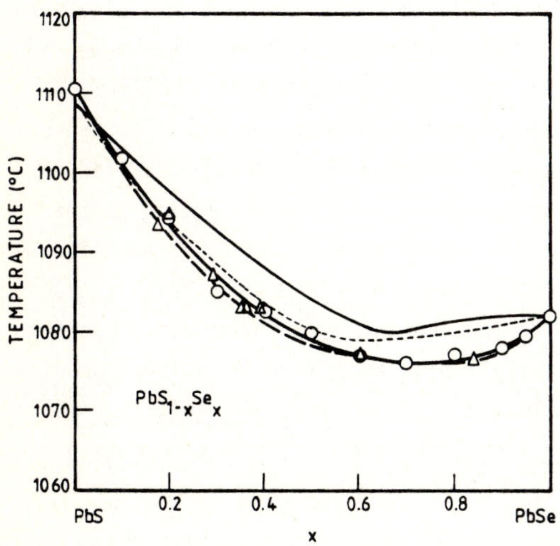

Fig. 2.14. The pseudobinary phase diagram of the PbS - PbSe system from various authors /73S1 , 74H1/

2.2.4 $PbTe_{1-x}S_x$

The pronounced mismatch in lattice constants of the binary compounds PbTe and PbS (Fig. 2.9) causes the alloy to be brittle. Accordingly, for this narrow-gap material there is not much device potential and the experimental data available are much less compared with the previously mentioned alloys.

2.2.5 Crystal Growth

Numerous techniques for the production of single crystals and thin films have been investigated during the last decade. Review articles were published in great number quite recently, devoted to this topic, e.g.: /70M1 , 74H1 , 78L1 , 79P , 80M/.

2.3 Imperfections

It was observed early that besides impurities deviations from stoichiometry introduce acceptors and donors in the IV-VI compounds and alloys /56B , 60B , 70M1 , 74H1 , 80H/. In general it was found that excess metal yields donor levels whereas excess nonmetal yields acceptor levels. The lattice point defects associated with excess metal are assumed to be nonmetal vacancies or metal interstitials while those which result from excess nonmetal are assumed to be metal vacancies or nonmetal interstitials. In the next two sections first native defects and then impurities are discussed.

2.3.1 Native Defects

As mentioned above deviations from stoichiometry introduce acceptors and donors. In Figs. 2.15-17 the metal-saturated and nonmetal-saturated solidus lines are shown for the alloys $Pb_{1-x}Sn_xTe$, $Pb_{1-x}Sn_xSe$, and $PbS_{1-x}Se_x$. Solidus lines are displayed graphically for several alloy compositions. The predominant native defect types in the IV-VI compounds and their alloys have not yet been established. For SnTe, BREBRICK /63B/ claimed evidence from density measurements that Sn vacancies present the predominant lattice point defects. Isothermal annealing procedures applied to $Pb_{1-x}Sn_xTe$ ($0.17 \lesssim x \lesssim 0.20$) yield electron or hole concentrations as low as the order of 10^{15} cm^{-3} /74H1/. However, isothermal annealing is a practical technique for significantly reducing the carrier concentration in IV-VI alloys only in cases where the solidus line comes close enough to the stoichiometric composition at temperatures well below the melting point but still high enough for equilibration with the vapour phase to occur in a reasonable length of time. When this condition is not satisfied, a two-temperature zone annealing technique can be used to obtain significantly lower carrier concentrations /74H1/.

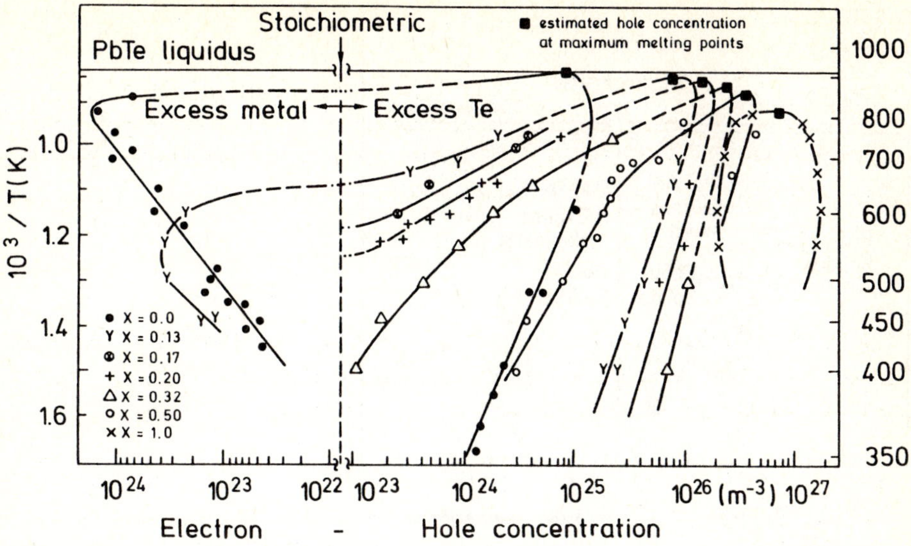

Fig. 2.15. Carrier concentrations at 77 K (300 K for SnTe) vs isothermal annealing temperature for $Pb_{1-x}Sn_xTe$ from various authors /80M/

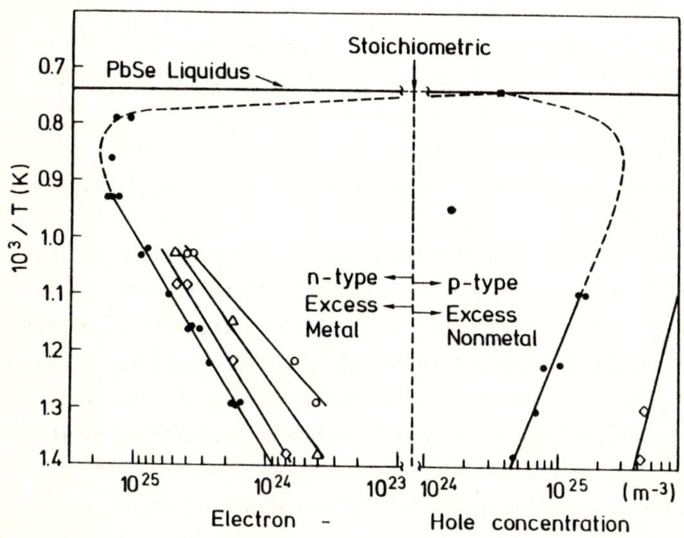

Fig. 2.16. Carrier concentrations at 77 K vs isothermal annealing temperature for $Pb_{1-x}Sn_xSe$. ▪ Estimated hole concentration of solid at maximum melting point. x = 0 (•); x = 0.07 (◇); x = 0.21 (▵); x = 0.25 (o) /73H1/

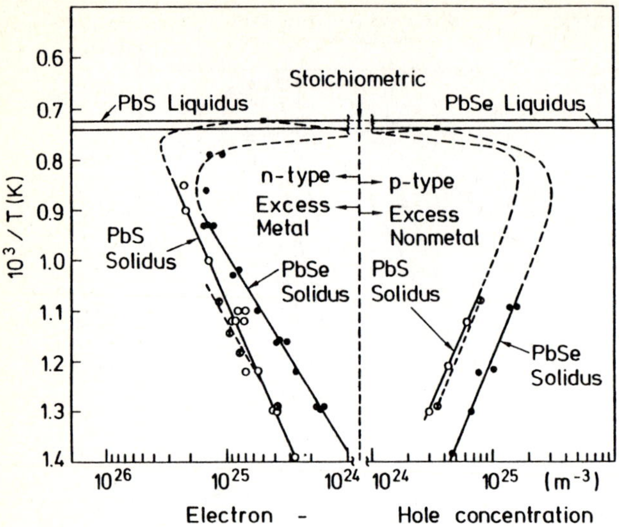

Fig. 2.17. Carrier concentrations at 77 K vs isothermal annealing temperature for $PbS_{1-x}Se_x$, x = 0 (o); x = 0.42 (⊗); x = 1 (•); ■ estimated concentration of solid at maximum melting point /74H1/

Numerous investigations were carried out in order to determine experimentally the energy levels associated with the acceptor and donor lattice defects.

However, neither thermal nor magnetic freeze-out of carriers was detected in crystals having predominantly extrinsic carriers introduced by stoichiometry deviations /58H, 76B1, 79N, 80H/. The ionisation energy of a hydrogen-like donor or acceptor, which usually represents the model for the shallow lattice defects, is expected to be very small in the IV-VI compounds.

Assuming for instance an effective mass of 0.05 m_0 and a static dielectric constant ε_s = 1000, an ionisation energy of about 10^{-6} eV is obtained from the hydrogen-like lattice defect model /70M2, 73S2/. According to this consideration a thermal freeze-out is expected at temperatures of the order of mK, and a magnetic freeze-out above 10 T at somewhat higher temperatures if the carrier density is of the order of 10^{16} cm^{-3} /69D2/. Higher carrier densities would lead to an appreciable overlap of electron wave functions associated with neighbouring defects. In this case a correspondingly higher magnetic field would be necessary in order to lift the overlap of the wave functions. The Hall coefficient of two PbTe samples is shown in Fig. 2.18. There is no carrier freeze-out to be seen even at strong fields up to 20 T /79T2/. According to the various experimental investigations, the extrinsic carriers introduced by vacancies and interstitials cannot be associated with the classical donor or acceptor model having levels in the energy gap between valence and conduction bands. PARADA and PRATT /69P, 71P, 73P/ pointed out that a vacancy potential differs significantly from a hydrogen-like impurity. They assume that a vacancy or

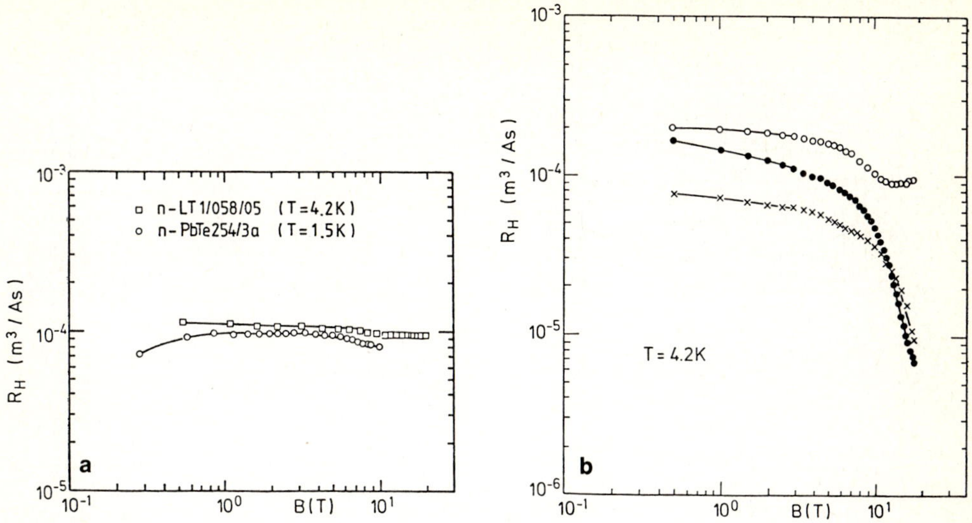

Fig. 2.18a,b. Hall coefficient vs magnetic field, (a) for two PbTe samples, and (b) for three $Pb_{0.8}Sn_{0.2}Te$ samples /79N , 79T2/

an interstitial represents a much stronger perturbation of the periodicity of the lattice than is the case for a chemical impurity. They think a vacancy to be qualitatively very similar to an F-centre in alkali halides. PARADA and PRATT /69P , 71P/ calculated the energy-band scheme for PbTe shown in Fig. 2.19 using the Green's function method. The unperturbed bands are presented by full lines and characterized as s or p states of the atoms. The mixing of these states is neglected. At the site of a vacancy the energy levels of the valence band are shifted up on the energy scale, some of them are even lifted above the conduction-band edge. The result of this energy state calculation is illustrated in Fig. 2.20. When a Pb atom is removed from the crystal leaving a vacancy, 4 valence electrons are lost to the crystal and only 1 level (= 2 states) is lost to the valence band. Consequently, 2 states are unoccupied, generating 2 holes in the valence band. Similarly a Te vacancy generates 2 electrons in the conduction band. According to this model, the vacancy states are located either deep in the valence band or high in the conduction band, i.e. they are resonant with the band states and a freeze-out of carriers cannot take place. The theoretical results are expected to be qualitatively valid also for other IV-VI compounds and alloys. Qualitatively the same conclusions were drawn by HEMSTREET /75H2 , 75H3/ from a cluster calculation which he carried out for the compounds PbTe, PbS, and SnTe. A cluster of 27 atoms was chosen with either an anion or cation or a corresponding vacancy in the centre. As can be expected for a molecule containing 27 atoms, the energy levels fall into groups of completely occupied or empty states.

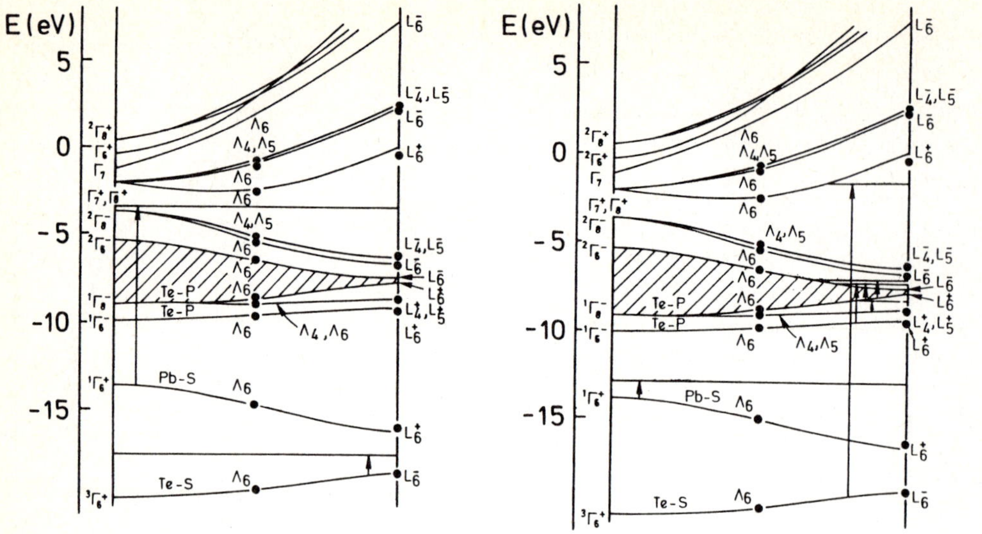

Fig. 2.19. Calculated energy band scheme for PbTe. The left-hand side shows a level due to a Pb vacancy split-off from the Pb s valence band. The right-hand side shows a level due to a Te vacancy split-off from the Te s valence band. The shaded region is the gap between the valence and conduction bands /69P/

The fundamental gap between the highest occupied and lowest unoccupied level was calculated to be about 2 eV. This large value has been interpreted by the poor convergence of the conduction-band energy value to the crystal value with increasing cluster size. The energy spectrum of the defect cluster is very similar to that of the complete cluster except that a two-fold degenerate state has disappeared from the valence states. This result confirms PARADA and PRATT's conclusions that a Pb vacancy generates two holes in the valence band.

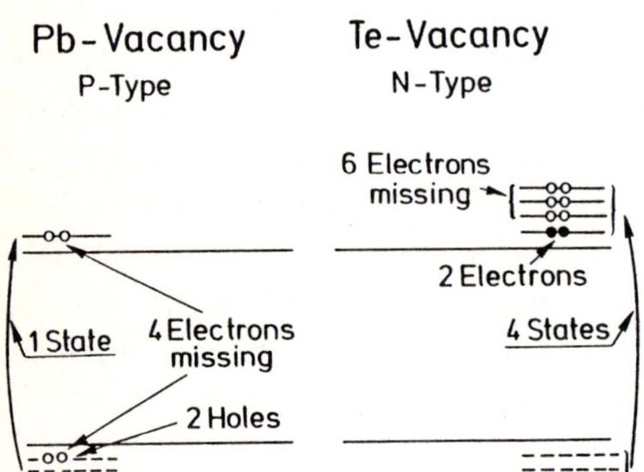

Fig. 2.20. Schematic representation of the effect of a Pb and a Te vacancy /71P/

The effect of the lead vacancy on the charge of the surrounding atoms turned out to be minimal. Most of the valence charge that is taken out of the cluster with the removal of the neutral Pb or Sn atom comes from either the interstitial region or the central site where the lead or tin atom originally resided. The charge density is not localized, it is very small and it is roughly constant over much of the volume of the vacancy sphere, rising slightly near the sphere boundary /69P/. This interesting result is a confirmation of PRATT's /73P/ arguments that a vacancy represents a small scattering cross-section for free carriers.

The results of another cluster calculation by CHOO as discussed in /73P/ yielded data for Pb and Te interstitials, presented in Fig. 2.21.

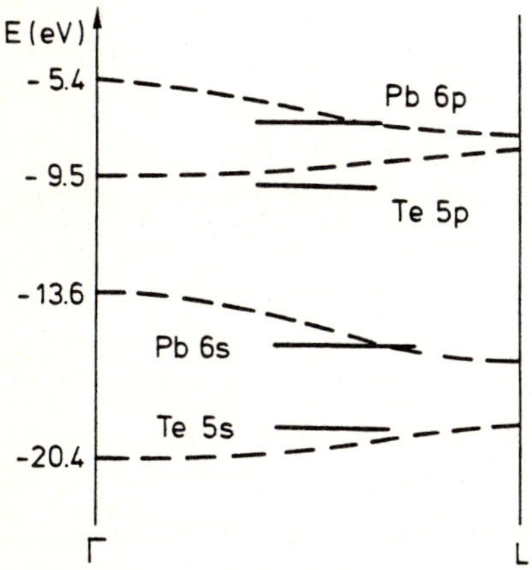

Fig. 2.21. Sketch showing estimate of the one electron Pb and Te interstitial s and p energies as found by CHOO by the cluster technique, from /73P/

According to PARADA and PRATT's /69P/ results a vacancy should be doubly ionised. Various experimental studies on this question were carried out to ascertain whether deviations from stoichiometry yield singly or doubly ionised defects, for instance for PbTe /63S , 66F , 70L , 74L1/, PbSe /63I/, PbS /56B/, and SnTe /63B , 77B1/.

Calculations show that the models for transport properties based on double and single ionisation give equally good fits to the experimental data. However, there is a tendency to assume that the defect-introduced acceptors and donors are doubly ionised /70L/.

Recently in III-V compounds antisite defects have become important /80K1/. This defect is formed when anions occupy cation sites, or vice versa. It was found that antisite defects introduce electronic states deep within the energy gap. As far as we know, this type of defect has not yet been considered in IV-VI compounds, however, it cannot be ruled out that it may play an important role in pure crystals. As mentioned at the end of the next section, recently a deep donor was found in PbTe and in PbSe which could not be identified until now.

2.3.2 Chemical Impurities

The properties of many impurities were studied, as is shown in Table 2.6, where most of the published data is compiled. The dopands were introduced during growth by diffusion and more recently by ion implantation. It was found that the acceptor or donor character of an impurity often depends on the annealing procedure applied to the crystal, i.e. whether this procedure was carried out under metal or nonmetal saturated conditions.

Many investigations were attributed to elements of the III group. For instance it was found that In has an energy level in $Pb_{1-x}Sn_xTe$ and $Pb_{1-x-y}Sn_xIn_yTe_{1-z}Se_z$ which is resonant with the conduction band for small x values but this level is situated in the band gap for higher mole fractions, as shown in Fig. 2.22 /75D2 , 79A2/.

It was suggested in /71A/ that the local levels of In form a narrow impurity band in which each In atom has two electron states. If $n_i < N_{In}$, the Fermi level is pinned in the impurity band and it escapes from this band only for $n_i > N_{In}$. A similar conclusion on the stabilisation of the Fermi level deduced from the assumption of variable valence of In in the $Pb_{1-x}Sn_xTe$ lattice was reached by ANDREEV et al. /76A1/. They assume that when In is dissolved in PbTe an In atom gives up two electrons and In^{2+} ions appear at Pb^{2+} sites. They conclude, however, that such a "neutral state" of In in the high permittivity of PbTe may be unstable since In^{2+} may give up its electron to another In ion according to the reaction $2\ In^{2+} \rightarrow In^{+} + In^{3+}$.

It is assumed by these authors that the energy necessary for this transfer may be compensated by the polarisation of the lattice around the charged In^{+} and In^{3+} centres located at the Pb^{2+} sites.

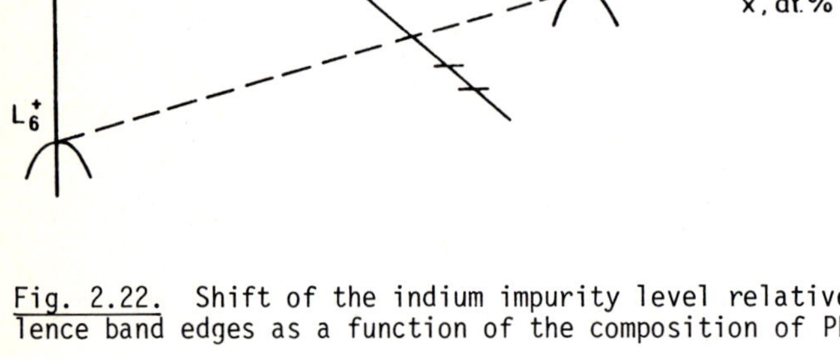

Fig. 2.22. Shift of the indium impurity level relative to the conduction and valence band edges as a function of the composition of $Pb_{1-x}Sn_xTe$ + 0.5 at.% In /79A2/

Table 2.6. Impurities in lead salts

Impurity	Compound	Remarks	Reference
H (donor)	PbTe	proton bombardement	71D1
Na (acceptor)	PbS	doping with NaS, transport properties	80P1
	PbTe	electrical transport, thermoelectric power	67C, 68C2
	PbTe	absorption spectra	79G3
	$Pb_{1-x}Sn_xSe$ ($x = 0 - 0.34$)	doping with NaSe, transport properties	80P1
	$PbS_{1-x}Se_x$ ($x = 0.5$)		79V4
Cr (donor)	PbTe	thermoelectric properties	79B2
Mn (donor)	$Pb_{1-x}Sn_xTe$ ($x = 0.18 - 0.23$)	paramagnetic properties, energy level 100 meV above conduction band at $T \to 0$	79L5
Fe	PbTe	magnetic properties, Fe replaces isovalently Pb	80A2
Cu (acceptor, donor)	PbTe	Cu in Te saturated crystals acceptor-like, in Pb saturated crystals donor-like, Hall coefficient (77 K, 300 K)	73S3
Ag (acceptor, donor)	PbTe	Ag in Te saturated and in Pb saturated crystals, Hall coefficient (77 K, 300 K)	73S3
	PbTe	thermoelectric properties	79B2
Cd (donor)	PbSe	Hall coefficient, lattice parameter, density	65R
	PbTe	Hall coefficient, lattice parameter, density	65R
	PbTe	Hall effect and solubility, absorption coefficient	76V1, 79V2
	$Pb_{1-x}Sn_xTe$ ($x = 0.2$)	Hall effect and solubility	76V2
	$Pb_{1-x}Sn_xTe$ ($x = 0.10 - 0.24$)	diffusion of Cd-In alloy, Hall effect	78L3
	$Pb_{1-x}Sn_xTe$ ($x = 0.2$)	spectral response of Cd diffused photodiodes	78L4

Table 2.6. (continued)

Impurity	Compound	Remarks	Reference
Cd (donor)	$Pb_{1-x}Sn_xTe$ (x = 0.18 - 0.22)	photoluminescence of epitaxial films	81A1
	$Pb_{1-x}Sn_xTe$	analysis of Hall effect data	81C
	$Pb_{1-x}Sn_xTe$ (x = 0 - 0.25)	LPE grown layers, electrical properties	82F
	$Pb_{1-x}Sn_xTe$ (x ≈ 0.2)	photoluminescence	82Z
Ga (acceptor, donor)	PbTe	Ga^+ donor ion state	77B2
		Ga^{3+} donor ion state (in the middle of the band gap)	77B2
		E_D = 0.21 eV (0.19 eV) above conduction band edge at T = 77 K (300 K)	73V
	$Pb_{1-x}Sn_xTe$ (x = 0.18 - 0.30)	electron transport under hydrostatic pressure	81A3
In (donor)	PbSe	Hall coefficient, lattice parameter, solubility, shallow or resonant level	65R
	PbTe	Hall coefficient, lattice parameter, solubility, shallow or resonant level	65R
	PbTe	Hall coefficient, electrical conductivity, magneto resistance, 70 meV above conduction band for T → 0, $dE_D/dT = -(2-4) \cdot 10^{-4}$ eV/K, $dE_D/dP > 7 \cdot 10^{-6}$ eV/bar	71A, 73K1
	PbTe	optical properties, 20 meV above conduction band	73D2, 75D2
	PbTe	electron transport	77L1
		absorption coefficient	81B
		electrical properties Hall mobility	79W, 80C
	$Pb_{1-x}Sn_xTe$	for dependence of energy level on x and on pressure see Figs. 2.22,23	79A2
	$Pb_{1-x}Sn_xTe$ (x = 0.10 - 0.24)	diffusion of Cd-In alloy, Hall effect	78L3

Table 2.6. (continued)

Impurity	Compound	Remarks	Reference
	$Pb_{1-x}Sn_xTe$ (x = 0 - 0.25)	transport coefficients (T = 77 - 900 K)	79G3
	$Pb_{1-x}Sn_xTe$ (x ≈ 0.2)	photoluminescence	82Z
	$Pb_{1-x}Sn_xTe$ (x = 0 - 0.25)	LPE grown layers, electrical properties	81Z, 82F
	$Pb_{1-x}Sn_xTe$ (x = 0.18 - 0.3)	discussion of an impurity fluctuation potential model	81V1
	$Pb_{1-x}Sn_xTe$ (0.2 ≲ 0.25 ; 1 at.% In)	activation energy from pressure dependent transport (P ⩽ 20 kbar, T = 4.2 - 100 K)	81A4
	$Pb_{1-x}Sn_xTe$ (x = 0.2)	laser annealing	79B1
	$PbSe_{1-x}Te_x$ (x = 0.2 , 0.5)	transport coefficients (T = 77 - 900 K)	79G3
	$PbSe_{1-x}Te_x$ (x = 0.92)	LPE grown layers, electrical properties	81Z, 82F
	$(PbSe)_{1-x}(SnTe)_x$ (x = 0 - 1)	electron transport	81G2
	$(PbSe)_{1-x}(SnTe)_x$ (x = 0 - 1, with 1 at.% In)	transport coefficients (T = 77 - 900 K)	81G2
Tl (acceptor)	PbS	Hall coefficient	80B1
	PbSe	Hall coefficient	77V1
	PbSe	Hall coefficient	80B1
	PbTe	resonant with valence band, about 240 meV below valence band edge, Hall coefficient (77 K, 300 K)	73S3
	PbTe	electrical and optical properties	79G2
	PbTe	absorption coefficient	79V1, 79V3

Table 2.6. (continued)

Impurity	Compound	Remarks	Reference
	PbTe	electrical and optical properties, quasi local impurity states inside valence band	81V2
	$Pb_{1-x}Sn_xTe$ (x = 0 - 0.25)	LPE grown layers, electrical properties	82F
	$PbS_{1-x}Se_x$		80V
Sn (donor)	$Pb_{1-x}Sn_xSe$ (x = 0 - 0.35)	study of the doping effect of tin	80P1
	$Pb_{1-x}Sn_xS$ (x = 0 - 0.05)	study of the doping effect of tin	80P1
As (acceptor)	PbTe	As^+-ion implantation 400 keV	75D1
	PbTe	in Pb saturated crystals	73S3
	$PbSe_{1-x}Te_x$ (x = 0.92)	electrical properties of LPE grown layers	82F
Sb (donor, acceptor)	PbTe	Sb^+-ion implantation created n-type layer in p-type films	73D1
	PbTe	in Te saturated crystals, amphoteric impurity appears to be located in two different lattice positions	73S3
	PbTe	thermoelectric properties	79B2
	$Pb_{1-x}Sn_xTe$ (x = 0.13 - 0.20)	Sb diffusion in p-type yielded n-type material	70A
	$Pb_{1-x}Sn_xTe$ (x = 0.2)	absorption edge shift due to Sb diffusion	78L4
Bi (donor, acceptor)	PbTe	Hall effect (77 K, 300 K)	73S3
	PbTe	thermoelectric properties	79B2
	PbTe	Hall data	79G1
	$Pb_{1-x}Sn_xTe$ (x = 0.1 - 0.22)	photoluminescence spectra	81A2
	$Pb_{1-x}Sn_xTe$ (x = 0 - 0.25)	LPE grown layers, electrical properties ($Pb_{0.8}Sn_{0.2}Te$: Bi always p-type)	82F
	$PbSe_{1-x}Te_x$ (x = 0.92)	LPE grown layers, electrical properties	82F

An earlier attempt to understand the behaviour of the elements of group III was done by ROSENBERG and WALD /65R/. They assume a localized state for In in PbSe, which is split off from the valence bands by the energy difference between the 5s orbital of In and the 6s orbital of Pb. The donor and acceptor behaviour of a group-III element in PbTe was studied by an ionic lattice approach by WEISER /81W1/. Generally speaking, theoretical knowledge of most of the experimentally investigated impurities is rather poor.

The hydrostatic pressure dependence of the In level in $Pb_{1-x}Sn_xTe$ (x = 0.22, 0.25) containing 0.5 at.% In was studied by AKIMOV et al. /77A1 , 79A2/. According to their results the pressure-induced shift of the conduction band makes the In level become resonant with the conduction band at a pressure of 1.3 kbar (T = 4.2 K). The impurity level relative to the band edges as a function of hydrostatic pressure is sketched in Fig. 2.23.

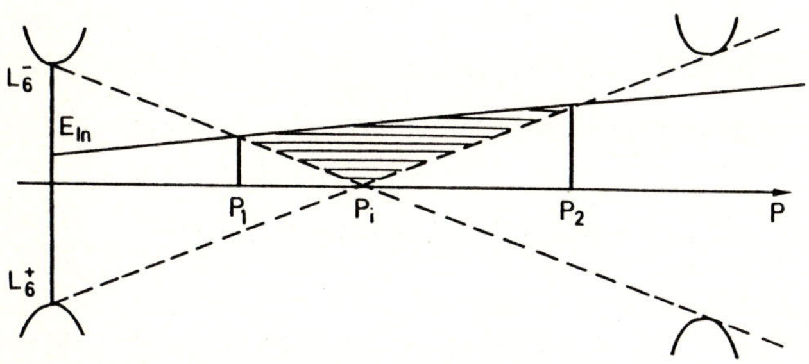

Fig. 2.23. Shift of the indium impurity level in the $Pb_{0.75}Sn_{0.25}Te$ + 0.5 at.% In alloy under pressure at 4.2 K /79A2/

As mentioned above, most of the impurities can be assumed to have shallow or even resonant levels. However, recently a deep donor level of an unidentified defect was found by SCHLICHT et al. /78S/ and by LISCHKA and HUBER /78L2/ in PbTe and by LISCHKA and PREIER /80L2/ in PbSe. In PbTe the observed donor causes a significant carrier freeze-out between 300 K and 100 K as is shown in Fig. 2.24. The analysis of the Hall data yielded a strong temperature dependence of this donor level /78S/ as plotted in Fig. 2.24 for three different samples. The temperature coefficient of the levels varies from sample to sample. The levels become resonant with the conduction band eventually between 170 and 200 K. A similar thermal freeze-out of electrons was observed in n-type $Pb_{0.82}Sn_{0.18}Te$ by SIZOV et al. /80S/. The measured Hall data was fitted with a donor level 20 meV below the conduction band. The authors relate the donor level to a Te vacancy. However, their conclusion is not substantiated by convincing arguments that rule out a chemical impurity.

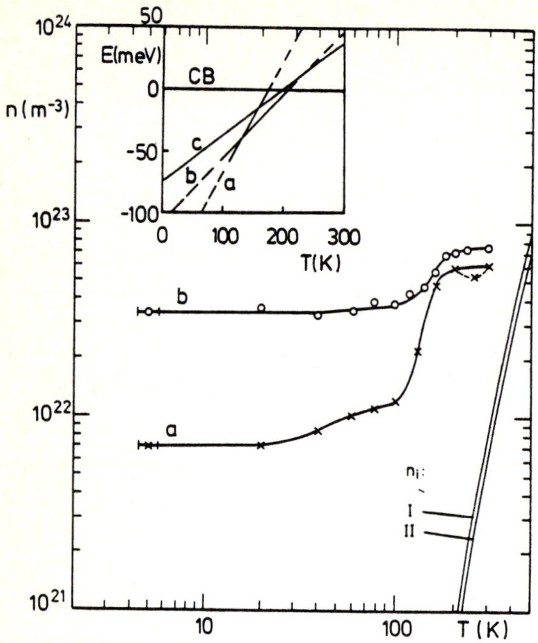

Fig. 2.24. Carrier concentration vs temperature for two PbTe samples. Here n_i represents the intrinsic carrier density calculated (I): with the six-band model with parameters from /73H3/, and (II): with the six-band model with parameters from /77F/. The insert shows the donor level energy as a function of temperature. (a,b) correspond to the presented Hall data respectively /78S/, (c) corresponds to data from /78L2/

3. Band Structure

In this chapter we are concerned with band-structure calculations and experimental results on band-edge characteristics of the lead salts. The first section outlines some general features of the overall band structure of the binary compounds which were obtained by different mathematical approaches such as empirical pseudopotential method (EPM), orthogonal-plane-wave (OPW), augmented-plane-wave (APW), and Green's function (KKR) calculations. The next section deals with the temperature, pressure, and composition dependence of the band gap. In the third section we consider the structure of the conduction and valence band near the L point of the Brillouin zone. Experimentally obtained band-edge parameters (e.g. effective masses, g factors) are compared with theoretical results of $\underline{k}\cdot\underline{p}$ models. In a subsection experimental and theoretical results on the energy levels in a magnetic field are discussed.

3.1 Band-Structure Calculations

3.1.1 Lead Chalcogenides

During the last two decades several experimental and theoretical investigations have been carried out to get an insight into the overall form of the band structures of the IV-VI semiconducting compounds. Concerning the lead salts PbS, PbSe, and PbTe most of the results obtained before 1973 were reviewed by DALVEN /73D3/. Therefore we emphasize the more recent results on these materials.

Relativistic effects are known to influence significantly the band structure of materials containing heavy atoms /63J/. A variety of computational techniques has been used to calculate the electronic band structure of the lead salts including relativistic effects.

A first-principles band calculation employing no adjustable parameter was carried out by HERMAN et al. /68H1/. They calculated energy eigenvalues at several key points of the Brillouin zone by a relativistic orthogonal-plane-wave (OPW) method. The remainder of the zone was then calculated by a (local) pseudopotential interpolation procedure.

It also seems to be an appropriate approach to calculate the bands in a (relativistic) muffin-tin potential. This has been done by CONKLIN et al. /65C/ for PbTe, and by RABII for PbS and PbSe /68R/, utilizing the augmented-plane-wave (APW) method. These calculations take relativistic effects into account by the use of "relativistic perturbations", which add the mass-velocity, Darwin, and spin-orbit interaction terms to the Hamiltonian.

OVERHOF and RÖSSLER /70 O/ utilized the relativistic Green's function method (KKR method) to calculate the band structure of PbS, PbSe, and PbTe. Inside the muffin-tin spheres they used the relativistic atomic potentials of LIBERMAN et al. /65L/ for Pb and nonrelativistic atomic potentials of HERMAN and SKILLMAN /63H/ for S, Se, and Te. These potentials were superimposed along the nearest-neighbour directions (it does not seem to be relevant whether one uses relativistic or nonrelativistic potentials for atomic numbers up to Z = 55).

The muffin-tin model calculations involve one fit parameter, the constant potential outside the muffin-tin spheres, which is determined by adjusting the theoretical smallest gap at the L point to the experimental one. The benefit of this type of calculation lies in the fact that it is possible to separate the contribution of potential and crystal structure, the latter having to be calculated only once ("structure constants"). For details of the APW method see for instance /67L/.

Several pseudopotential band-structure calculations have been published /73D3/. The band structure of PbS calculated by a local empirical pseudopotential method (EPM) /73K2/ is shown in Fig. 3.1. A more recent calculation has been carried out by MARTINEZ et al. /75M1/. Their Hamiltonian included a non-local and energy-dependent correction to the local atomic pseudopotential and a relativistic correction

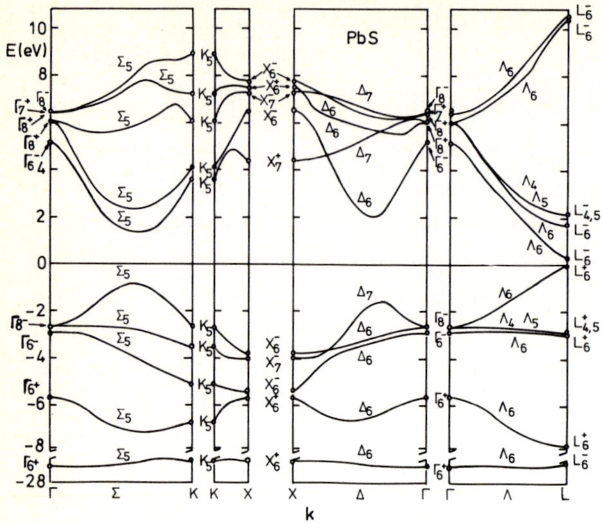

Fig. 3.1. Band structure of PbS calculated by a local empirical pseudopotential method /73K2/

due to the spin-orbit interaction. (Other relativistic terms have the full symmetry of the non-relativistic Hamiltonian; hence they can be considered as to be absorbed in the non-relativistic empirical pseudopotential.) Concerning the first correction s - p non-locality was simulated by an effective mass parameter whereas the explicit inclusion of a d-like non-local potential introduced two additional parameters. Two further parameters required for the spin-orbit interaction correction were chosen empirically by adjusting the $\Gamma_6^- - \Gamma_8^-$ splitting of the upper valence band to that found by the OPW method /68H1/. The results of this calculation are shown in Fig. 3.2 for PbSe and in Fig. 3.3 for PbTe.

Besides the $E(\underline{k})$ curves the densities of states are also displayed, which were calculated using the method of GILAT and DOLLING /64G/. Comparing the densities of states of the two compounds one can see the difference in covalency /80L1/. The valence bands are considerably broader in PbSe than in PbTe. In PbTe the upper (Pb) valence s band is clearly separated from the p bands whereas in PbSe there is already a small overlap between the s and p bands.

It should be mentioned, however, that it is just the large calculated bandwidth of the p valence bands in PbSe which is not in agreement with experiment /78G1/. The pronounced shoulder in the calculated density of valence states, which is mainly related to the lowest energy level with X_6^- symmetry, is not observed in XPS /73M1/ and UPS /73C1/ spectra of PbSe. The calculated binding energies at the X point are lower by 0.5 - 1.3 eV than those obtained experimentally by GRANDKE et al. /78G1/ from angle-resolved uv photoemission spectroscopy. For PbTe the agreement between non-local EPM calculation /75M1/ and experiment /78G1/ is much better.

The various band structures mentioned so far are all quite similar in their general features, although considerable differences may exist regarding the precise

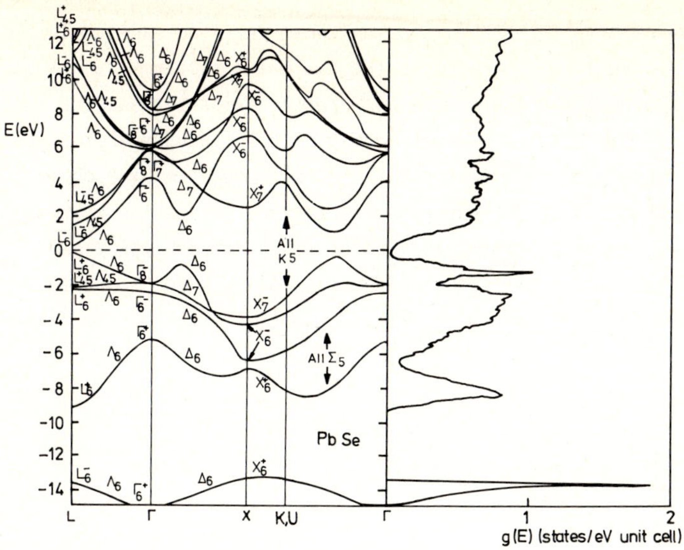

Fig. 3.2. Band structure of PbSe calculated by a non-local empirical pseudopotential method. The calculated density of states is also given /75M1/

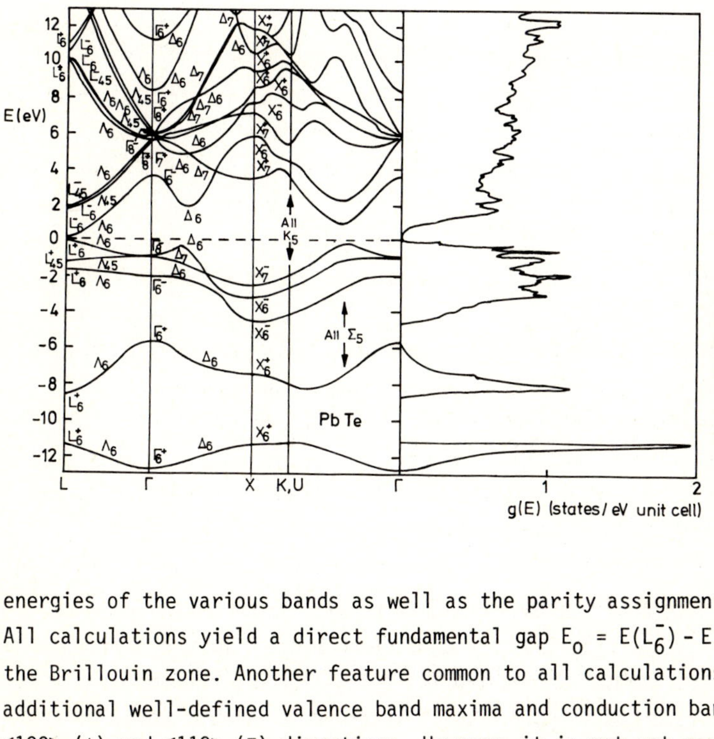

Fig. 3.3. Band structure of PbTe calculated by a non-local empirical pseudopotential method. The calculated density of states is also shown /75M1/

energies of the various bands as well as the parity assignments at critical points. All calculations yield a direct fundamental gap $E_0 = E(L_6^-) - E(L_6^+)$ at the L point of the Brillouin zone. Another feature common to all calculations is the existence of additional well-defined valence band maxima and conduction band minima along the <100> (Δ) and <110> (Σ) directions. However, it is not yet certain whether these represent subsidiary band edges (absolute extrema) or saddle points. There is experimental evidence for the existence of a second valence band maximum in PbTe, which becomes occupied by charge carriers at temperatures above room temperature (see e.g. /77S1/ and references therein). The Σ axis is a possible choice for this maximum. However, the exact location in k space of this extremum has not been verified experimentally as yet. This problem is treated in more detail in the following section.

Regarding the energy values of the bands some differences exist between the various calculations. As an example we have plotted in Fig. 3.4 the band energies at the L point for PbTe obtained by different computational methods. (We have chosen the L point because of its importance for experiment. Larger discrepancies between the various calculations may exist at other critical points.) The APW /65C/ and OPW /68H1/ results are quite similar whereas the KKR calculations /70 0/ provide rather individual values for the L(3) and L(4) valence bands.

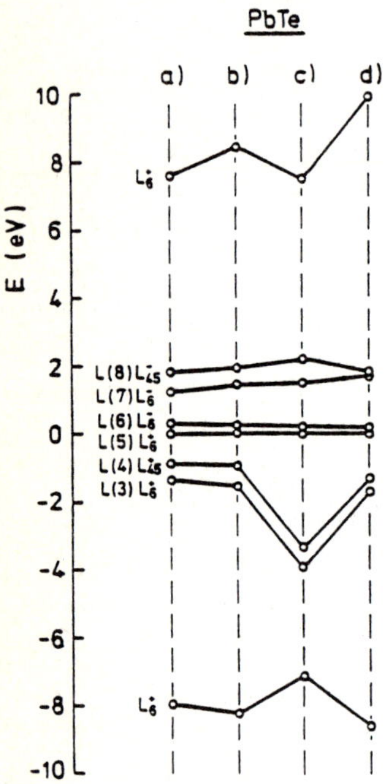

Fig. 3.4. Energy levels at the L point of the Brillouin zone for PbTe as obtained by different computational methods: a) APW /65C/, b) OPW /68H1/, c) KKR /70 0/, d) non-local EPM /75M1/

The most striking feature of the non-local EPM result /75M1/ is the very small energy separation between the L(7) and L(8) conduction bands. This results from the fact that MARTINEZ et al. /75M1/ assumed the two (single group) conduction band levels L_2^- and L_3^- to be reversed in PbTe from their order in PbS and PbSe. This inversion of conduction-band levels was first proposed by BERNICK and KLEINMAN /70B1/, and is illustrated in Fig. 3.5a. The spin-orbit coupling mixes states with L_6^- symmetry which originate from different single group functions. This means that the lowest conduction band L_6^- is a linear combination of L_6^- (L_2^-) and L_6^- (L_3^-). In PbSe the L_6^- (L_2^-) function dominates, whereas in PbTe the L_6^- (L_3^-) function determines the

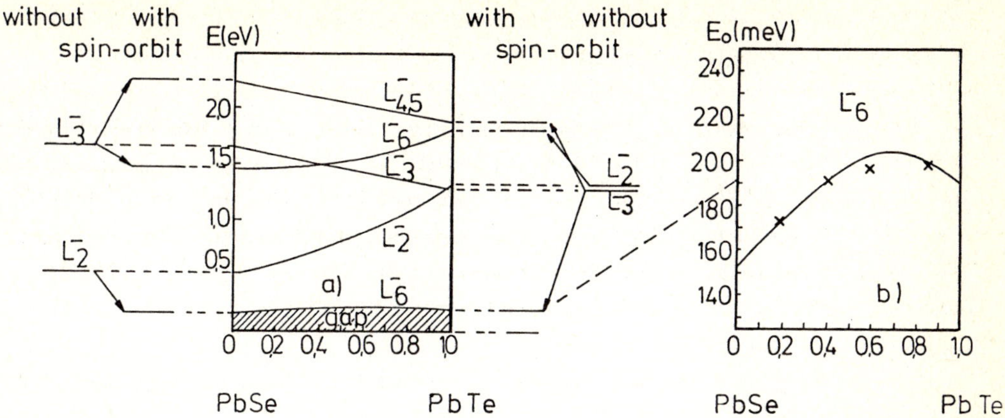

Fig. 3.5a,b. Conduction band level energies vs composition for $PbSe_{1-x}Te_x$.
a) assumed variations of the single group levels L_2^- and L_3^- and resulting variations of the double valued representations L_6^- and L_{45}^-
b) expanded representation of the fundamental gap, experimental data are also shown. Zero in energy is the highest valence band level /78J1/

lowest conduction band. MARTINEZ et al. /75M1/ stated that the inversion of conduction-band levels represents the only solution which can explain the very different experimentally observed anisotropies of the effective masses (see Sect. 3.3) in the lead salts.

This inconsistency in the L level scheme within the lead salt group is expected also from another point of view. There have been considerations that PbTe represents an anomalous member of this group: motivated by the fact that the lead salt series exhibits an unusual dependence of the gap energy on the atomic number of the chalcogen atom [E_0(PbS) > E_0(PbTe) > E_0(PbSe)] DALVEN /71D2/ examined the f-sum rule at the conduction-band and valence-band edges using experimental values of effective masses and energy gaps. He found that PbTe does not follow the same f sum as PbS and PbSe. He concluded that the irregular 5s level energy of the Te atom is responsible for this finding, and that one has to expect a different L_6^- band-edge state in PbTe if compared with PbS and PbSe.

The energy-gap anomaly has been reexamined by RABII /78R/ including the fourth member of the group: PbPo. According to his considerations the anomaly of PbTe is due to an anomalously large lattice constant which in turn is the result of a correspondingly large 5p orbital of the Te atom. Calculations show that the pressure necessary to reduce the lattice constant to a "normal" value within the group will almost remove the anomaly of the energy gap, if one takes into account the negative pressure coefficient of the gap (see next section).

To elucidate the problem of conduction-band inversion in PbSe and PbTe, JEDRZEJCZAK et al. /78J1/ investigated the composition dependence of the band gap in the ternary compound $PbSe_{1-x}Te_x$. Their experimental results obtained from measuring the photovoltaic effect in p-n junctions of this material are shown in Fig. 3.5b. In order to explain the composition dependence of the gap the authors used the following model: spin-orbit splitting is mainly due to the lead and hence can be assumed to be practically constant for all compositions. In a first approach the single group conduction-band levels L_2^- and L_3^- were assumed to vary linearly with composition showing a crossover at x = 0.97. Then for each composition the Hamiltonian incorporating three conduction band levels with spin-orbit interaction was diagonalised. The resulting gap obtained by this approach is qualitatively correct. To obtain a better quantitative fit in a second approach the authors assumed a bowing of the L_2^- level which is displayed in Fig. 3.5a and b.

The investigation /78J1/ was intended as a comment on the controversy about the band ordering around the optical gap in PbSe and PbTe brought up by GLOSSER et al. /74G1/, who had suggested that the currently accepted parities /75M1/ at the L point for PbTe were inconsistent with their measurements of band population effects in electroreflectance spectra. From the non-observation of a critical-point transition at L in PbTe (which appears in PbSe) they had concluded that in PbTe the bands L(5), L(7), and L(8) must have one parity while the bands L(3), L(4), and L(6) must have the other. (This was also concluded by NARITA and TAKAFUJI /76N1/ from SdH measurements.) However, as was pointed out in subsequent letters /75D3 , 75M2/, this conclusion is not mandatory and leads to properties which are difficult to reconcile with theoretical calculations and some experiments. In /78J1/ it was pointed out that the optical transitions in question ($L_6^+ \rightarrow L_6^-(L_2^-)$ for PbTe and $L_6^+ \rightarrow L_6^-(L_3^-)$ for PbSe according to the authors' model of conduction-band inversion) involve different momentum matrix elements with different symmetries. Although it is not quite certain whether this fact can account for the observability of the one transition and the non-observability of the other, JEDRZEJCZAK et al. /78J1/ concluded that the assumption of conduction-band inversion in PbSe and PbTe is compatible with experimental findings.

We have confined our discussion on band ordering to the L point because the ordering of states in this region determines most of the physical properties of the materials. Further discrepancies between the various calculations exist concerning the parity assignments at some other points, mainly at Γ and X. Within the framework of this article we cannot discuss all these discrepancies in detail. The most individual results were again obtained by the KKR calculation /70 O/. It appears to us that this calculation has not received wide acceptance.

As already mentioned, GRANDKE et al. /78G1/ have investigated the valence band structure of the lead salts by angle-resolved uv photoemission spectroscopy. They compared their experimentally obtained valence band energies with various band structure calculations. They found that the first-principles APW and OPW calculations provide good overall pictures of the (valence) band structures, whereas the agreement between empirically adjusted pseudopotential calculations and experiment is somewhat ambivalent. The local EPM calculation of KOHN et al. /73K2/ for PbS and PbSe is in reasonable agreement with experiment. The disagreement between experimental findings and the non-local EPM calculation of MARTINEZ et al. /75M1/ for PbSe has already been mentioned. Considerable discrepancies are also encountered when comparing the calculated binding energies of TUNG and COHEN /69T/ for PbTe with experimental values.

3.1.2 SnTe

Several band structure calculations are available for SnTe. An OPW calculation was carried out by HERMAN et al. /68H1/. RABII /69R/ utilized the APW method and TUNG and COHEN /69T/ a local empirical pseudopotential method. Another EPM calculation was done by BERNICK and KLEINMAN /70B1/.

A self-consistent relativistic APW calculation for SnTe has been carried out recently by MELVIN and HENDRY /79M1/. Their calculation included the most important relativistic effects exactly, namely mass-velocity and Darwin terms, whereas spin-orbit interactions were taken into account by a scheme based on degenerate perturbation theory of BOYER and KLEIN /75B/. The band structure obtained by this method is shown in Fig. 3.6.

While similarities exist in the results of these calculations, there are clear and important differences in detail. All calculations agree in that the conduction and valence-band symmetries at the L point are reversed from those in the lead salts, i.e. the valence band has L_6^- symmetry while the conduction band has L_6^+ symmetry. Moreover all calculations (except BERNICK and KLEINMAN's) agree in the complete band ordering at L. Three calculations /69R, 69T, 79M1/ yielded that the minimum gap in SnTe is not at the L point but that at L one finds a saddle point. The valence and conduction bands come closest together away from L in the LWU plane. This can be seen from Fig. 3.6b. The "hump structure" /69T/ shown in this figure can be explained as follows /71T1/. In a simple perturbation model, the L bands would cross at the position of the humps if the bands did not interact with each other. However, since the bands do repel each other, gaps are formed at these points. This means that in SnTe the band-structure symmetry within the humps (including L) is reversed from that in PbTe, and that outside the humps the PbTe and SnTe band-structure symmetry is the same. This fact has important consequences for the temperature dependence of the SnTe smallest gap (see next section).

A rough estimation shows that the band inversion taking place when going from PbTe to SnTe is mainly due to relativistic effects /66D1/. The band calculation of SCHLÜTER et al. /75S1/ yielded that 29% of the L_6^+ valence band charge comes from a Pb s state and 58% of the L_6^- conduction band charge comes from a Pb p state. From the tabulated values of the relativistic energy shifts of the valence states of atomic Pb and Sn /63H/, one can estimate an upward shift of the L_6^+ state by 0.80 eV and of the L_6^-

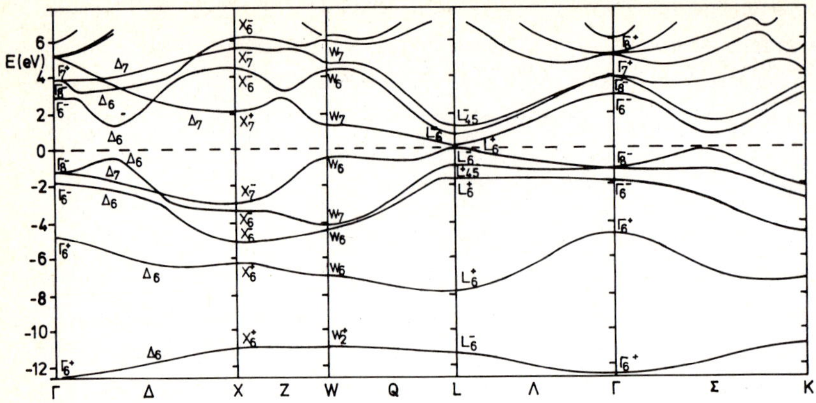

Fig. 3.6a. Band structure of SnTe obtained from a self-consistent relativistic APW calculation /89M1/

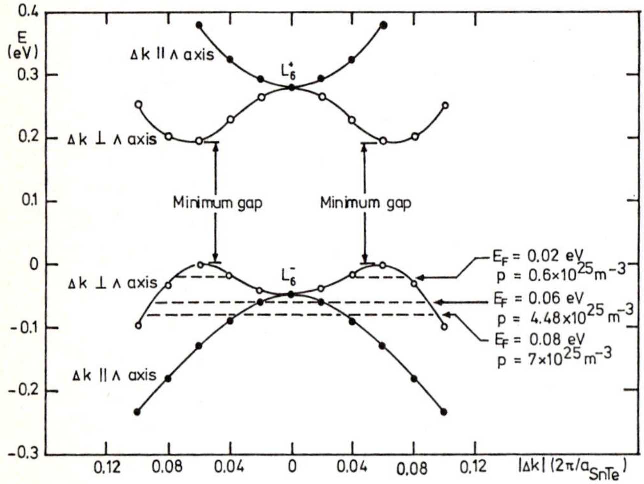

Fig. 3.6b. Band structure of SnTe near L for direction parallel and perpendicular to the Λ axis calculated by an empirical pseudopotential method. Fermi levels corresponding to different hole concentrations p are indicated /71T1/

state by 0.42 eV when replacing the Pb in PbTe by Sn. This results in a gap E_0 = $E(L_6^-) - E(L_6^+)$ = - 0.19 eV for SnTe (compared to E_0 = + 0.19 eV for PbTe), which is not too far from the measured value of - 0.3 eV /66E/.

The APW calculations of RABII /69R/ give a valence-band maximum along Σ which lies 0.3 eV above the L_6^- level thus indicating an indirect gap. RABII himself was sceptical about this finding and it has not been confirmed by other authors. All other calculations yield a second valence-band maximum along Σ slightly below the principal maximum at L and hence nearly degenerate with the L valence-band edge. A third valence-band maximum occurs along Δ and lies about 0.3 eV below the principal edge /79M1/.

3.2 Temperature, Pressure, and Composition Dependence of the Fundamental Gap

We will confine our discussion to the fundamental gap. As to higher interband energy gaps we mention only some publications. Room temperature reflectivity measurements over the energy range from 0.5 to 25 eV have been made by CARDONA and GREENAWAY /64C1/ on the lead salts and SnTe. ASPNES and CARDONA /68A2/ carried out electroabsorption and electroreflectance measurements on the lead salts. NISHINO et al. /71N/ measured electroreflectance and thermoreflectance of PbSe. Using the techniques of wavelength-modulation spectroscopy, KOHN et al. /73K2/ measured the reflectivity and the derivative of reflectivity of the lead salts. Transitions from core d levels to the conduction band were investigated by synchrotron-radiation reflectivity measurements /72C/. Most of these experimental results agree reasonably well with EPM band-structure calculations /75M4/. For a collection of experimental data see e.g. /83L/.

3.2.1 Lead Chalcogenides

As already mentioned in the previous section, the minimum energy gap in the lead salts is direct and occurs at the L point of the Brillouin zone. This conclusion has been established by band-structure calculations as well as by a variety of experimental investigations. In the case of PbTe the gap location was originally suggested by weak-field magnetoresistance measurements and subsequently confirmed by several other experiments /73D3/.

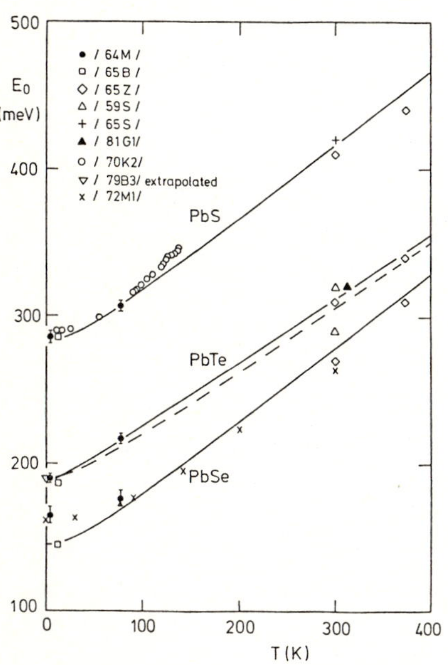

Fig. 3.7. Energy gap E_0 vs temperature for PbS, PbSe, and PbTe. For experimental methods see text. The dashed curve is calculated according to the relation given in /77A2/, the solid curves correspond to the relations of /79G5/

In Fig. 3.7 values of the fundamental gap of the three lead salts measured at various temperatures are displayed together with some empirical relations for the temperature dependence of the gap. The experimental values were obtained from magneto-absorption /64M/, magneto-emission /65B/, and magnetoreflectivity /79B3/ experiments, from the dependence of the refractive index on photon energy /65Z/, and from the analysis of optical absorption spectra in the region of the fundamental absorption edge /59S1, 65S, 70K2, 72M1, 81G1/. It should be noted that in the case of the lead salts (especially PbSe) the connection between optical data and fundamental gap may be quite complicated due to the Burstein shift and the existence of density of states tails in the energy gap /72M1, 73C2, 76T1/.

The energy-gap anomaly of the lead salt series, discussed in the preceding section, is evident from Fig. 3.7 at all temperatures. The data also show that $E_0(T)$ is linear with T above approximately 75 K, and non-linear at lower temperatures, the latter effect being due to the temperature dependence of the expansion coefficient in this temperature region (see below). This temperature dependence of the gap is usually described by empirical formulas of the VARSHNI type /67V/:

$$\text{(i)} \quad E_0(T) = E_0(0) + \alpha T^2/(T+\theta)$$

or according to GRISAR /79G5/ by:

$$\text{(ii)} \quad E_0(T) = E_1 + \sqrt{E_2^2 + \alpha^2(T+\theta')^2} \ .$$

The dashed curve in Fig. 3.7 corresponds to the type (i) relation given by ANDERSON /77A2/ while the solid curves were calculated by the type (ii) formulas of /79G5/. The parameters are given in Table 3.1.

The temperature coefficients $(\partial E_0/\partial T)_p$ can be seen from the linear portions of the curves in Fig. 3.7 to be:

$$(\partial E_0/\partial T)_p \cong \left\{ \begin{array}{ll} +\,0.51 \text{ meV/K} & \text{for PbS and PbSe} \\ +\,0.45 \text{ meV/K} & \text{for PbTe} \end{array} \right\}.$$

Table 3.1. Coefficients of the empirical relations describing the temperature dependence of the fundamental gap

	$E_0(0)$ (meV)	α (meV/K)	θ (K)	E_1 (meV)	E_2 (meV)	θ' (K)	
PbS		0.506		263	20	0	/79G5/
PbSe		0.506		125	20	0	/79G5/
PbTe		0.44		171.5	12.8	20	/79G5/
	190	0.45	50				/77A2/

Slightly lower values for $(\partial E_0/\partial T)_p$ were also reported: 0.44 meV/K for PbSe /64W, 72M1/ for temperatures between 150 K and 700 K and 0.41 meV/K for PbTe /66T/ in the temperature range from 80 to 350 K.

That $(\partial E_0/\partial T)_p$ is positive for the lead salts is unique among binary compound semiconductors and has been discussed by several authors (e.g. /71T1, 75S2/). The temperature dependence of the energy gap is determined by two contributions:

$$(\partial E_0/\partial T)_p = (\partial E_0/\partial T)_V - 3\alpha B\ (\partial E_0/\partial p)_T \quad , \tag{3.1}$$

with $\alpha = (1/L)(\partial L/\partial T)_p$ being the linear thermal expansion coefficient and B the bulk modulus. The first rhs term of (3.1) describes the electron-phonon interaction while the second term is the anharmonic lattice contribution. The latter term is known from experiment. Experimental values of the pressure coefficient $(\partial E_0/\partial p)_T$ of the energy gap were determined by measuring the shifts due to pressure of a) the peak position of spontaneous radiation /68B1/, b) current/voltage characteristics of diodes /71L1/, and c) the optical absorption edge /67P/. From the experimental data given in Table 3.2 one obtains for the dilatation term

$$(\partial E_0/\partial T)_{anh} \cong \begin{cases} +\ 0.34\ \text{meV/K} & \text{for PbS} \\ +\ 0.24\ \text{meV/K} & \text{for PbSe} \\ +\ 0.18\ \text{meV/K} & \text{for PbTe} \end{cases}$$

Table 3.2. Pressure coefficients of the gap together with values for the linear thermal expansion coefficient and the bulk modulus

	$(\partial E_0/\partial p)_T$ (10^{-11} eV/Pa)	α (10^{-5} K^{-1})	B (10^{10} N/m^2)
PbS	-9.1 (295 K /67P/)	2.03 (300 K /64N/)	6.22 (300 K /51B/)
PbSe	-8.6 (77 K, 300 K /73M2/)	1.94 (300 K /64N/)	4.8 /75S2/
	-9.1 (295 K /67P/)		4.1 (25 K /64N/)
	-9.2 (77 K /68B1/)		5.41 (298 K /71L2/)
	-10 (77 K /71L1/)		
PbTe	-7.4 (295 K /67P/)	1.98 (300 K /64N/)	3.98 (300 K /81M/)
		2.04 (303 K /68H2/)	4.11 (303 K /68H2/)
SnTe	8.85 ± 0.05 (300 K /71C1/)		3.78 (293 K /76S1/)
	(calc.)		3.60 (300 K /81M/)

In the cases of PbSe and PbTe the dilatation term contributes about one half to the total experimental temperature coefficient. These values have been reproduced theoretically by SCHLÜTER et al. /75S2/ on the basis of their EPM calculation /75M1/, and the value for PbTe was also obtained by TSANG and COHEN /71T1/. Detailed analysis of the deformation potential shows (/75S1, 75S2/) that the energy of s-like states rises faster with pressure than does the energy of p- or d-like states, provided the kinetic energies are approximately equal. Since in both compounds the valence states L(5) contain considerably more s-like charge than the conduction states L(6) (see Fig. 3 of /75S1/), the gap closes on the application of pressure.

Whereas the agreement between calculation and experiment is rather good for the lattice effect it seems to be much more difficult to explain the interaction of electrons with lattice vibrations in a quantitatively satisfying manner. According to BROOKS and YU in /71T1/ contributions of an electron-phonon interaction term containing atomic displacements only to first order (and thus describing the *instantaneous* emission and absorption of two phonons) can be accounted for by incorporating a temperature dependent Debye-Waller factor into the pseudopotential. Proceeding this way leads to reasonable results for PbTe /71T1, 75S2/ but predicts an opposite sign in the temperature coefficient for PbSe /75S2/. The different signs of the Debye-Waller temperature coefficients result from the different conduction band orderings at L /75S2/ which have been discussed in the preceding section. Preliminary calculations of SCHLÜTER et al. /75S2/ seem to indicate that the addition of a term containing atomic displacements in second order might help out of the dilemma. This term, introduced by FAN /50F/, corresponds to the scattering of an electron in state $|\underline{k},n\rangle$ (n being the band index) into another state $|\underline{k}',n'\rangle$ by emitting or absorbing a phonon of wave vector $\underline{q} = \underline{k}' - \underline{k} + \underline{G}^u$ followed by the reverse process ($\underline{G}^u$ is the reciprocal lattice vector of an umklapp process, which has to be considered if $\underline{k}' - \underline{k}$ falls outside the first Brillouin zone). As was pointed out by KEFFER et al. /70K1/, in the lead salts with their small band gaps the positive interband ($n \neq n'$) contributions might overcompensate the negative intraband ($n = n'$) terms thus opening the gap. The calculations of SCHLÜTER et al. /75S2/ for intravalley and intervalley scattering confirm this effect and also indicate that the effect is stronger in PbSe than in PbTe. However, quantitative calculations for the electron-phonon term in PbSe are still missing.

Investigations of the temperature dependence of the optical absorption edge /66T, 67P/ have shown that the energy gap becomes independent of temperature above 450 K and 700 K for PbTe and PbSe respectively. This effect can be explained by the existence of a second set of maxima in the valence band. As mentioned in the preceding section, possible locations for these additional extrema are the Σ directions of the Brillouin zone. In spite of the fact that these locations are not evident from experiment we will refer to the second set of maxima as Σ extrema.

The existence of a second valence band edge was first surmised by ALLGAIER /61A/ from analysing the temperature dependence of the Hall effect and subsequently confirmed by several investigations (see /68A1/ and references therein). The interpretation of the temperature dependence of the gap is as follows. The second valence band edge $\Sigma(5)$ lies energetically below the valence band edge L(5) at low temperatures. The indirect gap $E_0' = L(6) - \Sigma(5)$ is nearly temperature independent. Due to the positive temperature coefficient of the L(6) - L(5) gap $\Sigma(5)$ takes over the role of the valence band edge at high temperatures. This behaviour is sketched in Fig. 3.8. Values of the energy differences between non-equivalent valence band maxima are given in Table 3.3.

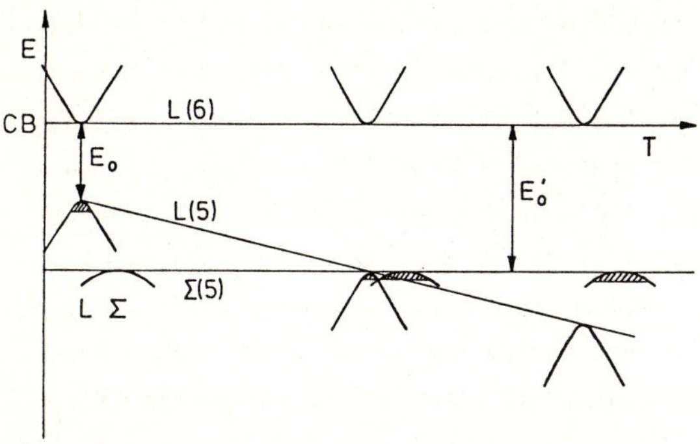

Fig. 3.8. Qualitative temperature dependence of the band extrema CB(L) = L(6), VB(L) = L(5), VB(Σ) = $\Sigma(5)$ of PbTe (after /67A1/)

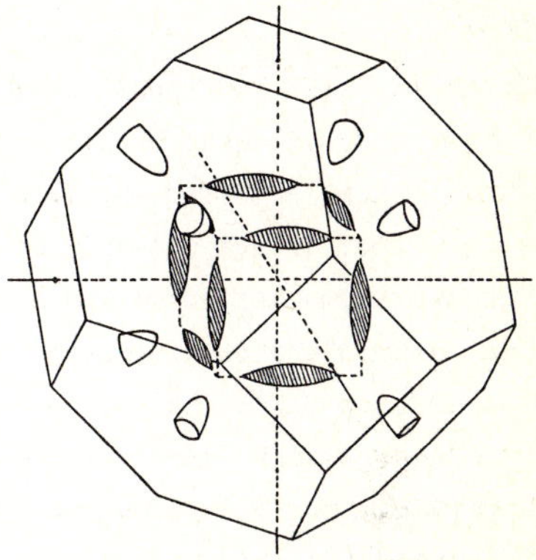

Fig. 3.9. Model for the constant-energy surfaces of the valence bands of PbTe. <111> oriented half-ellipsoids at the L points and <100> oriented ellipsoids at Σ points are shown /77S1/

Table 3.3. Values of the energy difference between non-equivalent valence band maxima $\Delta E_v = E(L_{6v}) - E(\Sigma_{5v})$ for $Pb_{1-x}Sn_xTe$

x	T (K)	ΔE_v (meV)	Ref.
0.00	0	170	67A1
0.28	10	295	74 0
0.28	300	190	74 0
0.47	10	340	74 0
0.47	300	240	74 0
0.71	77	420	78A1
0.78	77	410	78A1
0.88	77	400	78A1
1.00	77	360	78A1
1.00	0	300	67A2

SITTER et al. /77S1/ concluded from measuring the anisotropy of the conductivity of warm carriers and the magnetoresistance of p-type PbTe that the secondary valence band maxima have <100> orientation. Assuming them to be located at Σ led to the model for the constant-energy surfaces of the valence band shown in Fig. 3.9.

3.2.2 SnTe

Measurements of the fundamental gap in SnTe exhibit considerable scatter: MIS tunnelling experiments carried out on polycrystalline films with hole concentrations of 5 to $8 \cdot 10^{26}$ m^{-3} yielded a thermal gap of about 300 meV at 4.2 K and a negative temperature coefficient of - 0.2 meV/K between 4.2 and 100 K /66E, 68S/. From optical experiments carried out on a single-crystal film with $3.6 \cdot 10^{25}$ holes, small negative temperature coefficients (0 to - 0.15 meV/K) were obtained for photon energies below 400 meV and positive values for higher photon energies /69B1/. More recent investigations /79M2/ yielded the temperature dependence of the minimum optical gap shown in Fig. 3.10. The temperature coefficient of the gap in the high-temperature region changes its sign from negative to positive with increasing carrier concentration. The kinks in the $E_0(T)$ curves are attributed to a structural phase transition /79M2/ (Chap. 2).

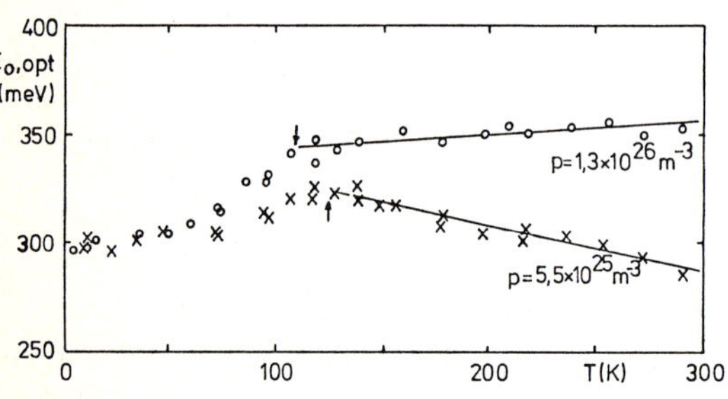

Fig. 3.10. Temperature dependence of the minimum optical gap for SnTe. Arrows indicate the temperatures at which the temperature coefficient of the gap changes discontinuously due to the structural phase transition. In the high temperature phase the temperature coefficient of the gap changes sign with increasing carrier concentration /79M2/

As mentioned before, the band structure near the fundamental gap in SnTe is more complicated if compared to the lead salts (Fig. 3.6b). Due to the different orderings of L_6^- - and L_6^+ - like bands inside and outside the hump structure, calculations yield different signs of the gap temperature coefficient for the two regions /71T1/. The temperature coefficient is positive (like in PbTe) outside the humps but it shows the opposite sign within the humps. This means that for large hole concentrations (for which the Fermi level lies below L_6^-) one should measure a positive temperature coefficient of the gap. This is in agreement with experiment /79M2/. For lower carrier concentrations ($E_F > E(L_6^-)$), however, the situation is much more complicated. In this case the Fermi surface includes regions of both positive and negative temperature coefficients. In an optical experiment all direct transitions from the Fermi surface contribute to the result, and it is not obvious which sign of the gap temperature coefficient will dominate at a given carrier concentration and temperature. From tunnelling experiments, on the other hand, one determines energy differences between empty states near band extrema. For this case the model is able to explain the observed negative temperature coefficient of the gap /66E/. The quantitative agreement between theoretical model and experiment, however, is very unsatisfactory up to now, also with respect to the structural phase transition.

To our knowledge the pressure dependence of the SnTe gap has not been measured. COHEN and TSANG /71C1/ calculated the value given in Table 3.2, which confirms PAUL's empirical rule /61P , 72M2/ that the pressure coefficient should be unique for the lead-tin chalcogenides with NaCl structure.

As in the lead salts there is experimental evidence for the existence of a second set of valence band maxima in SnTe, too. From Hall effect measurements at high temperatures, ANDREEV /67A2/ deduced for the energy difference between the main valence band maximum at L and the second valence band maximum the value $\Delta E_V(0\ K) = 0.3$ eV. ALADGADGYAN and TONEVA /78A1/ reported for the temperature coefficient $d\Delta E_V/dT = 2\cdot10^{-4}$ eV/K.

3.2.3 Ternary Compounds

3.2.3.1 $Pb_{1-x}Sn_xTe$

A model for the approximate variation with alloy composition of the energy band states at the L point in $Pb_{1-x}Sn_xTe$ has been proposed by DIMMOCK /71D3/ and is shown in Fig. 3.11. A linear variation with composition has been assumed. Starting from PbTe the fundamental gap (between the middle L_6^+ and L_6^- states) closes with increasing Sn content. At an intermediate composition the states cross over. With further increasing x value E_0 increases, the L_6^+ states now forming the conduction band and L_6^- the valence band. As mentioned in the preceding section this variation of the energy gap with composition can be understood qualitatively in terms of the difference between the relativistic effects in Pb and Sn. In Fig. 3.12 experimental values of the gap are displayed for different temperatures as a function of the mole fraction of SnTe. They were determined from optical absorption data at room temperature /65N , 66B1/, laser emission experiments /69B2/, photovoltaic cut-offs /66M1/, and tunnelling studies /77T1/. For the lead-rich compositions ($x \lesssim 0.2$), which are most important for device applications, the temperature and composition dependence of the

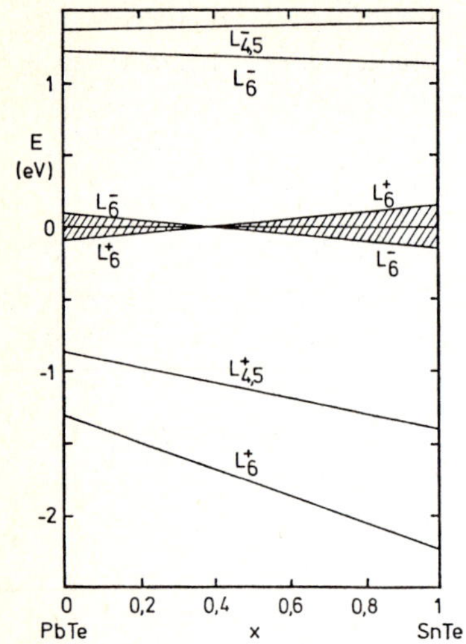

Fig. 3.11. Approximate position of the energy levels in the vicinity of the fundamental gap (shaded area) at the L point vs alloy composition in $Pb_{1-x}Sn_xTe$ /71D3/

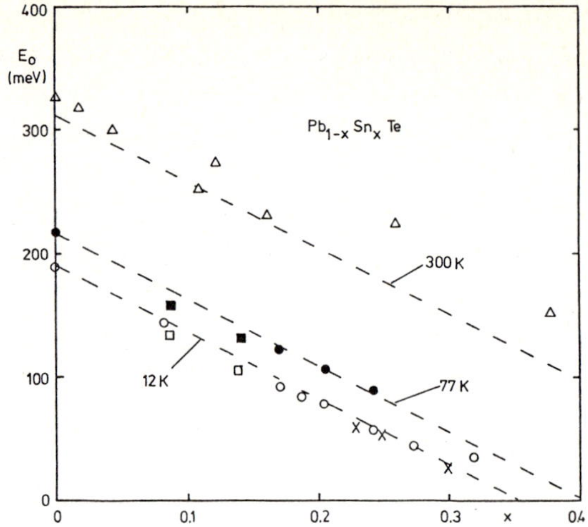

Fig. 3.12. Experimental data of the fundamental gap of $Pb_{1-x}Sn_xTe$ vs alloy composition:

△	optical absorption at 300 K	/65N, 66B/
●	laser emission at 77 K	/69B2/
○	laser emission at 12 K	/69B2/
■	photovaltaic cutoff at 77 K	/66M1/
□	photovoltaic cutoff at 12 K	/66M1/
×	tunnelling at 2 K or 4 K	/77T1/

The dashed lines were calculated for T = 12 K, 77 K, 300 K using the $E_0(x,T)$ expression of /79G5/

gap is often described by the empirical relations (E_0 in meV, T in K) from /77A2/

$$E_0(x,T) = 190 - 543x + \frac{0.45\ T^2}{T + 50}$$

or from /79G5/

$$E_0(x,T) = 171.5 - 535x + [(12.8)^2 + 0.19\ (T+20)^2]^{1/2} \quad .$$

Negative values of the gap correspond to a SnTe-like ordering of valence and conduction band. In Fig. 3.12 $E_0(x)|_T$ curves calculated according to the second formula /79G5/ for T = 12 K, 77 K, and 300 K are also shown.

For the temperature dependence of the photon energy emitted by laser diodes see Fig. 5.4a.

In accordance with PAUL's empirical rule /61P/ the pressure coefficient of the gap (Table 3.2) appears to be practically independent of the composition.

As has been discussed for the lead salts and SnTe, a second set of valence band maxima exists in these substances. Values of the energy differences ΔE_V between the main valence band maxima at L and the subsidiary ones are given in Table 3.3.

3.2.3.2 $Pb_{1-x}Sn_xSe$

As discussed in Sect. 2 PbSe and SnSe form a homogeneous phase with rocksalt structure only on the lead-rich side of compositions with $x \lesssim 0.4$. For these compositions the same considerations as those for $Pb_{1-x}Sn_xTe$ are valid. Fig. 3.13 shows experimental values of the gap for different compositions and temperatures. They were obtained from optical absorption /67S/ and laser emission /74Z1/ experiments.

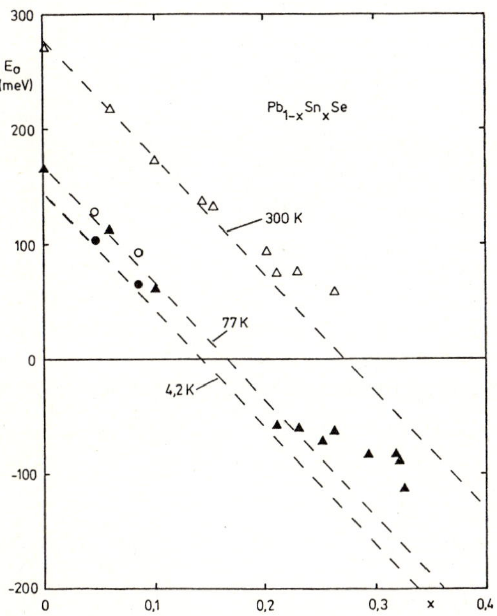

Fig. 3.13. Experimental data of the fundamental gap of $Pb_{1-x}Sn_xSe$ vs alloy composition:

△ optical absorption at 300 K /67S/

▲ optical absorption at 77 K /67S/

○ laser emission at 77 K /74Z1/

● laser emission at 4.2 K /74Z1/

The dashed curves were calculated for T = 4.2 K, 77 K, 300 K using the $E_0(x,T)$ expression of /79G5/

The temperature and composition dependence of the gap can be described by the relations (E_0 in meV, T in K) from /67S/

$$E_0(x,T) = 130 - 890x + 0.45\,T$$

or from /79G5/

$$E_0(x,T) = 125 - 1021x + [400 + 0.256\,T^2]^{1/2}$$

for $x \lesssim 0.2$. The $E_0(x)|_T$ curves shown in Fig. 3.13 were calculated according to /79G5/. Again negative E_0 values correspond to a band ordering inverted from that of PbSe. Figure 3.14 displays the temperature dependence of laser emission energy of $Pb_{1-x}Sn_xSe$ for different compositions /69H/. These results strongly support the band

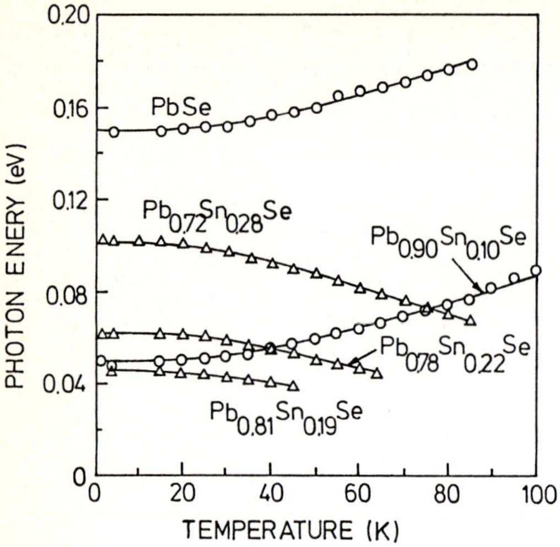

Fig. 3.14. Temperature dependence of the laser emission photon energy for diodes of $Pb_{1-x}Sn_xSe$ with $x = 0.0$, 0.10, 0.19, 0.22, and 0.28 /69H, 74H1/

crossing model discussed before (see also Fig. 5.4c). Several investigations have dealt with the pressure dependence of the gap /72M2, 73M2, 74Z1/. The pressure coefficient (Table 3.2) does not depend on composition (Paul's rule) and is also nearly temperature independent.

3.2.3.3 $PbS_{1-x}Se_x$

SCHOOLAR et al. /77S2/ have found from measuring photovoltaic cut-offs that in this pseudobinary system the gap varies like

$$E_0(x, T = 77K) \quad = \quad 305 - 130x \text{ (meV)} \quad .$$

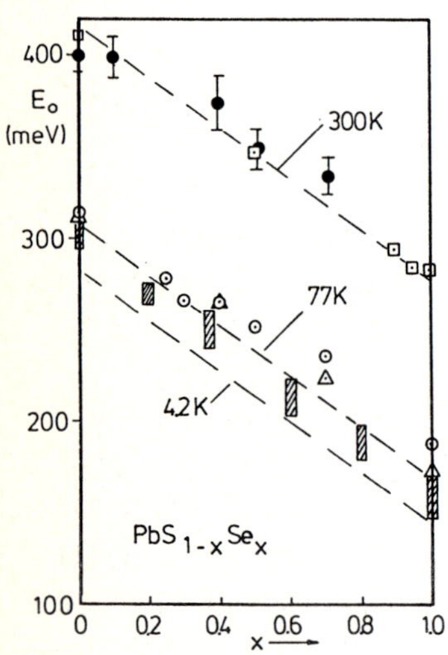

Fig. 3.15. Experimental data of the fundamental gap of $PbS_{1-x}Se_x$ vs alloy composition:
□ optical absorption at 300 K /59S1/
● optical absorption at 300 K /80V/
○ laser emission at 90 K /69K/
△ laser emission at 77 K /71C2/
▨ laser emission at 2 K /73M3/
The dashed lines were calculated for T = 4.2 K, 77 K, 300 K according to /79P/

PREIER and RIEDEL /74P2/ reported $\partial E_0/\partial T|_p = 0.53$ meV/K for $T \gtrsim 30$ K and $x = 0.6$. Further experimental data obtained from optical absorption /59S1, 80V/ and laser emission /69K, 71C2, 73M3/ experiments are displayed in Fig. 3.15. A suitable description of the energy gap variation is from /79P/ (E_0 in meV, T in K):

$$E_0(x,T) = 263 - 138x + [400 + 0.256\ T^2]^{1/2}.$$

$E_0(x)|_T$ curves calculated according to this equation are also shown in Fig. 3.15.

For the temperature dependence of the photon energy emitted by laser diodes see Fig. 5.4b.

3.3 Band-Edge Characteristics

As mentioned in Sect. 3.1, both the conduction and valence-band edges are at or near the L point of the Brillouin zone. The magneto-resistance studies (e.g. /58A2, 64A, 64C2/) which lead to this result also delineated the constant energy surfaces near L. For the three lead salts the surfaces of constant energy for both electrons and holes are (for sufficiently low carrier densities) prolate spheroids of revolution, their major axes being <111> or equivalent directions and their centres being at L (Fig. 3.9). At higher carrier concentrations ($\gtrsim 10^{25}$ m^{-3}) the constant energy surfaces become non-ellipsoidal. For SnTe the surfaces of constant energy are much more complicated due to the inversion of band ordering. Each pocket of carriers centred about L consists of multiply connected surfaces /71C1/.

Due to the small energy gaps of these materials the dispersion relation $E(\underline{k})$ may be strongly non-parabolic.

3.3.1 $\underline{k}\cdot\underline{p}$ Models

The most accurate model presently used to describe the band structure near the L point is the six-band $\underline{k}\cdot\underline{p}$ model of MITCHELL and WALLIS /66M2/ and DIMMOCK /71D3/. This model includes the coupling between the highest valence band and the far bands and between the lowest conduction band and the far bands (Fig. 3.11) in a second-order perturbation approximation and treats the resulting two-band problem exactly. The corresponding matrix Hamiltonian /71D3/ is given in (3.2) where P_t and P_ℓ represent the direct transverse and longitudinal momentum matrix elements taken between the extremal valence and conduction band states; $m_t^\pm$, $m_\ell^\pm$, $g_t^\pm$ and $g_\ell^\pm$ /73A/ are the far-band contributions to the transverse and longitudinal band-edge effective masses and g factors (see Table 3.4). General expressions for the parabolic effective masses and g factors in terms of five momentum matrix elements and two spin-orbit mixing coefficients have been given in /66M2/. It should be noted, however, that MITCHELL and WALLIS /66M2/ used different algebraic signs for the masses and g factors of the valence band if compared with the work of DIMMOCK /71D3/. The

$$\mathcal{H} = \begin{array}{c} \\ \end{array}\begin{array}{cccc} L_6^-(\uparrow) & L_6^-(\downarrow) & L_6^+(\uparrow) & L_6^+(\downarrow) \end{array}$$

$$\mathcal{H} = \left[\begin{array}{c|c|c|c}
\frac{E_o}{2} + \frac{\hbar^2}{2m_t^-}(k_x^2 + k_y^2) + \frac{\hbar^2}{2m_\ell^-} k_z^2 + \frac{1}{2} g_\ell^- \mu_B B_z & \frac{1}{2} g_t^- \mu_B (B_x - i B_y) & \frac{\hbar}{m_o} P_\ell k_z & \frac{\hbar}{m_o} P_t (k_x - i k_y) \\
\hline
\frac{1}{2} g_t^- \mu_B (B_x + i B_y) & \frac{E_o}{2} + \frac{\hbar^2}{2m_t^-}(k_x^2 + k_y^2) + \frac{\hbar^2}{2m_\ell^-} k_z^2 - \frac{1}{2} g_\ell^- \mu_B B_z & \frac{\hbar}{m_o} P_t (k_x + i k_y) & - \frac{\hbar}{m_o} P_\ell k_z \\
\hline
\frac{\hbar}{m_o} P_\ell k_z & \frac{\hbar}{m_o} P_t (k_x - i k_y) & - \frac{E_o}{2} - \frac{\hbar^2}{2m_t^+}(k_x^2 + k_y^2) - \frac{\hbar^2}{2m_\ell^+} k_z^2 - \frac{1}{2} g_\ell^+ \mu_B B_z & - \frac{1}{2} g_t^+ \mu_B (B_x - i B_y) \\
\hline
\frac{\hbar}{m_o} P_t (k_x + i k_y) & - \frac{\hbar}{m_o} P_\ell k_z & - \frac{1}{2} g_t^+ \mu_B (B_x + i B_y) & - \frac{E_o}{2} - \frac{\hbar^2}{2m_t^+}(k_x^2 + k_y^2) - \frac{\hbar^2}{2m_\ell^+} k_z^2 + \frac{1}{2} g_\ell^+ \mu_B B_z
\end{array}\right] \quad (3.2)$$

with $\underline{k} = \frac{1}{\hbar}(\underline{p} + e\underline{A})$, $\underline{p} = -i\hbar\underline{\nabla}$, $\underline{A} = \frac{1}{2}(\underline{B} \times \underline{r})$.

Table 3.4. Band-edge masses and g factors in terms of $\underline{k} \cdot \underline{p}$ parameters (for $E_o > 0$)

	Total	=	Two-band contribution	+	Far-band and free-electron contribution
Masses	$\frac{m_o}{m_t^c}$	=	$\frac{2P_t^2}{E_o m_o}$	+	$\frac{m_o}{m_t^-}$
	$\frac{m_o}{m_\ell^c}$	=	$\frac{2P_\ell^2}{E_o m_o}$	+	$\frac{m_o}{m_\ell^-}$
	$\frac{m_o}{m_t^v}$	=	$\frac{2P_t^2}{E_o m_o}$	+	$\frac{m_o}{m_t^+}$
	$\frac{m_o}{m_\ell^v}$	=	$\frac{2P_\ell^2}{E_o m_o}$	+	$\frac{m_o}{m_\ell^+}$
g factors	$\lvert g_t^c \rvert$	=	$\frac{4P_t P_\ell}{E_o m_o}$	+	g_t^-
	$\lvert g_\ell^c \rvert$	=	$\frac{4P_t^2}{E_o m_o}$	+	g_ℓ^-
	$\lvert g_t^v \rvert$	=	$\frac{4P_t P_\ell}{E_o m_o}$	+	g_t^+
	$\lvert g_\ell^v \rvert$	=	$\frac{4P_t^2}{E_o m_o}$	+	g_ℓ^+

work of ADLER et al. /73A/ differs from that of DIMMOCK /71D3/ in that the algebraic signs of $g_{t,\ell}^+$ are reversed.

Several sets of band parameters have been published (/71D3, 73H3, 77F2, 78G4, 79A1, 79B3, 79Z, 81I/ for $Pb_{1-x}Sn_xTe$, /78B1, 78K1, 78M, 79P, 80B2, 81K1/ for $Pb_{1-x}Sn_xSe$, and /79P/ for $PbS_{1-x}Se_x$). A selection of data is given in Table 3.5. For Pb(Sn)Te most of the published data agree reasonably well. The best overall agreement with experimental data, such as density of states and effective masses, seems to be provided by the data set of HEWES et al. /73H3/.

Table 3.5. L point band parameters of the lead salts

Compound	x	$\frac{2P_t^2}{m_o}$ (eV)	$\frac{2P_\ell^2}{m_o}$ (eV)	$\frac{m_o}{m_t^-}$	$\frac{m_o}{m_t^+}$	$\frac{m_o}{m_\ell^-}$	$\frac{m_o}{m_\ell^+}$	g_t^-	g_t^+	g_ℓ^-	g_ℓ^+	Ref.
$Pb_{1-x}Sn_xTe$	0	5.6	0.52	11.6	10.0	1.2	0.7	3.4	-7.7	-1.8	-3.8	73H3
	0.17	↓	↓	11.6	12.0	↓	0.9	3.4	-9.1	↓	0.18	↓
	0.30	↓	↓	11.6	10.0	↓	0.7	3.4	-8.1	↓	-0.82	↓
	0.40	↓	↓	11.6	9.5	↓	0.7	3.4	-8.1	↓	-0.82	↓
	0.50	↓	↓	10.6	7.5	↓	0.7	3.3	-7.5	↓	-0.82	↓
	0.60	↓	↓	9.6	7.5	↓	0.7	3.2	-7.5	↓	-0.82	↓
	0	8.2	0.70	17	3.9	2.9	0.35					77F2
	0	5.9	0.58	14	10	2.4	1.3	-1.3	-0.35	-3	-1	79A1
	0.05	↓	0.56	↓	↓	↓	↓	↓	↓	↓	↓	↓
	0.10	↓	0.54	↓	↓	↓	↓	↓	↓	↓	↓	↓
	0.16	↓	0.52	↓	↓	↓	↓	↓	↓	↓	↓	↓
$Pb_{1-x}Sn_xSe$	0	3.0	1.7	4.3	8.7	3.1	3.3	-3.5	-0.8	-5.1	-3.3	80B2
$PbS_{1-x}Se_x$	0-1	3.0	1.6	1.9 +2.4x	2.7 +6.0x	3.7 -0.8x	3.7 -0.4x					79P

For the magnetic field free cases (B = 0) (3.2) leads to the following dispersion relation for the conduction and valence band (neglecting spin)

$$\left[\frac{E_0}{2} + \frac{\hbar^2}{2m_t^-} k_t^2 + \frac{\hbar^2}{2m_\ell^-} k_\ell^2 - E\right] \cdot \left[-\frac{E_0}{2} - \frac{\hbar^2}{2m_t^+} k_t^2 - \frac{\hbar^2}{2m_\ell^+} k_\ell^2 - E\right] = \frac{\hbar^2}{m_0^2}\left[P_t^2 k_t^2 + P_\ell^2 k_\ell^2\right] \quad (3.3)$$

with $k_t^2 = k_x^2 + k_y^2$, $k_\ell^2 = k_z^2$.

This relation is non-parabolic and describes non-ellipsoidal energy surfaces. If the carrier density is not too high ($\lesssim 10^{25}$ m^{-3}) one can drop the terms quartic in k. Then the constant energy surfaces are ellipsoidal.

If one neglects the far band contributions ($m_o/m_{t,\ell}^{+,-} \to 0$) the usual ellipsoidal non-parabolic two-band KANE dispersion relation /56K/ is obtained:

$$(E - \frac{E_0}{2})\,(\frac{1}{2} + \frac{E}{E_0}) = \frac{\hbar^2}{2m_t} k_t^2 + \frac{\hbar^2}{2m_\ell} k_\ell^2 \quad , \qquad (3.4)$$

with $m_{t,\ell} = m_{t,\ell}^c = m_{t,\ell}^v$, which is equivalent to assuming exact mirror symmetry of the conduction and valence band.

Another model often used to analyse experimental data of the lead salts is the two-band model of COHEN /61C/:

$$E = \frac{E_0}{2} + \frac{\hbar^2 k_t^2}{2m_t\left(\frac{1}{2} + \frac{E}{E_0} + \frac{\hbar^2 k_\ell^2}{2m_\ell^+ E_0}\right)} + \frac{\hbar^2}{2m_\ell^-} k_\ell^2 \qquad (3.5)$$

which follows from (3.2) by assuming m_o/m_t^+, $m_o/m_t^- \to 0$ and $P_\ell \to 0$ /77F2/.

In Fig. 3.16 we have plotted the two dispersion relations (3.3) and (3.4) for x = 0.2 and E_0 = 0.1 eV. For the multi-band model (MBM) we used the parameters of APPOLD et al. /79A1/ (see Table 3.5). The two-band model parameters were chosen to fit the average MBM band-edge masses: $2P_t^2/m_o$ = 7.1 eV , $2P_\ell^2/m_o$ = 0.68 eV. For comparison the parabolic approximation as well as some Fermi energies (calculated for the case of complete degeneracy) are also shown in the figure. Obviously the deviations from parabolicity are strong at carrier concentrations $\gtrsim 10^{23}$ m^{-3}. The asymmetry between conduction band and valence band depends on the band parameter set used (3.11).

3.3.2 Statistics

If the surfaces of constant energy are ellipsoids with a major (rotational) axis a and minor axes b the volume in $\underline{k}$ space containing states up to an energy E is simply $V(E) = (4\pi/3)\cdot a(E)\cdot b^2(E)$. Therefore the density of states in one ellipsoid is given by

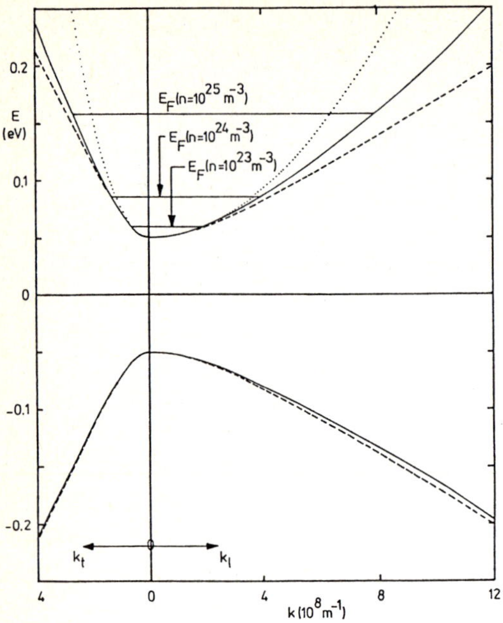

Fig. 3.16. Dispersion relation for $Pb_{0.8}Sn_{0.2}Te$ with $E_o = 0.1$ eV. Solid lines were calculated according to the multi-band model (MBM) with the band parameters of /79A1/. Dashed lines correspond to the two-band model with the parameters chosen to fit the average MBM band-edge masses (see text). Dotted lines represent the parabolic approximations. Fermi energies for three different electron concentrations calculated for the case of complete degeneracy are also shown

$$g(E)\,dE = \frac{1}{3\pi^2}\left(2ab\,\frac{\partial b}{\partial E} + b^2\,\frac{\partial a}{\partial E}\right) dE \quad . \tag{3.6}$$

If N (= 4) denotes the number of equivalent ellipsoids in the first Brillouin zone,

$$m_d^{c,v} = N^{2/3}\,[m_\ell^{c,v}(m_t^{c,v})^2]^{1/3} \tag{3.7}$$

the density of states effective mass, and

$$\mathcal{N}^{c,v} = 2\left(\frac{m_d^{c,v}\,kT}{2\pi\hbar^2}\right)^{3/2} \tag{3.8}$$

the conduction or valence band effective density of states, the density of states can be written as

$$g^{c,v}(z)\,dz = \frac{2}{\sqrt{\pi}}\,\mathcal{N}^{c,v}\,\tilde{g}^{c,v}(z)\,dz \tag{3.9}$$

with

$$\tilde{g}(z) = [z(1+\beta z)]^{1/2}\,(1+2\beta z) \tag{3.10a}$$

for the two-band model, and

$$\tilde{g}^{c,v}(z) = \left[\frac{z(1+\beta z)}{(1+\Delta_\ell^{c,v}\beta z)(1+\Delta_t^{c,v}\beta z)^2}\right]^{1/2}$$

$$\cdot\left[1 + 2\beta z - \frac{1}{3}\beta z(1+\beta z)\left(\frac{2\Delta_t^{c,v}}{1+\Delta_t^{c,v}\beta z} + \frac{\Delta_\ell^{c,v}}{1+\Delta_\ell^{c,v}\beta z}\right)\right] \tag{3.10b}$$

for the ellipsoidal truncation of the multi-band model (MBM) where $z = E/(kT)$ and $\beta = kT/E_0$. The quantities

$$\Delta_{\ell,t}^{c,v} = m_{\ell,t}^{c,v}\left(\frac{1}{m_{\ell,t}^{-,+}} - \frac{1}{m_{\ell,t}^{+,-}}\right) \tag{3.11}$$

defined according to /77F2/ measure the asymmetry between the conduction and valence band edge masses.

In Figs. 3.17a,b we have plotted the density of states of the conduction band (a) and of the valence band (b) according to (3.9,10b) using different band parameter sets of Table 3.5. The experimental data of DUBROVSKAYA and RAVICH /66D2/ for the conduction band and of DUBROVSKAYA et al. /67D/ for the valence band are also shown.

As can be seen from the figure, none of the theoretical curves can account for the experimentally observed rapid increase of the density of states at higher energies. However, as has been pointed out by ZIEP and GENZOW /79Z/, an extended multi-

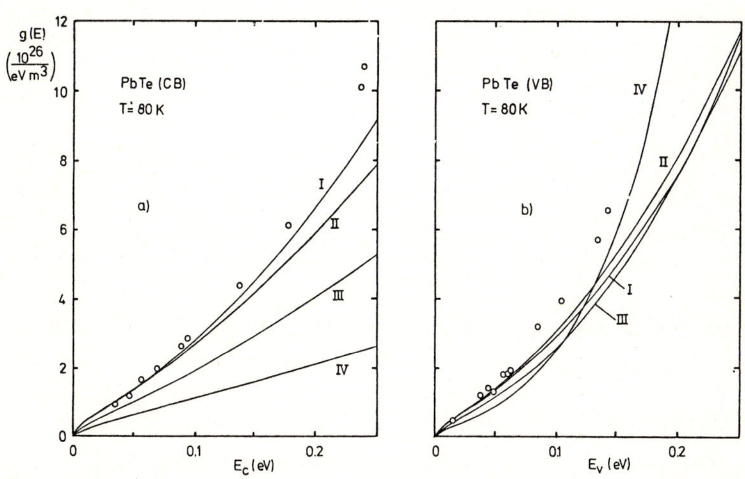

Fig. 3.17a,b. Density of states of the conduction band (a) and of the valence band (b) of PbTe. The curves were calculated on the basis of the ellipsoidal truncation of the multi-band model using different sets of band parameters: I /71D3/, II /73H3/, III /79A1/, IV /77F2/. Experimental data of /66D2/ for the conduction band and of /67D/ for the valence band are also shown

band model, which also takes into account the interaction of the higher conduction bands with lower valence bands, could possibly describe the observed behaviour satisfactorily.

The electron concentration in the conduction band is given by

$$n = \frac{2}{\sqrt{\pi}} \mathcal{N}^c \, G^c(\eta) \quad , \tag{3.12}$$

where $\eta = E_F/(kT)$ is the reduced Fermi energy and

$$G^c(\eta) = \int_0^\infty \frac{\tilde{g}^c(z)\, dz}{1 + \exp(z - \eta)} \quad . \tag{3.13}$$

Obviously, if $\tilde{g}^c(z)$ is given by (3.10a), $G^c(\eta)$ is equal to the generalized Fermi integral $\frac{2}{3}\, {}^0L_0^{2/3}(\eta,\beta)$ (see /74Z2/). In the case of strong degeneracy (3.12) can be evaluated as

$$n = \frac{4}{3\sqrt{\pi}} \mathcal{N}^c \, [\eta(1+\beta\eta)]^{3/2} \tag{3.14a}$$

for the two-band model, and

$$n = \frac{4}{3\sqrt{\pi}} \mathcal{N}^c \, \frac{[\eta(1+\beta\eta)]^{3/2}}{(1 + \Delta_t^c \beta\eta)\,(1 + \Delta_\ell^c \beta\eta)^{1/2}} \tag{3.14b}$$

in the case of the multi-band model (MBM). Similar relations are obtained for holes.

In the MBM case the intrinsic carrier density n_i can in general be obtained by determining the intrinsic Fermi level η_i from the relation

$$\frac{\mathcal{N}^c}{\mathcal{N}^v} = \frac{G^v(-\eta_i - \beta^{-1})}{G^c(\eta_i)} \quad . \tag{3.15}$$

Under non-degenerate conditions $([1 + \exp(z - \eta)]^{-1} \approx \exp(\eta - z))$ n_i is given by

$$n_i = \frac{2}{\sqrt{\pi}} \sqrt{\mathcal{N}^c \mathcal{N}^v} \exp\left(-\frac{\beta^{-1}}{2}\right) \left[\left(\int_0^\infty e^{-z}\, \tilde{g}^c(z)\, dz\right) \left(\int_0^\infty e^{-z}\, \tilde{g}^v(z)\, dz\right)\right]^{1/2} \quad . \tag{3.16}$$

In Figs. 3.18a-c we have plotted the intrinsic carrier densities for the three lead compounds $Pb_{1-x}Sn_xTe$, $Pb_{1-x}Sn_xSe$, and $PbS_{1-x}Se_x$ as a function of temperature. The curves were calculated using the ellipsoidal truncation of the multi-band model with band parameters given in Table 3.5 (/79A1/ for $Pb_{1-x}Sn_xTe$, /80B2/ for $Pb_{1-x}Sn_xSe$, and /79P/ for $PbS_{1-x}Se_x$) and the $E_0(x,T)$ relations of /79G5/ and /79P/. The curves of Figs. 3.18 can be compared to those calculated by KASZUBA and ROGALSKI /81K2/,

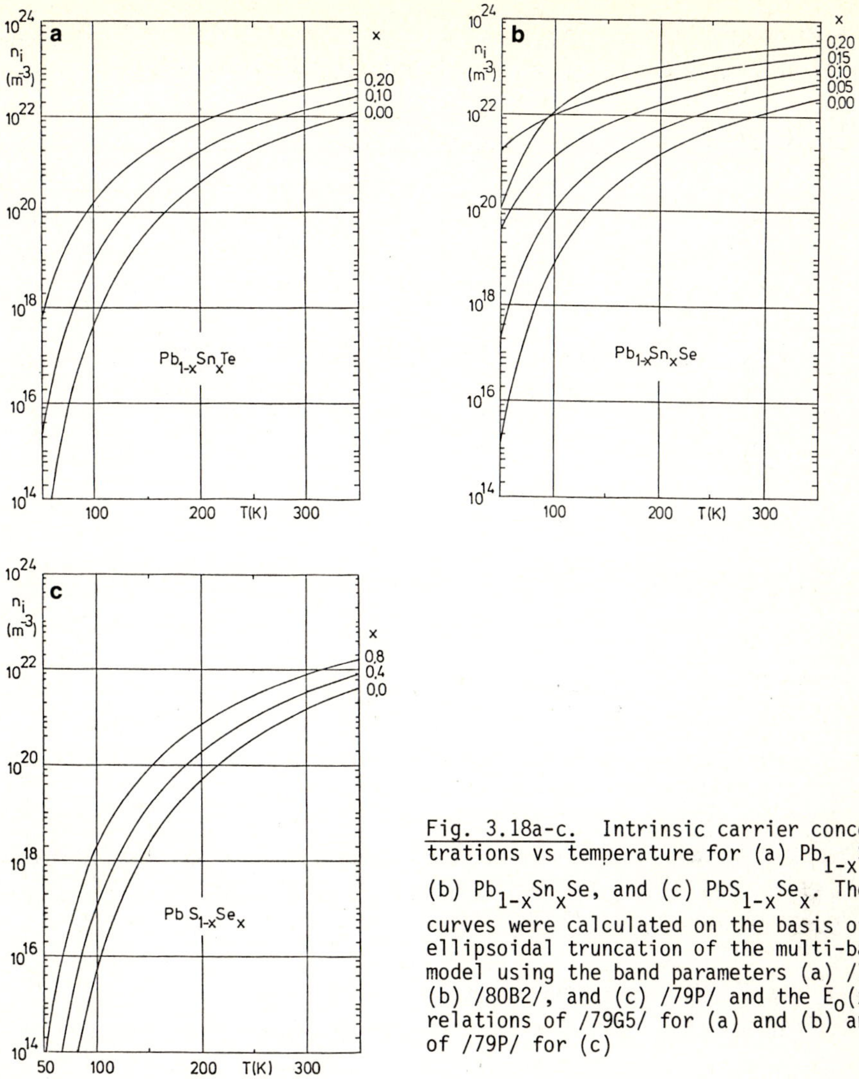

Fig. 3.18a-c. Intrinsic carrier concentrations vs temperature for (a) $Pb_{1-x}Sn_xTe$, (b) $Pb_{1-x}Sn_xSe$, and (c) $PbS_{1-x}Se_x$. The curves were calculated on the basis of the ellipsoidal truncation of the multi-band model using the band parameters (a) /79A1/, (b) /80B2/, and (c) /79P/ and the $E_0(x,T)$ relations of /79G5/ for (a) and (b) and of /79P/ for (c)

who used relation (3.16), but neglected the asymmetry between $\tilde{g}^c(z)$ and $\tilde{g}^v(z)$. As can be seem from this comparison, neglecting terms $\Delta^{c,v}_{\ell,t}$ has very little effect on n_i for the lead-rich compositions of the compounds. On the other hand, the n_i values depend on the choice of band parameters. In the case of PbTe the parameter sets I /71D3/ and II /73H3/ yield n_i values which are about 25% higher than those displayed in Fig. 3.18a. Parameter set IV /77F2/ yields n_i values which are somewhat smaller than those of Fig. 3.18a.

For the tin-rich compounds with $E_0 < 0$ (3.10) are not generally valid. Since in this case the band-edge masses are given by

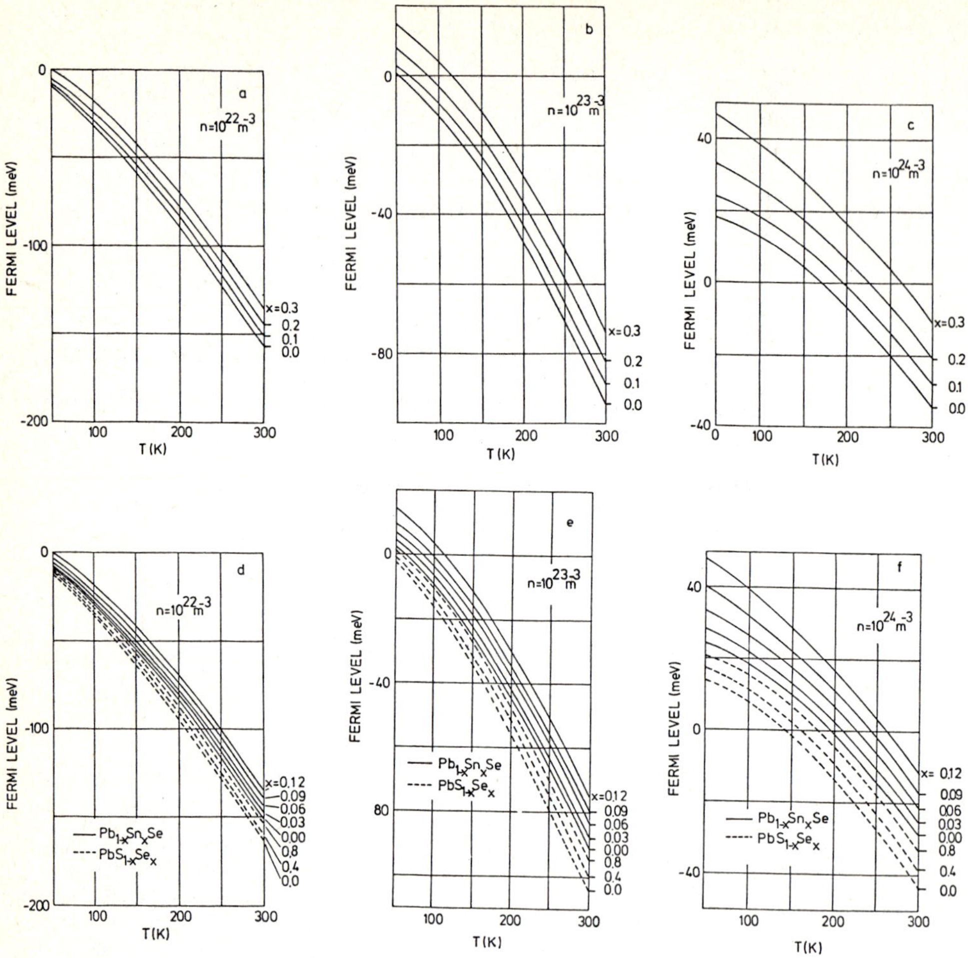

Fig. 3.19a-f. Calculated temperature dependence of the Fermi level of $Pb_{1-x}Sn_xTe$ (a-c) and $Pb_{1-x}Sn_xSe$ / $PbS_{1-x}Se_x$ (d-f) for different electron concentrations: (a,d) $n = 10^{22}\ m^{-3}$, (b,e) $n = 10^{23}\ m^{-3}$, (c,f) $n = 10^{24}\ m^{-3}$ /80G1/

$$\frac{m_0}{m_{t,\ell}^{c,v}} = \frac{2P_{t,\ell}^2}{m_0|E_0|} - \frac{m_0}{m_{t,\ell}^{+,-}} , \qquad (3.17)$$

it may happen that some masses become negative, the surfaces of constant energy being no longer ellipsoidal.

Figures 3.19a-f show Fermi energies as functions of temperature for the three lead compounds for three electron concentrations /80G1/, calculated from the two-band modification of (3.12) and using the parameters of /79P/.

3.3.3 Effective Masses

3.3.3.1 $Pb_{1-x}Sn_xTe$

As mentioned previously, the energy bands in the lead salts are non-parabolic, resulting in effective masses which are functions of the carrier concentration. This fact has been confirmed by several experiments. The hole masses in PbTe have been studied especially thoroughly. Figure 3.20 shows the <111> cyclotron hole mass (which, if the constant-energy surface is an ellipsoid of revolution, equals the transverse effective mass m_t^v) determined from SdH experiments /78J2/ plotted as a function of hole concentration. The transverse hole mass increases as the carrier concentration increases, imaging the non-parabolicity of the valence band of PbTe.

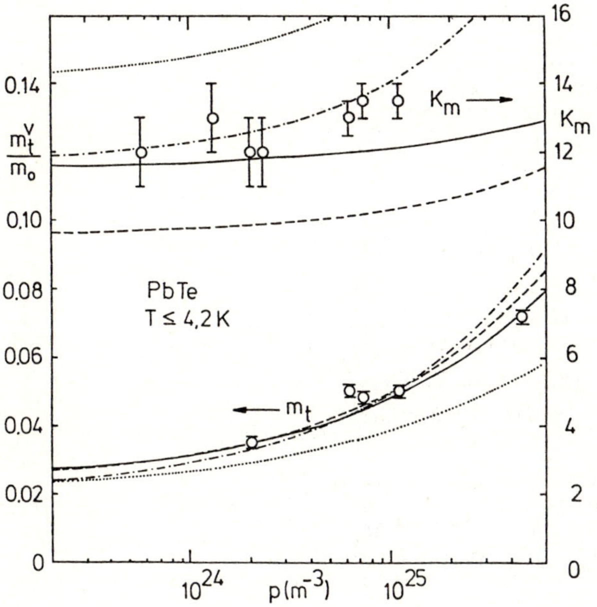

Fig. 3.20. Anisotropy K_m and cyclotron mass m_t^v vs hole concentration for PbTe at $T \leq 4.2$ K. Experimental data were taken from /78J2/. The curves were calculated according to (3.18,19) using different sets of band parameters: solid lines /73H3/, dashed lines /79A1/, dotted lines /71D3/, and chain-dotted lines /77F2/

These recent experiments /78J2/ showed that the anisotropy factor K_m, which is defined as the square of the maximum-to-minimum cyclotron cross-section ratio

$$K_m = \left(\frac{A_{max}}{A_{min}}\right)^2 = \frac{m_\ell}{m_t}\,\frac{1 + \Delta_t \beta z}{1 + \Delta_\ell \beta z} \quad , \qquad (3.18)$$

is practically independent of carrier concentration with a value of 13 ± 1. This result is in contrast to previous investigations /64C2, 71T2/ which yielded K_m decreasing from about 13 for $p = 4 \cdot 10^{23}$ m^{-3} to about 6 for $p = 4 \cdot 10^{24}$ m^{-3}. The very large value of the mass anisotropy in PbTe can be understood theoretically on the basis of the band level ordering discussed in Sect. 3.1. In Fig. 3.20 values of m_t^v and K_m calculated by utilizing the ellipsoidal truncation of the multi-band model with different sets of band parameters (Table 3.5) are also shown:

$$m_t^{cyc} = m_t(k=0)\,\frac{1 + 2\beta\eta + \Delta_t(\beta\eta)^2}{(1 + \Delta_t\,\beta\eta)^2} \quad . \tag{3.19}$$

Experimental data of effective masses and mass anisotropy for n-type and p-type PbTe obtained prior to /78J2/ have been analysed extensively on the basis of the multi-band model by FOLEY and LANGENBERG /77F2/. Measured low-temperature data of the band-edge masses of PbTe are given in Table 3.6.

Table 3.6. Low temperature ($T \leqslant 8$ K) band edge masses of the binary compounds

Compound	m_t^c/m_o	m_t^v/m_o	m_ℓ^c/m_o	m_ℓ^v/m_o	Ref.
PbTe	0.024	0.022	0.24	0.31	64C2
		0.026			77R1
	0.0202		0.214		81I
PbSe	0.040	0.034	0.07	0.068	64C2
PbS	0.080	0.075	0.105	0.105	64C2

The temperature dependence of the band-edge masses of the mixed crystal $Pb_{1-x}Sn_xTe$ calculated according to the relations given in Table 3.4 and using the band parameters of HEWES et al. /73H3/ and the $E_0(x,T)$ relation of GRISAR /79G5/ is displayed in Figs. 3.21a,b. Experimental values of effective masses have been reviewed by HEWES et al. /73H3/. Several further effective mass investigations have been carried out. Without attempting to be exhaustive we mention the measurements of Faraday rotation /77S3/, magneto-photoluminescence /78G3/, thermoelectric power oscillations /81D/, and SdH effect /77A3 , 77D2/. The pressure dependence of the cyclotron mass has been measured by AKIMOV et al. /78A2/ on $Pb_{1-x}Sn_xTe$ ($x = 0.18$ and 0.23) with respect to the transition to zero-gap state.

From Hall-effect measurements as a function of temperature the effective mass of holes in the second valence band (Σ) extrema has been determined. The values given in literature range from $0.6\ m_o$ to $1.4\ m_o$ /65D , 66A1 , 66A2 , 67A1/. SITTER et al. /77S1/ reported a heavy hole mass anisotropy ratio of approximately 10. This value derived from magnetoresistance experiments is based on the assumption that polar optical scattering is dominant.

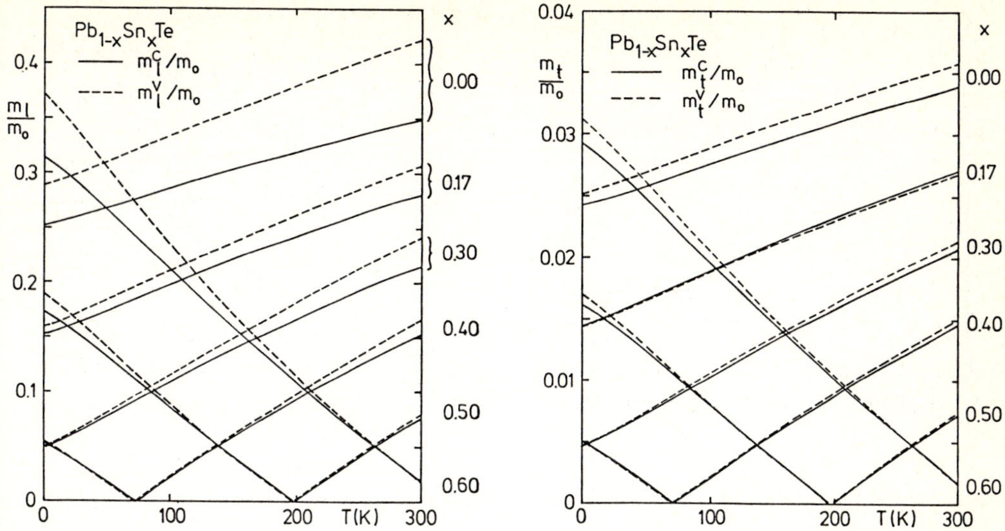

Fig. 3.21a,b. Transverse (a) and longitudinal (b) effective band edge masses vs temperature for various alloy compositions of $Pb_{1-x}Sn_xTe$. The curves were calculated using the band parameters of /73H3/ and the $E_0(x,T)$ relation of /79G5/

3.3.3.2 $Pb_{1-x}Sn_xSe$

Experimental values of effective masses have been determined from measurements of four transport coefficients /80G2/, the thermoelectric power in magnetic fields /78K1, 80F/, SdH oscillations /72M3, 77K3, 78B2/, magnetoplasma cyclotron absorption /67B/, Burstein-Moss effect /79L6/, magneto-photoluminescence /79G4/, Faraday rotation /79S2/, IR reflectivity and magnetoreflectivity /72M4/, and plasma reflection /77K2/. The dependence of the effective mass on hydrostatic pressure below and above the band-inversion point has also been studied /78M/. Values of the band-edge masses of PbSe obtained from SdH measurements at 4 K /64C2/ are given in Table 3.6.

The temperature dependence of the band edge masses calculated according to /80B2/ and /79G5/ (Table 3.5) is plotted in Figs. 3.22a,b.

The mass anisotropy in this mixed crystal is much smaller than in $Pb_{1-x}Sn_xTe$. For PbSe $K_m^c \approx 1.7$ and $K_m^v \approx 1.9 - 2.0$ /64C2, 72K2/ have been reported. These values decrease with increasing tin content, reaching approximately $K_m^{c,v} \approx 1.2$ for $x = 0.2$ /77V2/.

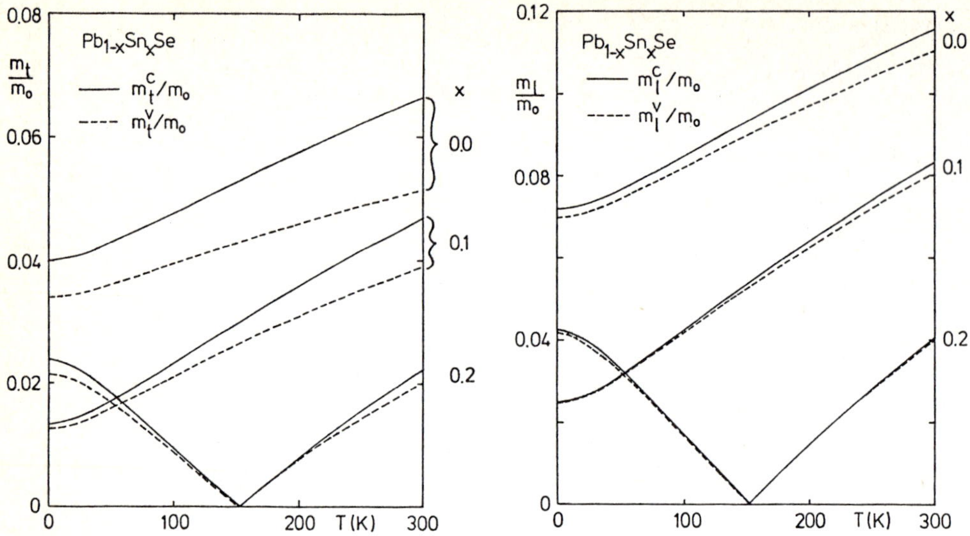

Fig. 3.22a,b. Transverse (a) and longitudinal (b) effective band-edge masses vs temperature for three alloy compositions of $Pb_{1-x}Sn_xSe$. The curves were calculated using the band parameters of /80B2/ and the $E_0(x,T)$ relations of /79G5/

3.3.3.3 $PbS_{1-x}Se_x$

Experimental low-temperature values of the band-edge effective masses of PbS /64C2/ are given in Table 3.6. VEIS et al. /79V5/ investigated the dependence of the dielectric susceptibility effective mass on the hole density at 300 K and found that it could be described by the two-band Kane model with the parameters $m_t = (0.156 \pm 0.014)m_0$ and $E_0 = 0.41$ eV. The density-of-states effective mass was determined from Seebeck, Hall, and transverse Nernst-Ettingshausen coefficient, and was found to be $(0.44 \pm 0.04)m_0$ at 300 K and $(0.23 \pm 0.03)m_0$ at 120 K. The temperature and composition dependence of the effective masses of the mixed crystal can be calculated according to /79P/ (Table 3.5). Studies of the valence band structure of $PbSe_{0.5}S_{0.5}$ /79V4/ confirmed that the valence band parameters of the mixed crystal are close to the arithmetic means of the values reported for the binary compounds.

3.3.4 Energy Levels in a Magnetic Field; g Factors

Equation (3.2) can be solved exactly for the eigenenergies in the case that the magnetic field is in the <111> direction by relating the operators $k_x \pm i k_y$ to the raising and lowering operators of harmonic oscillator functions /66M2 , 71D3/. The procedure has been outlined by MITCHELL and WALLIS /66M2/. The resulting valence-conduction-band Hamiltonian in renormalized form is given by the matrix

$$
\begin{array}{c|cccc}
 & C^{1/2}F_n & V^{-1/2}F_{n+1} & C^{-1/2}F_{n+1} & V^{1/2}F_n \\
\hline
C^{1/2}F_n & a_n & e_n & 0 & f \\
V^{-1/2}F_{n+1} & e_n & b_n & -f & 0 \\
C^{-1/2}F_{n+1} & 0 & -f & c_n & e_n \\
V^{1/2}F_n & f & 0 & e_n & d_n
\end{array}
$$

with

$$a_n = \frac{E_0}{2} + (n+\frac{1}{2})\hbar\omega_\ell^- + \frac{1}{2} g_\ell^- \mu_B B$$

$$b_n = -\frac{E_0}{2} - (n+\frac{3}{2})\hbar\omega_\ell^+ + \frac{1}{2} g_\ell^+ \mu_B B$$

$$c_n = \frac{E_0}{2} + (n+\frac{3}{2})\hbar\omega_\ell^- - \frac{1}{2} g_\ell^- \mu_B B$$

$$d_n = -\frac{E_0}{2} - (n+\frac{1}{2})\hbar\omega_\ell^+ - \frac{1}{2} g_\ell^+ \mu_B B$$

$$e_n = \left[4\frac{P_t^2}{m_0}(n+1)\mu_B B\right]^{1/2}$$

$$f = \frac{\hbar}{m_0} P_\ell k_z$$

with $\hbar\omega_\ell^\pm = 2\frac{m_0}{m_t^\pm}\mu_B B$.

Here $C^{\pm 1/2}$ and $V^{\pm 1/2}$ represent the renormalized conduction and valence-band functions /66M2/, F_n are the harmonic-oscillator functions. The Landau energies at $k_z = 0$ are obtained by diagonalising the two decoupled 2×2 matrices:

$$
\begin{aligned}
E^{c,n,+}_{v,n,-} &= \frac{1}{2}(a_n+b_n) \pm \frac{1}{2} E_0/|E_0| \cdot \left[(a_n-b_n)^2 + 4e_n^2\right]^{1/2} \\
E^{c,n,-}_{v,n,+} &= \frac{1}{2}(c_n+d_n) \pm \frac{1}{2} E_0/|E_0| \cdot \left[(c_n-d_n)^2 + 4e_n^2\right]^{1/2} \quad .
\end{aligned}
\tag{3.20}
$$

(In this notation the n-th |-> state corresponds to the (n + 1)-st |+> state.) The highest valence-band Landau level $E_{v,-1,-} = b_{-1}$ and the lowest conduction-band Landau level $E_{c,-1,-} = c_{-1}$ are linear functions of B and depend only on the far-band contributions. In a two-band model (without the far-band contributions) the spin splitting would be exactly equal to the orbital splitting. The interaction with the four distant levels removes this degeneracy.

For an arbitrary direction of $\underline{B}$ with respect to the <111> axis, (3.2) cannot be solved exactly. Three approximate solutions have been reviewed in /80B2/. Here we cite the result of DIMMOCK /71D3/ only. Provided that the contributions due to interaction with the more distant bands are small compared to the direct valence-conduction band interactions, the Landau levels are given by (3.20) making the following replacements:

$$g_\ell^\pm \to \bar{g}^\pm = (P_\ell g_t^\pm \sin^2\theta + P_t g_\ell^\pm \cos^2\theta)/P_\theta$$

$$\omega_\ell^\pm \to \bar{\omega}^\pm = \left[\frac{1}{2}(P_\ell^2 \omega_\ell^{\pm 2} + P_t^2 \omega_t^{\pm 2})\sin^2\theta + P_t^2 \omega_\ell^{\pm 2}\cos^2\theta\right]/(\omega_\ell^\pm P_t P_\theta) \qquad (3.21)$$

$$e_n^2 \to \bar{e}_n^2 = 4\,(n+1)\,\mu_B B\, P_t P_\theta/m_0 \quad ,$$

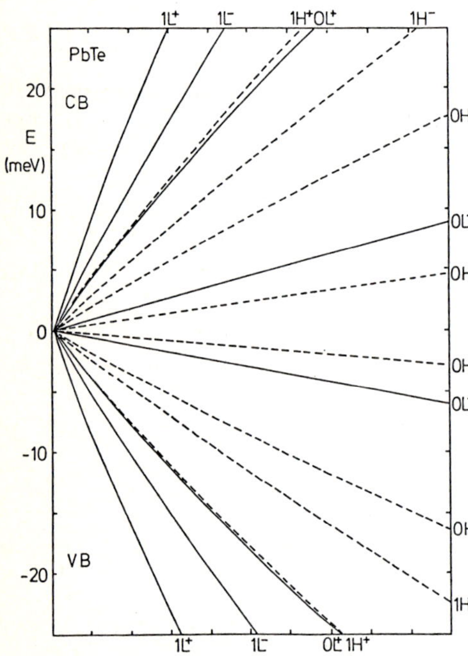

Fig. 3.23. Energies of the lowest Landau levels vs magnetic field for PbTe (B ∥ <111>). L denotes light carriers ($\theta = 0^0$), H represents heavy carriers ($\theta = 70.53^0$). The curves were calculated according to (3.20,21) using the band parameters of /79A1/. Zero in energy denotes the respective band edge (E_0 = 189 meV)

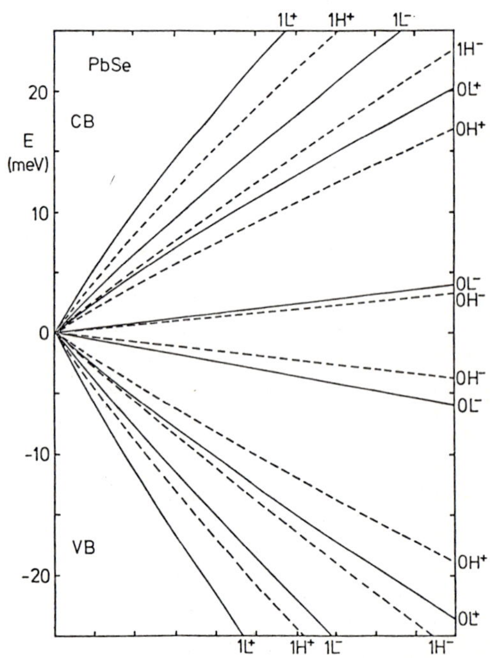

Fig. 3.24. Energies of the lowest Landau levels vs magnetic field for PbSe (B ∥ <111>). L denotes light carriers ($\theta = 0^0$), H represents heavy carriers ($\theta = 70.53^0$). The curves were calculated according to (3.20,21) using the band parameters of /80B2/. Zero in energy denotes the respective band edge (E_0 = 145 meV)

where $\omega_t^{\pm} = eB/(m_t^{\pm} m_\ell^{\pm})^{1/2}$, $\theta = \sphericalangle(\underline{B}, \langle 111\rangle)$, and $P_\theta = (P_\ell^2 \sin^2\theta + P_t^2 \cos^2\theta)^{1/2}$.

Landau energies calculated according to (3.20,21) ($\underline{B} \parallel \langle 111\rangle$) for the light (L) carriers ($\theta = 0^0$) in the <111> ellipsoid and the heavy (H) carriers ($\theta = 70.53^0$) in the oblique $\langle\bar{1}11\rangle$, $\langle 1\bar{1}1\rangle$, and $\langle\bar{1}\bar{1}1\rangle$ ellipsoids are plotted for PbTe and PbSe in Figs. 3.23,24. The band parameters used were those given in /79A1/ (PbTe) and /80B2/ (PbSe). (The Landau quantum numbers for the $|-\rangle$ states have been raised by one.) As has been pointed out by MITCHELL and WALLIS /66M2/ and DIMMOCK /71D3/, the Landau-level spacing is larger than the spin splitting if $E_0 > 0$ while the reverse is true if $E_0 < 0$. This result is independent of the magnetic field direction /71D3/. Expressions for the effective masses and g factors can be obtained from (3.20,21) in the low field limit in terms of the Landau level spacing $\Delta E(n,\pm)/\Delta n$ and the spin splitting $E(n,+) - E(n-1,-)$. In the limit $B \to 0$, the g factors at the band edges are given by the expressions of Table 3.4. Calculated band-edge g factors for $Pb_{1-x}Sn_xTe$ as a function of temperature are plotted in Figs. 3.25a,b. Due to the non-parabolicity of the bands the g factor decreases with both the magnetic field and the carrier density.

Experimental values of the g factors in $Pb_{1-x}Sn_xTe$ have been reviewed by HEWES et al. /73H3/. More recent results for PbTe obtained by SCHABER and DOEZEMA /79S1/ from FIR spin-flip resonance measurements are given in Table 3.7 together with some

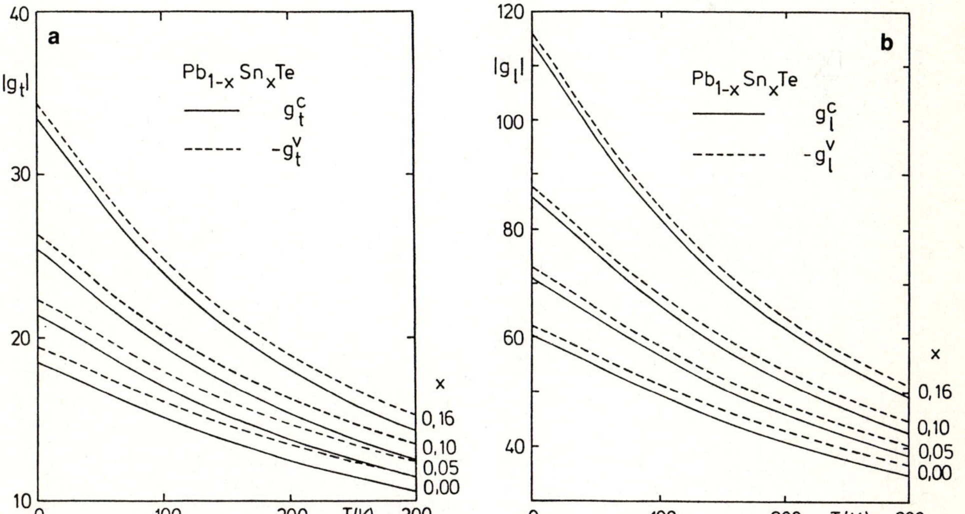

Fig. 3.25a,b. Calculated transverse (a) and longitudinal (b) g factor at conduction band and valence band edge as a function of temperature for different alloy compositions of $Pb_{1-x}Sn_xTe$. The band parameters of /79A1/ and the $E_0(x,T)$ relation of /79G5/ have been used

Table 3.7. g Factors of the binary compounds ($T \leq 4.2$ K)

Compound	n,p (m^{-3})	g_ℓ^c	g_ℓ^v	g_t^c	g_t^v	Method	Ref.
PbTe	band edge	45 ± 8	51 ± 8			SdH effect	64C2
	$5 \cdot 10^{22}$	59.5 ± 0.7	58.3 ± 1.0	15.3 ± 1.5	18.8 ± 1.5	FIR spin flip resonance	79S1
PbSe	band edge	27 ± 7	32 ± 7			SdH effect	64C2
	$6 \cdot 10^{23}$			19.6		magnetoacoustic quantum oscillations	72K2 , 74B1
	$1.3 \cdot 10^{24}$			17.6			
	$1.2 \cdot 10^{24}$				17.1		
PbS	band edge	12 ± 3	13 ± 3			SdH effect	64C2

values for PbSe obtained from investigations of magnetoacoustic quantum oscillations /74B1/. Results for the mixed crystal $Pb_{1-x}Sn_xTe$ can e.g. be found in /78G3, 79Y2/, for $Pb_{1-x}Sn_xSe$ in /79G4/. The dependence of the (near band edge) g factor along the <100> direction on the band gap E_0 determined from magneto-photoluminescence spectra /78G3 , 79G4/ is displayed in Figs. 3.26,27. The values were obtained assuming mirror symmetry of the bands ($|g^c| = |g^v|$).

For the magnetic field dependence of the low-temperature effective g factors of the conduction band of PbTe epitaxial films see Fig. 4.21 (Section 4.3.4).

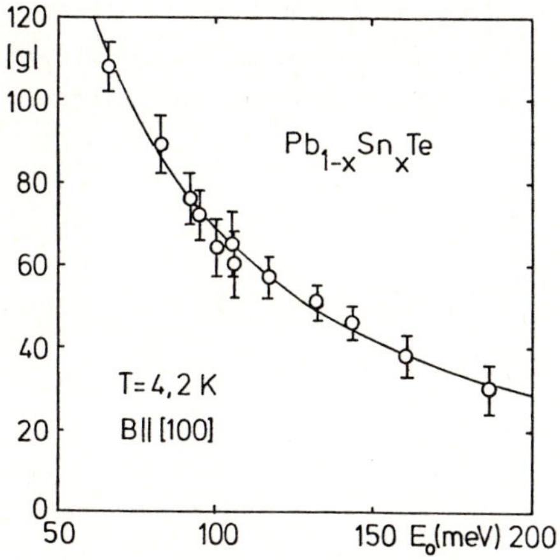

Fig. 3.26. Near-band-edge g factor along the <100> direction for $Pb_{1-x}Sn_xTe$ as a function of gap energy. Experimental values were obtained from magneto-photoluminescence spectra /78G3/ (B ∥ <100>, T = 4.2 K). The solid line is a least-squares fit to the data: $g = 8.1/E_0\,(eV) - 12$

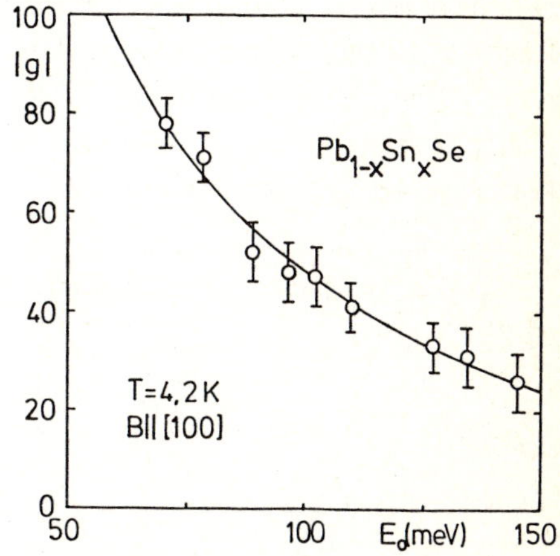

Fig. 3.27. Near-band-edge g factor along the <100> direction for $Pb_{1-x}Sn_xSe$ as a function of gap energy. Experimental data were obtained from magneto-photoluminescence spectra /79G4/ (B ∥ <100>, T = 4.2 K). The solid line is a least-squares fit to the data: $g = 7.1/E_0\,(eV) - 23$

4. Transport Properties

This chapter deals with various transport properties of the lead chalcogenides and their alloys. Data on band-edge characteristics such as gap energy, effective masses, g factors, etc. determined from transport measurements have already been included in the preceding chapter and will not be repeated here.

4.1 Hall Coefficient, Carrier Concentration

Calculated values of the intrinsic carrier density of the alloys have already been presented in Sect. 3.3 (see Figs. 3.18). The measured temperature dependence of the intrinsic carrier density of $Pb_{0.82}Sn_{0.18}Te$ is plotted in Fig. 4.1 /75 O/. (For $Pb_{0.72}Sn_{0.28}Te$ see /77K1/.) As has been discussed in Sect. 2.3 the main source of extrinsic carriers is deviations from stoichiometry. The energy levels associated with these defects are resonant with either the valence band or the conduction band. Hence a freeze-out of carriers cannot take place, the lead salts and their alloys behave like metals at low temperatures (see Figs. 2.24 , 4.2). The type and the concentration of defects can be controlled by the applied conditions of crystal growth and annealing. As an example Fig. 4.3 shows the hole concentration of bulk crystals and liquid phase epitaxial layers of $Pb_{0.8}Sn_{0.2}Te$ as a function of the reciprocal annealing or growth temperature /77P/.

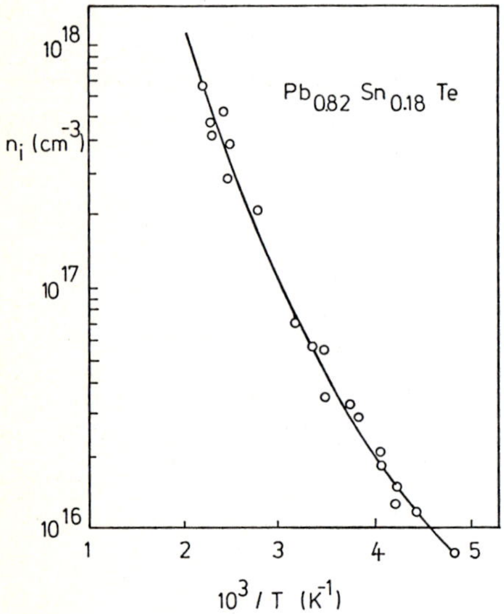

Fig. 4.1. $Pb_{0.82}Sn_{0.18}Te$. Temperature dependence of the intrinsic carrier density /75 O/

Some uncertainties exist concerning the degree of ionization of the defects. From theoretical considerations /69P , 73P , 75H3/ one expects metal vacancies to act as doubly ionized acceptors and chalcogen vacancies as doubly ionized donors. Ion implantation experiments (for PbS /56B/, for PbSe /63I/, and for PbTe /63S , 66F , 70L/) have been carried out to obtain information about interstitials. In the case of $Pb_{1-x}Sn_xTe$ metal interstitials seem to act as singly ionized donors whereas Te interstitials seem to be neutral /80G3/. KUCHERENKO et al. /81K1/ interpreted their transport measurements on $Pb_{1-x}Sn_xSe$ ($0.31 \leq x \leq 0.37$) by invoking a model of two singly charged levels of non- stoichiometric defects.

Fig. 4.2

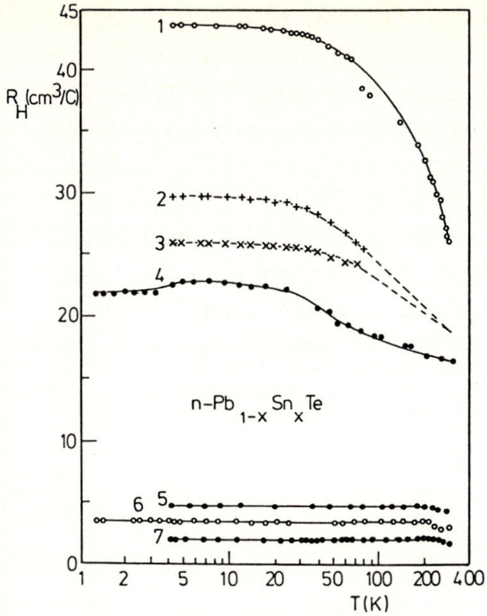

Fig. 4.3

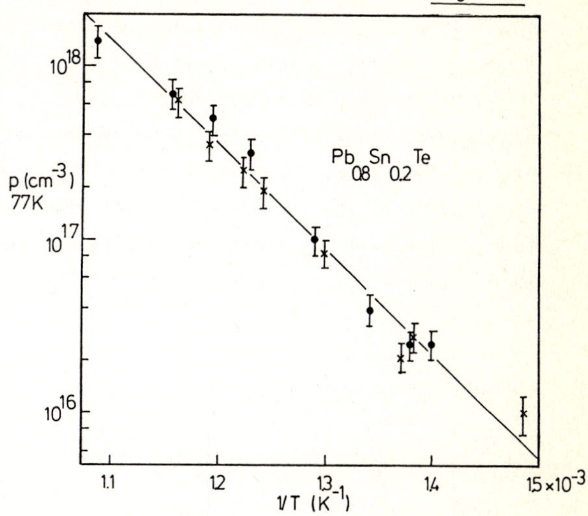

Fig. 4.2. $Pb_{1-x}Sn_xTe$. Temperature dependence of the Hall coefficients measured at B = 0.1 T for various samples (data taken at T = 4.2 K) /77T2/:

	x	n	μ_H
(1)	x = 0.13 ± 0.015 ,	n = $1.4 \cdot 10^{17}$ cm^{-3} ,	μ_H = $7.8 \cdot 10^5$ cm^2/(Vs)
(2)	0.09 ± 0.005	2.0	6.1
(3)	0.09 ± 0.005	2.4	5.4
(4)	0.12 ± 0.01	2.7	5.2
(5)	0.20 ± 0.02	10	1.5
(6)	0.16 ± 0.0015	16	1.2
(7)	0.15 ± 0.015	28	0.90

Fig. 4.3. $Pb_{0.8}Sn_{0.2}Te$. Hole concentration measured at 77 K as a function of the reciprocal annealing or growth temperature for annealed bulk crystals (×) and isothermal LPE layers (•) /77P/

The effect of electron irradiation on the electrical properties of p-type $Pb_{1-x}Sn_xTe$ ($x \gtrsim 0.2$) has been reported in /74H2/.

Hall effect measurements have also been carried out to find the energy levels associated with chemical impurities, see Sect. 2.3.2 for results.

Several investigations have been done on p-type material to get information about the valence band structure:

for $Pb_{1-x}Sn_xTe$ /67A1 , 68C2 , 74 0 , 75 0 , 78A1/ ,

for $Pb_{1-x}Sn_xSe$ /75V , 78V2/ ,

for $PbSe_{0.5}S_{0.5}$ /79V4/ .

From these investigations it became evident that in the lead salts a second set of valence band maxima exists with a higher effective hole mass and a lower mobility, separated from the main extrema by a few tenths of an eV and possibly located along the Σ direction of the Brillouin zone. Some parameters of this second set of extrema are given in Table 3.3 (see Sect. 3.2)

Investigations on p-type $Pb_{1-x}Sn_xTe$ ($0.18 \leq x \leq 0.26$) /75Z, 78A2/ and n-type $Pb_{1-x}Sn_xSe$ ($x = 0.125$) /78B1/ have shown that the Hall coefficient is practically pressure independent at liquid helium temperatures. The experimentally observed minimum in the pressure dependence of the resistivity (see next section) is therefore entirely due to a maximum in mobility as a result of band inversion under pressure.

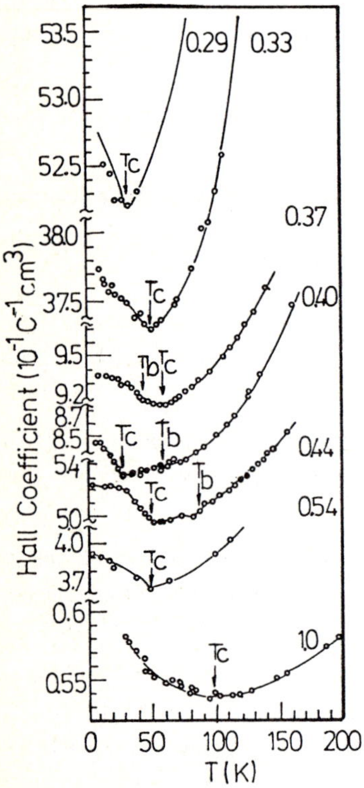

Fig. 4.4

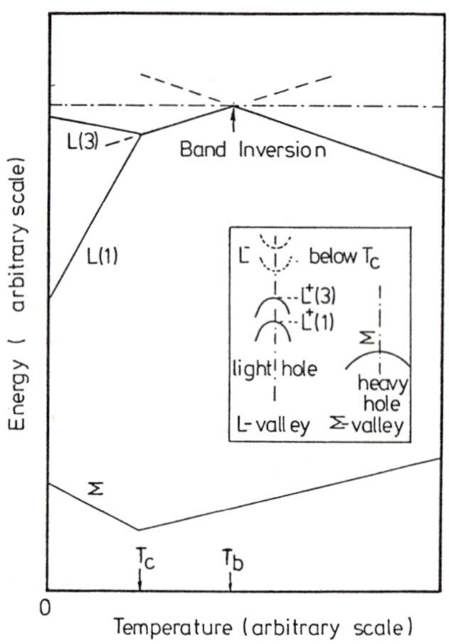

Fig. 4.5

Fig. 4.4. $Pb_{1-x}Sn_xTe$. Temperature dependence of the Hall coefficient for various samples with different alloy compositions /79K2/. T_c denotes the temperature of the structural phase transition, T_b the band inversion temperature

Fig. 4.5. $Pb_{1-x}Sn_xTe$. Schematic temperature dependence of the band-edge structure /79K2/. T_c and T_b have the same meaning as in Fig. 4.4

No magnetic freeze-out of carriers has been observed in the lead salts (see Fig. 2.18).

The observation of weak field oscillations in the Hall coefficient of $Pb_{0.82}Sn_{0.18}Te$ has been reported by LASHKAREV et al. /76L, 77L5/. These oscillations observed in the field range between 1 and 100 mT are attributed to the presence of carriers with very low effective masses of the order of $10^{-4}m_0$.

KOBAYASHI et al. /79K2/ have carried out galvanomagnetic measurements with respect to the structural phase transition in $Pb_{1-x}Sn_xTe$ ($0.29 \leq x \leq 1.0$). Their Hall effect results, shown in Fig. 4.4, have been interpreted on the basis of the band structure model sketched in Fig. 4.5. According to this model the increase of the Hall coefficient above and below the transition temperature T_c is caused by a redistribution of carriers between the main L valleys and the subsidiary Σ valleys. However, the data given in /79K2/ are not sufficient to allow the conclusions to be checked. For resistivity measurements with respect to the structural phase transition see the following section.

4.2 Carrier Mobility, Scattering Mechanisms

Much effort has been directed towards the understanding of the dominant scattering mechanisms in the lead chalcogenides (see e.g. the review articles by RAVICH et al. /71R/ and ZAWADZKI /74Z2/, and publications /58A1, 64K, 71K, 73L3, 73S5, 82P2/). At low temperatures the mobility seems to be dominated by ionized-defect scattering /73L3, 82P2/. Due to the large dielectric permittivities of the lead salts the carrier scattering by the Coulomb potentials of ionized impurities is relatively weak, which results in very high mobilities at low temperatures. The inefficiency of the Coulomb scattering makes the scattering by the (short-range) ion core potentials in samples with high carrier concentrations substantial. RAVICH et al. /71R/ calculated that above carrier concentrations of $0.8\cdot10^{19}$ cm^{-3} in PbTe, $2\cdot10^{19}$ cm^{-3} in PbSe, and $6\cdot10^{19}$ cm^{-3} in PbS the scattering by the ion core is the dominant mechanism at 4 K. In very pure PbTe crystals with carrier concentrations $\lesssim 10^{17}$ cm^{-3} unusually high mobilities up to $4\cdot10^{6}$ cm^2/Vs /82P2/ (even higher /82T/) have been observed at liquid helium temperatures (Fig. 4.6). To explain these results PALMETSHOFER et al. /82P2/ used a model potential which seems to be more adequate to describe the intrinsic defects of the lead salts than the usual screened Coulomb potential (Fig. 4.7a).

At higher temperatures the important scattering mechanisms are the interactions between carriers and acoustical and optical phonons. Typically the mobility varies with temperature as $T^{-\nu}$ at higher temperatures, where ν is approximately 2.5 for pure crystals. According to RAVICH et al. /71R/ analysis of the temperature and carrier concentration dependence of the mobility has to account for the following effects:

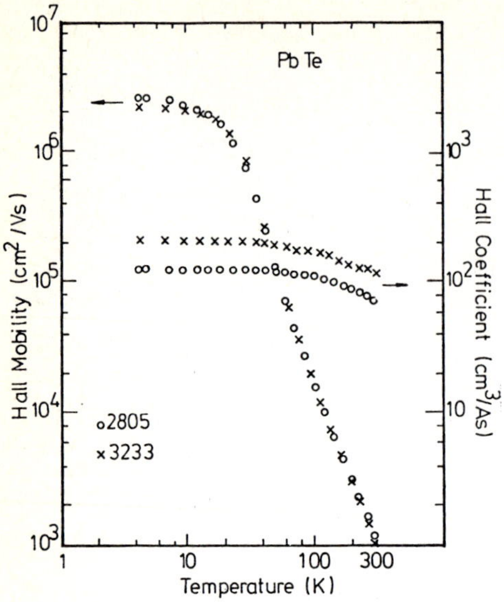

Fig. 4.6. PbTe. Temperature dependence of the Hall mobility and the Hall coefficient for two epitaxial films /82P2/

1. Collisions between carriers; they redistribute momenta and energies of different carriers, thus affecting other scattering modes. Even under degenerate conditions they are of importance for thermal effects.

2. Polar scattering by optical phonons; since in the lead salts the static dielectric constant is much larger than the optical one, one has to deal with a pronounced dispersion of the optical phonon branch due to screening effects of free carriers.

3. The energy dependence of the matrix element of carrier interaction with acoustical phonons in the non-parabolic energy band. The non-parabolicity of the energy bands manifests itself not only via the energy-dependent effective mass and hence the density of states but also through the electronic wave-function admixture and hence the scattering probability.

Figures 4.7b,c display the electron concentration dependence of the mobility of PbTe at 77 K and room temperature, respectively. Theoretical curves based on the considerations mentioned above are also shown. As can be seen from these figures polar optical phonon scattering determines the mobility in samples with low carrier concentrations at both temperatures. At higher carrier concentrations ($\gtrsim 5\cdot10^{18}$ cm^{-3}) acoustical phonon scattering becomes dominant. RAVICH et al. /71R/ considered impurity scattering to be negligible in the lead salts at room temperature and to be of importance at 77 K only for carrier concentrations $\gtrsim 10^{20}$ cm^{-3}. DAS and NAG /78D4/ have re-examined experimental data of the mobility of PbTe at 77 and 300 K taking into account the anisotropy of the band structure.

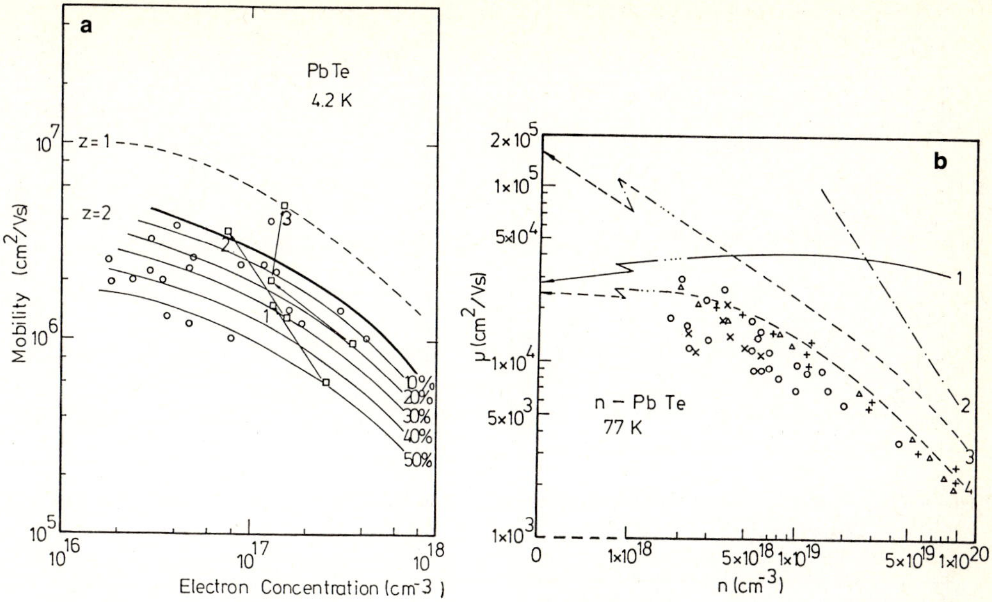

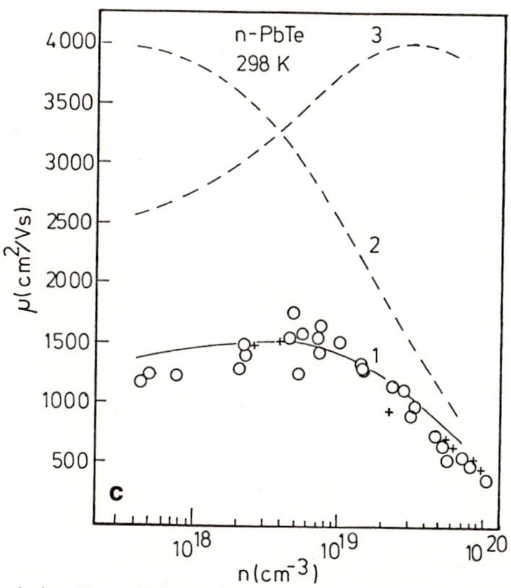

Fig. 4.7a-c. n-PbTe. Electron concentration dependence of the mobility for three different temperatures:
(a) T = 4.2 K. The curves are calculated for ionized defect scattering on singly (z = 1) and doubly (z = 2) ionized defects; the percentages denote compensation. Full dots represent experimental data of samples grown by the hot-wall technique, open squares indicate high mobility bulk samples. The arrows give the change due to annealing (several weeks at 400° C) /82P2/
(b) T = 77 K. The curves are calculated for (1) polar optical, (2) charged impurity, (3) acoustic, (4) mixed scattering mode. The arrows indicate theoretical values at low concentrations. Experimental data from various authors /71R/
(c) T = 298 K. The curves are calculated for (3) polar optical, (2) acoustic, (1) mixed scattering mode. Experimental data from various authors /71R/

Figures 4.8 and 4.9 display the temperature and carrier concentration dependence of the mobility of PbSe /73S5/. The electrical conductivity of PbSe has been measured under strong unaxial pressure by VOLOTSKOI and KAIDANOV /76V5/.

Several authors have investigated the mobilities in the lead salt based alloys. The system $Pb_{1-x}Sn_xTe$ has been studied especially thoroughly /71B , 74G2 , 74 O , 76G2 , 76S2 , 76S3 , 77B3 , 77I , 77T2 , 77T3 , 79T3 , 81S2/. Most of the considerations

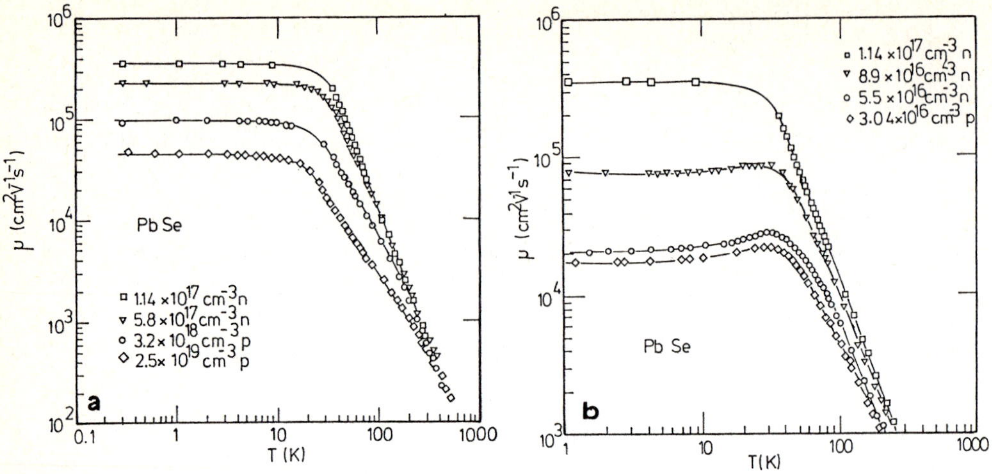

Fig. 4.8a,b. PbSe. Temperature dependence of the mobility: (a) for high carrier concentrations, (b) for low carrier concentrations /73S5/

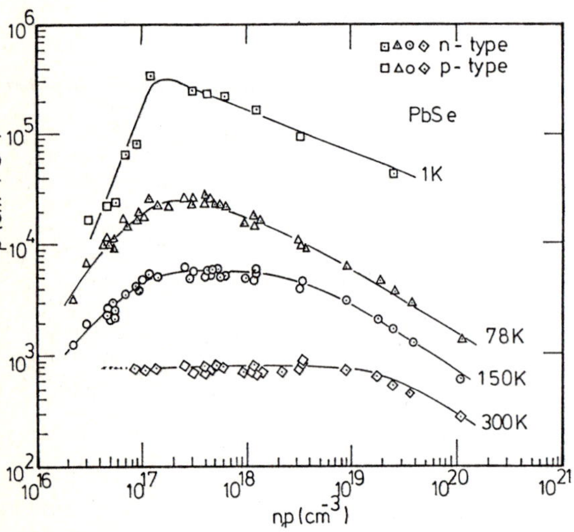

Fig. 4.9. PbSe. Carrier-concentration dependence of the mobility at four different temperatures /73S5/

stated above for the binary alloys are applicable to the mixed crystals, too. However, since the mobility due to acoustical phonon scattering is proportional to the reciprocal of the density of states (which means $\mu_{ac} \propto E_0^{-3/2}$ in the two-band model) its importance is diminished for the small gap mixtures. In these materials scattering by impurity atoms and vacancies is important even at room temperature /74 0, 76S3/.

Another scattering mechanism is reported to be important in the mixed crystals with comparatively low carrier concentrations at low temperatures - the disorder or alloy scattering /76S3, 77T2/ which is due to the chemical disorder of the component atoms. From Fig. 4.10 it can be seen that this type of scattering is dominant in $Pb_{1-x}Sn_xTe$ ($0.17 \leqslant x \leqslant 0.22$) with carrier concentrations below approximately

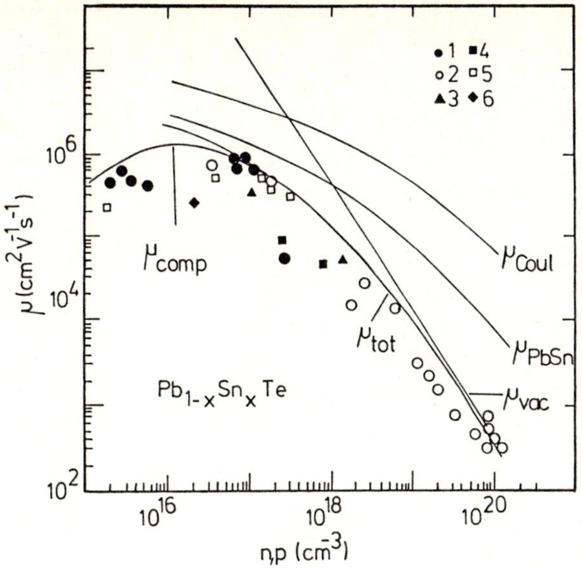

Fig. 4.10. $Pb_{1-x}Sn_xTe$. Carrier-concentration dependence of the mobility at 4.2 K. The curves are calculated for scattering of carriers by the long-range Coulomb part of the vacancy potential (μ_{Coul}) and its internal part (μ_{vac}), and by the difference between the potentials of Pb and Sn (μ_{Pb-Sn})$\cdot\mu_{comp}$ accounts for compensation in samples with low carrier density. Experimental data are taken from various authors: (1) x = 0.18 , n; (2) x = 0.18 , p; (3) x = 0.17 , n; (4) x = 0.216 , p; (5) x = 0.17 - 0.20 , n; (6) x = 0.17 , p /76S3/

$10^{18}cm^{-3}$ at 4.2 K. SIERANSKI and SZATKOWSKI /81S2/ argued that it is necessary to consider both intravalley and intervalley disorder scattering when interpreting the experimentally observed mobilities in order to remain consistent with other experimental findings (i.e. the composition dependence of the gap).

Mobility changes in $Pb_{1-x}Sn_xTe$ ($x \approx 0.2$) due to electron irradiation at 10 K have been found by GREEN /76G2/. The mobility increased in n-type material whereas it decreased in p-type samples.

The temperature dependence of the mobility ratio $b = \mu_n/\mu_p$ of $Pb_{0.82}Sn_{0.18}Te$ has been investigated by ORLETSKII et al. /75 0/ in the temperature range 200 to 450 K. They found

$$b = 0.76 + 6.1\ 10^{-3}\ T/K \qquad \text{for } 220 \leqslant T/K \leqslant 400 \quad .$$

Above 400 K b increases superlinearly /75 0/. The temperature dependence of the mobility ratio of $Pb_{0.72}Sn_{0.28}Te$ has been measured by KONDRATENKO et al. /77K1/ (Fig. 4.11). Measurements of the pressure dependence of the resistivity of p-type $Pb_{1-x}Sn_xTe$ with $x \approx 0.2$ have been carried out at 77 K /75Z/ and at 4.2 K /78A2/. Resistivity vs pressure exhibits a minimum at the point of band inversion. Since the carrier concentration was measured to be practically independent of pressure, the minimum in resistivity can be solely attributed to a change in mobility under pressure. In Fig. 4.12 the pressure dependencies of the relative changes in the Hall mobility $\mu_H(p)/\mu_H(0)$, in the momentum relaxation time $\tau(p)/\tau(0)$, and in the reciprocal of the cyclotron mass $m^*_{<111>}(0)/m^*_{<111>}(p)$ are plotted /78A2/. They all exhibit a maximum at the point of band inversion. AKIMOV et al. /78A2/ concluded from their

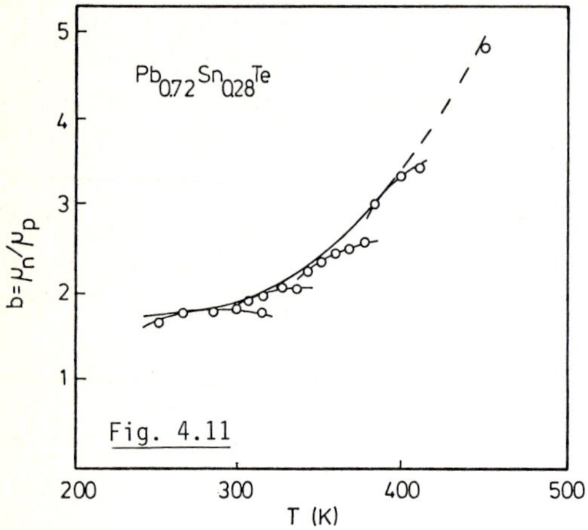

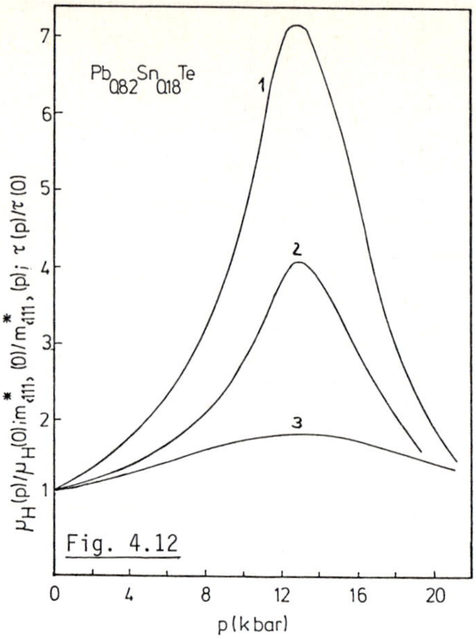

Fig. 4.11. $Pb_{0.72}Sn_{0.28}Te$. Temperature dependence of the mobility ratio $b = \mu_n/\mu_p$ /77K1/

Fig. 4.12. $Pb_{0.82}Sn_{0.18}Te$. Pressure dependencies of the relative changes in (1) the Hall mobility $\mu_H(p)/\mu_H(0)$, (2) the momentum relaxation time $\tau(p)/\tau(0)$, and (3) the reciprocal of the cyclotron mass $m^*_{<111>}(0)/m^*_{<111>}(p)$ at 4.2 K /78A2/

measurements that in $Pb_{1-x}Sn_xTe$ alloys near the transition to the zero-gap state at liquid helium temperatures the dominant hole relaxation mechanism is scattering by charged centres.

Resistance anomalies have been found in $Pb_{1-x}Sn_xTe$ $(0.29 \leqslant x \leqslant 1.0)$ caused by the structural phase transition /75K, 76K2, 79G6, 79K2/. The electrical resistivity shows an anomalous increase near the transition temperature which is interpreted in terms of an increasing carrier scattering by the soft TO phonon (see Sect. 5.4). The temperature dependencies of the resistivity are shown in Fig. 4.13 for various alloy compositions /79K2/.

The mobility of the system $Pb_{1-x}Sn_xSe$ has not been investigated as thoroughly as that of $Pb_{1-x}Sn_xTe$. One does not expect substantial differences between these two alloys since the anisotropy of the effective masses seems to be of little importance for the scattering processes /77T2/. We mention only a few publications on $Pb_{1-x}Sn_xSe$. The pressure dependence of the resistivity has been measured for $x = 0.17$ and 0.20 material by MELNGAILIS et al. /73M4/ and for $x = 0.125$ crystals by BRANDT et al. /78B1/. The temperature dependence of the electrical conductivity

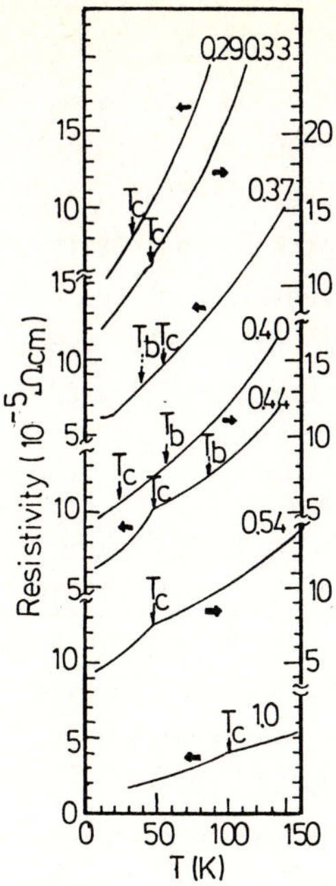

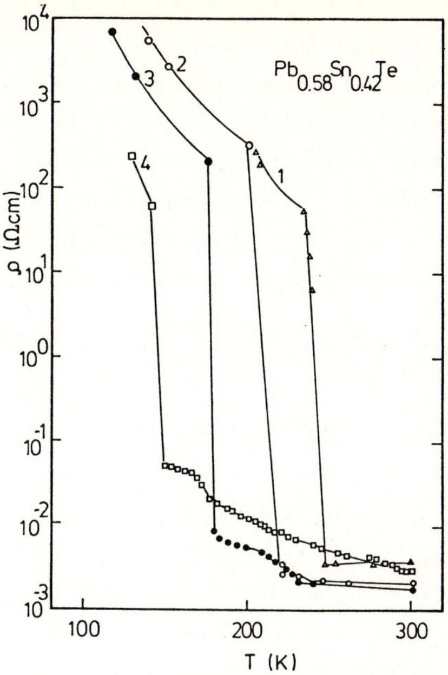

Fig. 4.13. $Pb_{1-x}Sn_xTe$. Temperature dependence of the resistivity for various samples with different alloy compositions /79K2/ (cf. Fig. 4.4)

Fig. 4.14. $Pb_{0.58}Sn_{0.42}Se$. Temperature dependence of the resistivity for four samples with different hole concentrations: (1) $3\cdot10^{18}$ cm^{-3}, (2) $4.2\cdot10^{18}$ cm^{-3}, (3) $5.6\cdot10^{18}$ cm^{-3}, and (4) $7.2\cdot10^{18}$ cm^{-3} /79K3/

of polycrystalline $Pb_{0.65}Sn_{0.35}Se$ has been investigated by VINOGRADOVA et al. /77V3/ up to 900 K.

Some recent investigations have dealt with compositions near the miscibility gap ($x \approx 0.42$) /78V3, 79K3/. Figure 4.14 shows the temperature dependence of the resistivity of four $Pb_{0.58}Sn_{0.42}Se$ samples with different hole concentrations. The anomalous increases in resistivity (4 to 5 orders of magnitude) are attributed to a structural phase transition from the rock salt structure to the orthorhombic structure /78V3, 79K3/.

4.3 Magnetotransport

This section is mainly dedicated to magnetic quantum effects, such as Shubnikov-de Haas effect, magneto-phonon effect, cyclotron resonance, etc. We will, however, start with a short discussion of weak- and strong-field magnetoresistance.

4.3.1 Magnetoresistance

For a many-valley semiconductor galvanomagnetic measurements in a non-quantising magnetic field are of interest because they yield information about the mobility anisotropy $K' = \mu_t/\mu_\ell = K_m/K_\tau$ ($K_m = m_\ell/m_t$, $K_\tau = \tau_\ell/\tau_t$) /70G/, whereas studies of the magnetic quantum effects give directly the effective-mass anisotropy. Weak-field magnetoresistance data offered strong evidence of the <111> orientation of the constant-energy ellipsoids in the lead salts /57S, 58A2, 59S2, 60A, 61A/. Under degenerate conditions the magnetoresistance ratio Y can be expressed as /58A2, 77T2/:

$$Y = \lim_{B\to 0} \Delta\rho_\ell(B)/\Delta\rho_t(B) = \frac{2(2K'+1)}{K'+2} \qquad \text{with } \Delta\rho_\ell(B) = \rho_\ell(B) - \rho_0 \tag{4.1}$$

where $\Delta\rho_\ell(B)/\rho_0$ is the longitudinal magnetoresistance for $\underline{j} \parallel \underline{B} \parallel$ <100> and $\Delta\rho_t(B)/\rho_0$ is the transverse magnetoresistance for $\underline{j} \parallel$ <100> and $\underline{B} \parallel$ <010>. Calculating the K' values from the measured weak-field magnetoresistance ratios Y and comparing them with the mass-anisotropy factors K_m obtained from Shubnikov - de Haas experiments made TAKITA et al. /77T2/ conclude that the scattering of carriers in n-type $Pb_{1-x}Sn_xTe$ ($0.09 \leq x \leq 0.25$) is almost isotropic at low temperatures in spite of the large anisotropy of the effective masses.

At higher temperatures (77, 90, and 300 K) the *mobility* anisotropy in p-type PbTe has been found to be smaller than the *effective-mass* anisotropy at low temperatures /59S2, 60A, 77G/. Values between approximately 2.5 and 5.6 have been reported. There seems to be a tendency for K' to decrease with increasing carrier concentration. To our knowledge no reliable data on the carrier concentration dependence of the *effective-mass* anisotropy at higher temperatures have been published. Therefore it seems difficult to check the theoretical conclusions of GRYAZNOV and RAVICH /70G/, who predicted a nearly isotropic relaxation time in the case of acoustical phonon scattering. The scattering due to optical phonons, however, should be strongly anisotropic and mainly determined by K_m /70G/.

The weak-field magnetoresistance results depend sensitively on the crystal perfection /74H3, 78A3/. In a series of papers (/80A1, 81R/ and references therein) ALLGAIER et al. showed that weak-field galvanomagnetic measurements can be used to detect symmetry-lowering effects due to substrate-induced strain in thin layers of PbTe and SnTe /76N4/.

Recently ALLGAIER et al. /82A/ reported on weak- and strong-field magnetoresistance measurements on n-type PbTe epitaxial film, covering a wide range of temperatures (1.8 to 300 K) and magnetic field intensities. Figure 4.15 shows one result as an example. In this figure the magnetoresistance curves obtained at nine different temperatures with the magnetic field oriented perpendicular to the film plane are plotted together versus the product of Hall mobility and magnetic induction ($0.03\ T \lesssim B \lesssim 8\ T$). The magnetoresistance appears to be quite insensitive to the

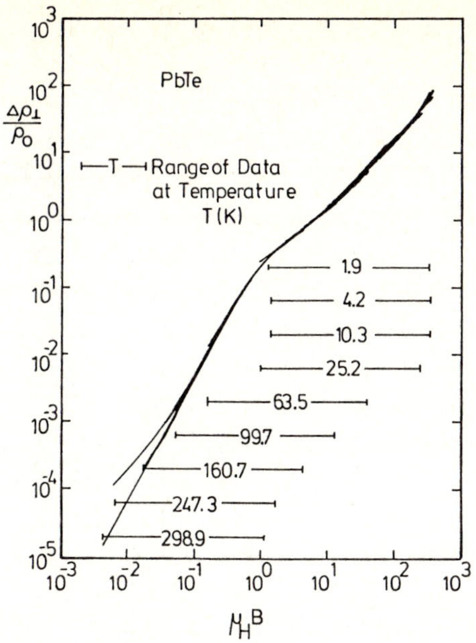

Fig. 4.15. PbTe. Magnetoresistance $\Delta\rho/\rho_0$ as a function of the mobility magnetic-field intensity product $\mu_H B$ for the transverse ($\underline{B}$ normal to the film plane) configuration. Data were taken at nine different temperatures in a magnetic field ranging from ≈ 0.03 T to ≈ 8 T /82A/

temperature and hence the scattering mechanisms. It rather seems to be a universal function of $\mu_H B$. Moreover, even the sample-to-sample scatter was found to be weak. This insensitivity to various parameters is in contrast to essentially all existing theories on magnetotransport. A comprehensive citation of publications on this subject is given in /82A/.

FINLAYSON and JOHNSON /75F/ have measured a negative longitudinal magnetoresistance in pure n-type PbS crystals at low temperatures which they attribute to scattering by localized spins.

4.3.2 Shubnikov - de Haas Effect

Several investigations of magnetic quantum oscillations in the magnetoresistance (Shubnikov - de Haas (SdH)effect) have been carried out on the lead and tin salts:

PbTe /70B4 , 78B4 , 78J2 , 79C/
PbSe /78B2/
SnTe /66B2/

as well as on the mixed crystals:

$Pb_{1-x}Sn_xTe$ /71B , 71M , 76N1 , 77A3 , 77D2 , 77T2/
$Pb_{1-x}Sn_xSe$ /72M3 , 75I2 , 77K3 , 77V2 , 78B1 , 78B2 , 78M , 81K3/
$PbSe_{1-x}Te_x$ /81A5/ .

Most of the information on band structure obtained from these measurements has already been presented in the preceding chapter and will not be repeated here. We will, however, discuss some peculiar results.

BURKE and CARVER /78B4/ observed SdH oscillations in the magnetoresistance of epitaxial films of n-PbTe on BaF_2 substrates. For a film with $n = 6\cdot10^{16}$ cm^{-3} all electrons were found to be in a *single* <111> valley at 4.2 K. This result is a consequence of a thermally induced shear strain which in turn occurs because of the difference between the thermal coefficients of expansion of PbTe and BaF_2. Due to the mismatch in lattice constants at low temperatures the PbTe films undergo an "antibismuth" distortion /80A1/, i.e. they are stretched in the film plane and compressed along the trigonal axis. This causes the energy of the <111> ellipsoid to be lowered relative to the others. The estimated energy shift is more than sufficient to transfer all the carriers into this single ellipsoid /78B4/.

The degree of strain in PbTe films on BaF_2 is significantly affected by the choice of growth parameters in the hot-wall technique. A comparison of four films with thicknesses between 2.8 and 11.1 μm grown at NAVSWC (Naval Surface Weapons Center, White Oak Laboratory) and one film (thickness 16 μm) grown at Linz University revealed that in the Linz sample no strain effects were observable /79C/. The latter sample did not show any noticeable carrier transfer. This can be seen from Fig. 4.16,

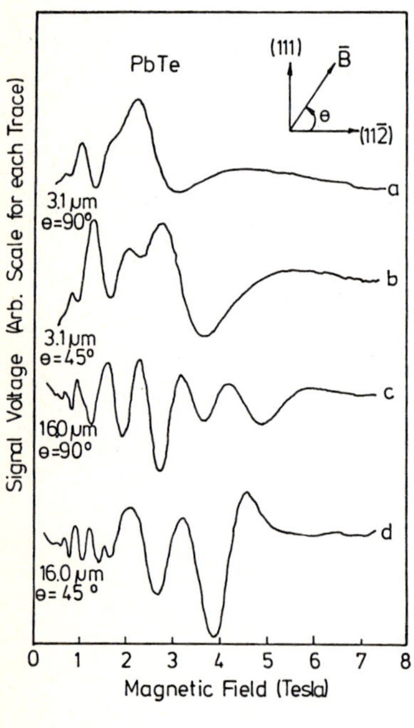

◄ Fig. 4.16. PbTe. Shubnikov - de Haas oscillations at 4.2 K for two epitaxial layers on BaF_2: (a),(b) thickness 3.1 μm, grown at NAVSWC; (c),(d) thickness 16.0 μm, grown at Linz University. Data were taken for two orientations of the magnetic field /79C/

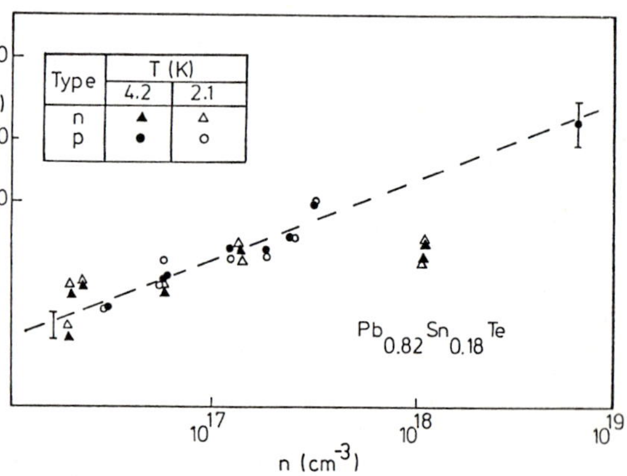

Fig. 4.17. $Pb_{0.82}Sn_{0.18}Te$. Carrier concentration dependence of the Dingle temperature T_D. For all samples (except one) the T_D values have been determined at 2.1 K and 4.2 K. The dashed line represents the theoretical dependence $T_D \sim \sqrt{n}$ /78B5/

where the SdH oscillations obtained for two orientations of the magnetic field are plotted for a thin (d = 3.1 μm) White Oak sample and the Linz sample /79C/. Whereas the White Oak sample shows only a single SdH frequency, the Linz sample yields a pattern corresponding to at least two SdH frequencies.

The Landau level broadening in $Pb_{0.82}Sn_{0.18}Te$ has been studied by determining the Dingle temperature T_D for a variety of samples /77A3 , 77D2/. An analysis of the data was given in /78B5/. In Fig. 4.17 the carrier concentration dependence of T_D is displayed. BAGINSKII and DMITRIEV /78B5/ concluded that the level broadening and the attenuation of the oscillation amplitude is connected with the presence of random fluctuating potentials in the crystal due to the disorder in the distribution of Pb and Sn atoms, foreign impurities, and native defects.

NARITA and TAKAFUJI /76N1/ investigated spin-splitting in the transverse (TM) and longitudinal (LM) magnetoresistance of $Pb_{1-x}Sn_xTe$ ($0 \leqslant x \leqslant 0.28$). They found that the SdH peaks corresponding to a crossing of the Fermi level and a spin-down level are observable in TM but not in LM. This anomaly can be explained by the selection rules, theoretically derived in a study of a similar effect in $Hg_{1-x}Cd_xTe$ /75N/.

Some SdH measurements have been carried out on $Pb_{1-x}Sn_xSe$ under hydrostatic pressure /78B1 , 78M/. The energy-gap dependence of the cyclotron masses was found to be in agreement with the band-inversion model of DIMMOCK /71D3/.

AKOPYAN et al. /81A5/ measured the SdH effect in n-type $PbSe_{1-x}Te_x$ crystals with various compositions. The effective electron-mass anisotropy K_m as a function of the alloy composition is displayed in Fig. 4.18. It was found to be practically independent of the electron density.

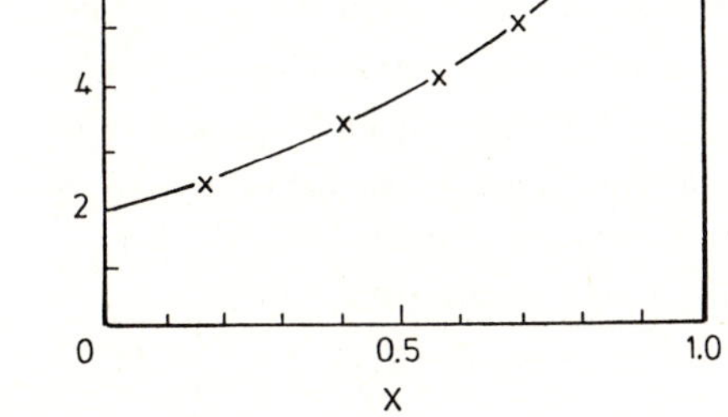

Fig. 4.18. $PbSe_{1-x}Te_x$. Anisotropy coefficient of the effective mass as a function of alloy composition /81A5/

4.3.3 Magnetophonon Effect

Oscillations in the magnetoresistance at elevated temperatures are mainly due to resonant inelastic scattering of electrons between Landau levels by longitudinal optical phonons, but interactions with other phonons have also been observed /79C/. Measurements of this effect yield valuable information on the electron-phonon interaction. Magnetophonon (MP) oscillations in the transverse ($\underline{j} \perp \underline{B}$) magnetoresistance have been measured on PbTe films /77M, 77R3, 78B4, 79C/, $Pb_{0.8}Sn_{0.2}Te$ films /77M/, $Pb_{0.8}Sn_{0.2}Te$ bulk crystals /77R3/, and PbS bulk crystals /76A2/.

Investigations on n-PbTe films on (111) BaF_2 substrates have shown that the amplitudes of the MP oscillations from the <111> ellipsoid, whose symmetry axis is normal to the film plane, increase monotonically with increasing film thickness /79C/. This effect could be due to either a reduction of strain with increasing film thickness and hence a reduction of the number and degeneracy of the <111> carriers or a reduction of the scattering from interface defects /79C/.

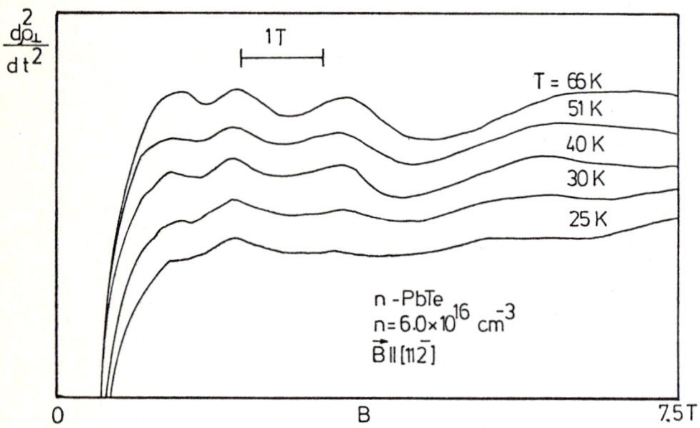

Fig. 4.19. PbTe, <111>-oriented film. Temperature dependence of the magnetophonon oscillations for $\underline{B} \parallel \langle 11\bar{2} \rangle$, $\underline{j} \parallel \langle 1\bar{1}0 \rangle$. The second derivative $d^2\rho_\perp/dt^2$ of the magnetoresistance $\rho_\perp$ is plotted vs magnetic field /78B4/

The temperature dependence of the MP oscillations of an n-PbTe film with n = $6\cdot10^{16}$ electrons/cm^3 is displayed in Fig. 4.19 /78B4/. The decrease of the amplitudes with decreasing temperature is expected when the decrease in the number of phonons is dominant over the accompanying increase in the relaxation time of the carriers /78B4/.

RAMAGE et al. /77R3/ deduced from their MP data on p-type PbTe and $Pb_{0.8}Sn_{0.2}Te$ that the frequency of the phonon which interacts most strongly with the free holes via intravalley scattering is approximately equal to the unscreened LO phonon frequency (114 cm^{-1}) and this value is independent of the carrier concentration in the sample.

4.3.4 Magnetooptical Properties

The magnetooptical properties of the lead salts and the mixed crystals $Pb_{1-x}Sn_xTe$ and $Pb_{1-x}Sn_xSe$ have been reviewed recently by BAUER /80B2/. Intraband transitions (which involve quantum states in a single band, e.g. cyclotron resonance, spin resonance) as well as interband transitions (involving states in the valence and conduction bands) have been discussed. In view of this review article we will mention only a few additional results.

Optical four-wave mixing of the beams of two CO_2 lasers has been observed in PbTe epitaxial layers by PASCHER /82P3/. Four kinds of resonances have been found: a) quantum oscillations due to mixing induced by non-parabolicity, b) spin resonances, c) cyclotron resonances, and d) combined cyclotron and spin resonances of the non-linear susceptibility. In Figs. 4.20a,b the functional relations between the observed resonance positions in the magnetic field and the frequency differences $\Delta\omega$ between the two CO_2 lasers (i.e. the intraband transition energies) are displayed for two orientations of the magnetic field. The identifications of the transitions are noted in the inserts of these figures. From these results the effective g factors

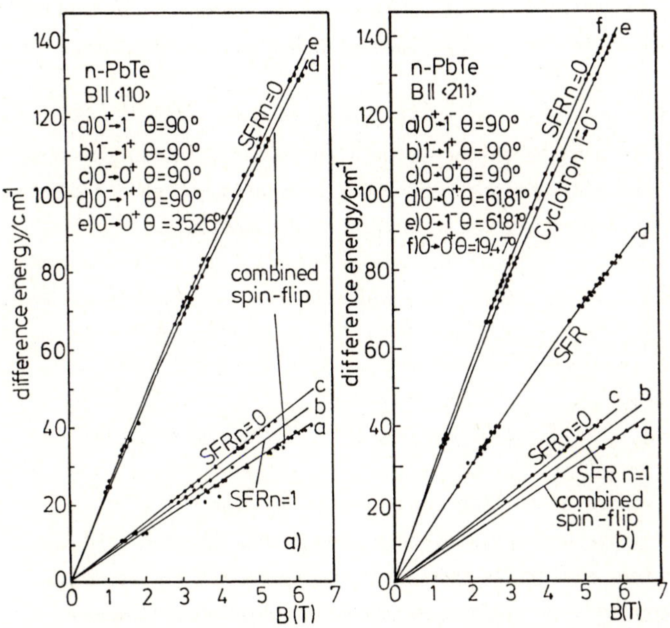

Fig. 4.20a,b. PbTe. Magnetic-field dependence of the intraband transition energies for (a) $\underline{B} \parallel$ <110> and (b) $\underline{B} \parallel$ <211>. The full dots denote the magnetic field positions at which resonances occurred for a variety of frequency differences $\Delta\omega$ between the two CO_2 lasers. θ denotes the angle between the magnetic field direction and the axis of the constant-energy ellipsoid /82P3/

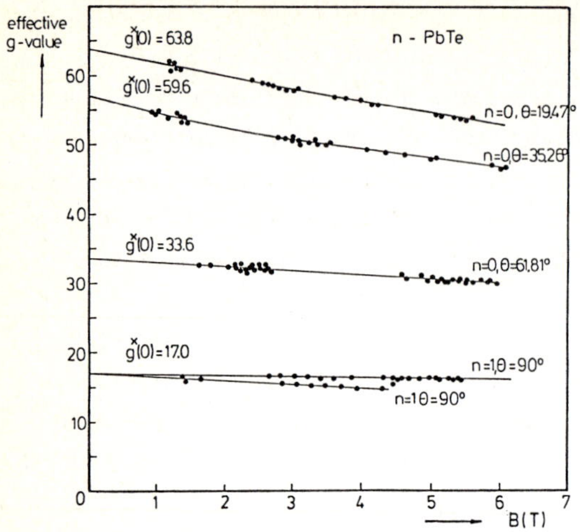

Fig. 4.21. PbTe. Magnetic-field dependence of the effective g factors for four directions /82P3/

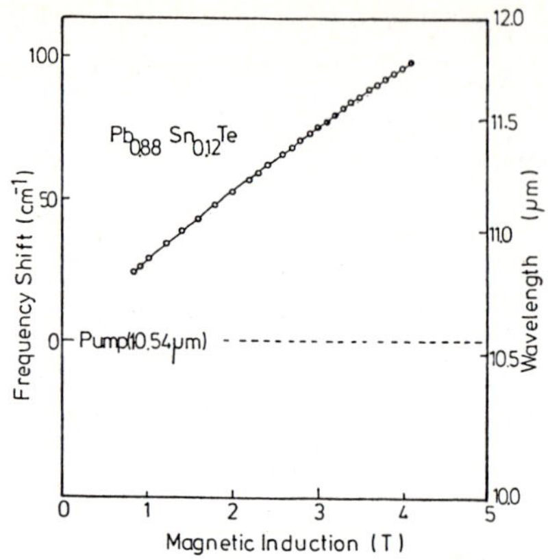

Fig. 4.22. $Pb_{0.88}Sn_{0.12}Te$. Wavelength of the spin-flip Raman emission and frequency shift from the pump wavelength at 10.54 μm as a function of magnetic field (T = 10 K) /79Y2/

have been calculated for four different angles θ between the magnetic field and the axis of the ellipsoid. The magnetic field dependence of the g values is plotted in Fig. 4.21.

The decrease in the effective g factor with increasing magnetic field has also been observed in n-$Pb_{0.88}Sn_{0.12}Te$ by measuring stimulated spin-flip Raman emission /79Y2/. The frequency shift and the wavelength of the Stokes emission obtained using the 10.54 μm CO_2 laser line as a pump are shown in Fig. 4.22. The value of the effective g factor estimated from the slope of the curve in Fig. 4.22 varies from 50 to 64 depending on the magnetic field /79Y2/. The magnetic-field dependence of the laser emission from $Pb_{1-x}Sn_xSe$, $Pb_{1-x}Sn_xTe$, and $PbS_{1-x}Se_x$ devices is discussed in Sect. 5.2.

4.4 Warm and Hot Carrier Effects, Impact Ionization

A couple of investigations is devoted to transport effects in high electric fields. Depending on how far their temperature deviates from thermal equilibrium with the lattice temperature the carriers are called warm or hot. The mobility μ of warm carriers is usually characterized by a coefficient β /73S2/ which describes the deviation from Ohm's law according to the relation

$$\mu(E) = \mu_0(1 + \beta E^2) \quad , \qquad (4.2)$$

where μ_0 is the weak field mobility and E is the electric field. In general β is dependent on the crystallographic direction along which the field is applied. SHENDEROVSKII and GORLEY /79S4/ have calculated the β coefficient for PbSe, PbTe and $Pb_{1-x}Sn_xTe$ (x ≈ 0.2). Data of β(E ∥ <001>) as a function of temperature are shown in Fig. 4.23. The data are representative for 10^{17} impurities/cm^3. The authors considered scattering with impurities and optical phonons and studied the effect of non-parabolicity on β. At temperatures below the Debye temperature ($\hbar\omega_{LO}$), β is negative (Fig. 4.23), at higher temperatures β becomes positive, i.e. the mobility increases with increasing electric field. This property is characteristic for dominant polar optical phonon scattering.

The mobility of hot carriers was studied for PbSe in /72S , 73S6/, for PbTe in /72S , 73S6 , 75H5 , 76C , 78J3/, and for $Pb_{1-x}Sn_xTe$ in /72S , 73S6 , 75G2/. Data ob-

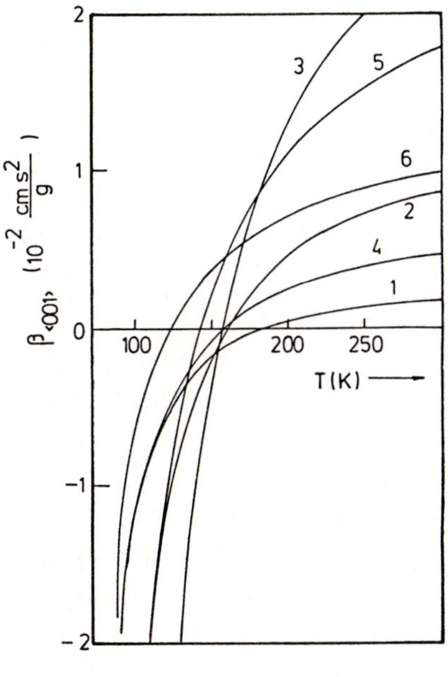

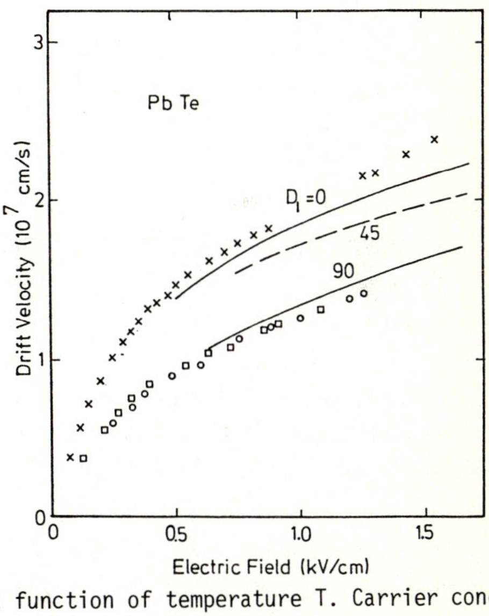

Fig. 4.23. Coefficient β(E ∥ <001>) as a function of temperature T. Carrier concentration is 10^{17} cm^{-3}. (1,4) PbSe, (2,5) PbTe, (3,6) $Pb_{1-x}Sn_xTe$. (1 - 3) and (4 - 6) are calculated assuming non-parabolicity and parabolicity of conduction band, respectively. (The scale for (6) is 10^{-1} cm s^2/g ≅ 1.11·10^{-6} cm^2/V^2) /79S4/

Fig. 4.24. PbTe. Drift velocity versus electric field of electrons at 77 K for an electric field oriented along a <100> axis of the crystal. Experimental data are from □ /72S , 73S6/, × /75H5/, ○ /76F/. Solid and broken lines are calculated including scattering by polar-optical phonons, acoustic phonons, and first-order-coupled intervalley phonons. Parameter is the deformation potential coupling constant D_1 (in eV) /76F/

tained from experiments with PbTe are compared with calculated data in Fig. 4.24. FERRY /76F/ tried to elucidate the discrepancy between the experimental values by a theoretical analysis using the deformation potential coupling constant D_1 as a parameter as shown in Fig. 4.24. However, the difference in the experimental data obtained by HEINRICH et al. /75H5/ and by CARVER et al. /76C/ are still not understood. In both cases similar substrates and preparation procedures were used. It was speculated that the investigated PbTe films may be strained. If the strain were large enough and the conduction band in the samples of HEINRICH et al. were still strain split at 77 K, the equivalent intervalley scattering would be reduced. This result would be in line with the calculated data for $D_1 = 0$.

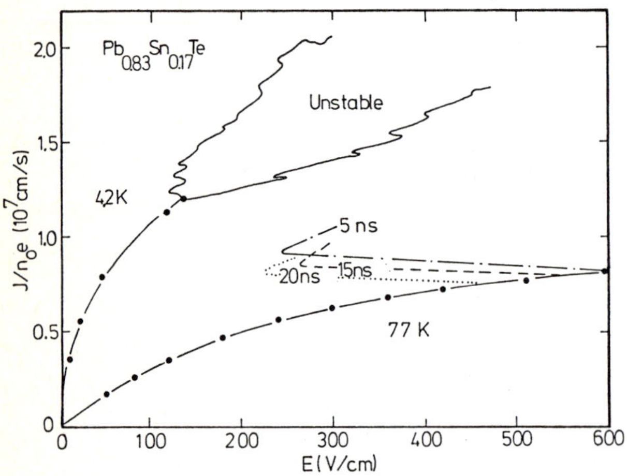

Fig. 4.25. $Pb_{0.83}Sn_{0.17}Te$. Drift velocity versus electric field at 4.2 K and 77 K. Electrical instabilities due to impact ionization are found at elevated fields. The 77 K curves were measured 5, 15, and 20 ns after the electric field was switched on /72S/

At still higher fields carrier generation due to band-band impact ionization takes place. As shown in Fig. 4.25 the current-voltage characteristic exhibits electric instabilities. In this field range ST-ONGE et al. /72S, 73S6/ observed oscillations at several hundred MHz. ST-ONGE et al. concluded that the current transport is governed simultaneously by impact ionization and by self-pinching of an electron-hole plasma. However, HEINRICH et al. /75H5, 78J3/ arrived at another interpretation for the transport properties under conditions of impact ionization. They observed travelling high field domains and claimed that they cause the electric instability rather than a self-pinching of the electron-hole plasma. In conclusion, the hot carrier data available at present do not allow a satisfactory theoretical description of the observed high field transport properties.

The Hall effect of hot electrons in PbTe was investigated in electric fields up to 1.1 kV/cm at 80 K /77R2/. The analysis of the data gave evidence that a strong electron transfer between equivalent valleys takes place. In addition it was observed that the crystallographic orientation had no significant influence on the threshold field for impact ionization nor on the appearance of the electric instability /77R2/.

4.5 Thermal Conductivity and Thermoelectric Effects

The thermal conductivity of various IV-VI compounds and their alloys was studied by ALEKSEEVA et al. /75A/. The experimental values are shown in Fig. 4.26. The analysis of the data has revealed that thermal conductivity is governed by the contribution from the crystal lattice, the electronic contribution being always smaller than 10% of the total thermal conductivity. The comparison of the data obtained from mixed

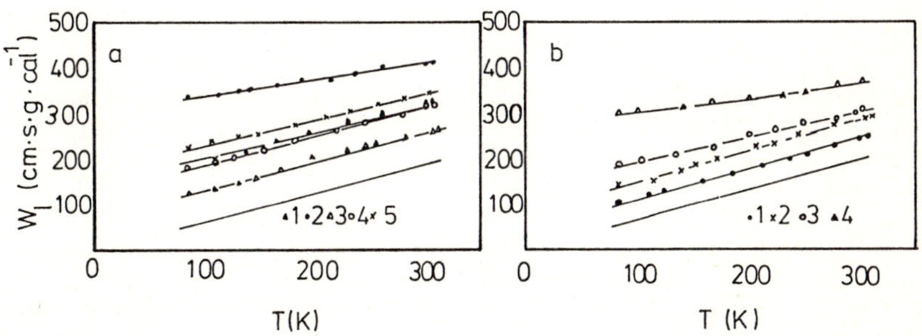

Fig. 4.26a,b. a) Temperature dependencies of the lattice thermal resistivity of PbTe-PbS and PbTe-PbSe solutions (all compositions are given in mole %): 1) 95PbTe-5PbS; 2) 90PbTe-10PbS; 3) 95PbTe-5PbSe; 4) 90PbTe-10PbSe; 5) 85PbTe-15PbSe; the lowest line represents pure PbTe.
b) Temperature dependencies of the lattice thermal resistivity of PbTe-SnTe and PbTe-GeTe solutions (mole %): 1) 95PbTe-5SnTe; 2) 90PbTe-10SnTe; 3) 95PbTe-5GeTe; 4) 90PbTe-10GeTe; the lowest line represents pure PbTe. Carrier concentrations are between $7 - 34 \cdot 10^{17}$ cm^{-3} /75A/

crystals with those of the binary compound PbTe shows (see Fig. 4.26) an increase of thermal resistivity in the mixed crystals. This increase is related to the additional scattering of phonons by impurities in the mixed crystals /75A/.

KNITTEL and GOLDSMID /79K4/ measured the thermal conductivity of $Pb_{1-x}Sn_xTe$ in a transverse magnetic field. The magnetic field induced changes of thermal conductivity are related partly to a removal of the electronic contribution and partly to a phonon-drag contribution.

To get information on band structure thermoelectric effects with and without magnetic fields were studied by various groups recently. Some of the band parameters obtained from such experiments are presented in Chap. 3. Some of these studies are: PbTe: /66D2, 67D, 75G3/; $Pb_{1-x}Sn_xTe$: /76B4, 78Y, 79T4, 81D/; PbSe: /78V2/; $Pb_{1-x}Sn_xSe$: /73K4, 74K3, 75V, 76K4, 77V3, 78K1, 80F, 80P2/; $PbS_{1-x}Se_x$: /79V4, 80V/.

4.6 Magnetic Susceptibility

Magnetic susceptibility has been measured in $Pb_{1-x}Sn_xTe$ ($0.11 \leq x \leq 0.31$, $x = 0.85$, 1.0) /74L3, 74T, 75L1, 78L6/ and $Pb_{1-x}Sn_xSe$ ($x = 0.06$, 0.20) /74T, 75L2/ utilizing the relative Faraday method.

Figure 4.27 shows the temperature dependencies of the experimental susceptibility χ_{exp} of nine samples with different electron and hole densities /78L6/. The susceptibility is diamagnetic throughout the investigated range of carrier densities.

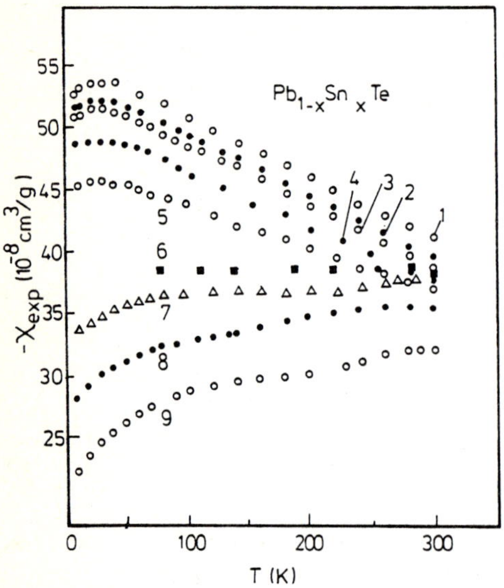

Fig. 4.27. $Pb_{1-x}Sn_xTe$. Temperature dependence of the experimental magnetic susceptibility χ_{exp} for nine samples (data taken at 77 K) /78L6/:

(1) $x = 0.18$, $n = 3.0 \cdot 10^{16}$ cm^{-3}
(2) $x = 0.18$, $n = 1.2 \cdot 10^{17}$ cm^{-3}
(3) $x = 0.18$, $p = 3.5 \cdot 10^{17}$ cm^{-3}
(4) $x = 0.18$, $p = 5.2 \cdot 10^{17}$ cm^{-3}
(5) $x = 0.18$, $n = 1.2 \cdot 10^{18}$ cm^{-3}
(6) $x = 0.18$, $p = 4.9 \cdot 10^{18}$ cm^{-3}
(7) $x = 0.18$, $p = 3.0 \cdot 10^{19}$ cm^{-3}
(8) $x = 0.18$, $p = 8.9 \cdot 10^{19}$ cm^{-3}
(9) $x = 0.20$, $p = 6.0 \cdot 10^{20}$ cm^{-3}

The experimental susceptibility consists of various components: lattice part χ_{latt}, carrier contribution χ_c, and defect contribution χ_d: $\chi_{exp} = \chi_{latt} + \chi_c + \chi_d$.

Although a theory of the magnetic susceptibility of $Pb_{1-x}Sn_xTe$ taking into account the characteristic features of its band structure is missing up to now, the experimental results /78L6/ could be described (at least qualitatively) on the basis of ZAWADZKI's theory /63Z/. It was assumed that $\chi_{latt} + \chi_d \approx -54 \cdot 10^{-8}$ cm^3/g (temperature independent).

The carrier contribution χ_c is strongly affected by the carrier redistribution between light- and heavy-hole valence band. The contribution of the heavy holes was concluded to be diamagnetic. This conclusion is in agreement with experimental results on samples with high hole concentrations, which show a decrease in the paramagnetism of light holes with rising temperature.

5. Optical Properties

5.1 General Remarks

Optical experiments as absorption, reflectivity, photoemission, and various modulation techniques of transmission and reflexion are useful to get information on band parameters. Most data presented in Chap. 3 were obtained by one of these experimental methods. On the other hand, optical investigations particularly in the far-infrared region of the spectrum yield information on lattice properties, which are very interesting in this class of compounds because of the structural phase transition. As discussed in Sect. 2.1 this phase transition may be provoked by different parameters such as temperature, pressure and composition ratio of the alloy in question. Another field of infrared investigations deals with free carriers. Here the experiments concern properties such as effective masses, g factor, carrier concentration, and last but not least the coupling of free carriers with excitations of the crystal. In the following sections of this chapter we shall present some experimental results of optical investigations carried out to obtain quantitative information about the basic properties of the narrow-gap lead salts.

5.2 Absorption, Reflexion and Emission near the Fundamental Energy Gap

Data on the absorption coefficient of PbTe for photon energies near the fundamental band gap from various authors were recently analysed by GELMONT et al. /81G1/. The absorption coefficient shown in Fig. 5.1 was obtained from transmission and reflexion experiments. In their analysis GELMONT et al. tried to determine the contributions of the different physical processes to the total optical response. The absorption coefficient is mainly determined by an allowed direct transition. This is shown in Fig. 5.1 by comparing the experimental values with the theoretical absorption data (Curve 2) for an allowed transition.

For the remaining part of the absorption (Curve 3) the following interpretation is proposed /81G1/: there are contributions a) by an indirect transition effective between a valence band Σ extremum and the L minimum of the conduction band, the theoretical data for this process are presented by Curve 4, b) by an indirect transition between a valence band Λ extremum and the L minimum of the conduction band, the theoretical values being presented by Curve 6, c) the remaining structures (a - d) are assumed to be caused by phonon-assisted transitions between bound energy levels of lattice defects. The two symmetrical sets of peaks (e - h , i - n) are assumed to correspond to transitions from the same level to two levels related to different conduction band extrema and separated by the energy 0.195 eV. The peaks e, f, i, j, k include the emission of an optical phonon, whereas the peaks g, h, l, m, n include

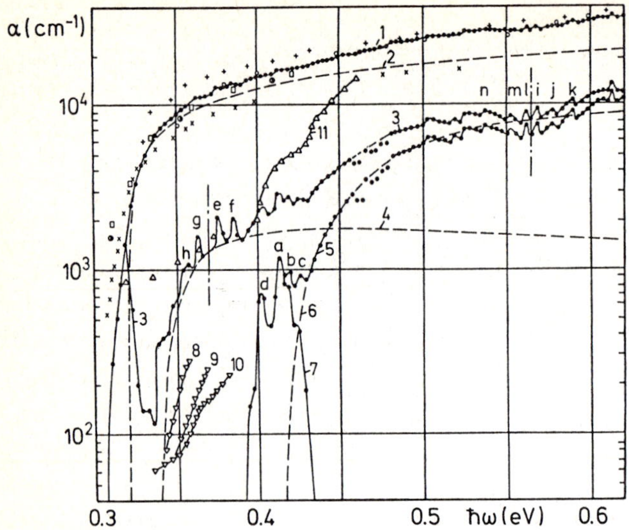

Fig. 5.1. PbTe. Absorption coefficient α vs the radiation energy $\hbar\omega$ at 313 K. 1) Experimental plot of $\alpha(\hbar\omega)$; 2) calculated curve of the fundamental absorption α_1 for an allowed direct transition; 3) $(\alpha - \alpha_1)$; 4) theoretical values $\alpha_2(\omega)$ for an indirect transition VB(Σ) - CB(L); 5) $(\alpha - \alpha_1 - \alpha_2)$; 6) theoretical values α_3 for an indirect transition VB(Λ) - CB(L); 7) the impurity absorption. Experimental data at 8) 420 K /66T/; 9) 459 K /66T/; 10) 515 K /66T/; 11) 685 K /76V4/. From various authors: × /59S1/, ○ /73D4/, □ /74P1/, + /79Z/, ◑ /68M/, ● /81G1/

the absorption of the same phonons. Further experimental studies concerning the absorption and reflection are presented for $Pb_{1-x}Sn_xTe$ by /59S1, 61S, 66T, 67T, 72D, 73D4, 76G1, 77D1, 78G2, 78L/, for $Pb_{1-x}Sn_xSe$ by /59S1, 61S, 75G1, 76G1, 78V1, 79L4, 79L6/, and for $PbS_{1-x}Se_x$ by /59S1, 61S, 80V/.

The variation of the absorption coefficient with composition ratio in $Pb_{1-x}Sn_xTe$ is shown in Fig. 5.2 for x = 0, 0.1, 0.2 /73D4/. The data were obtained from epitaxial films at 300 K. The solid lines represent calculated data from a two-band Kane model.

Absorption spectra of the long-wavelength edge of $Pb_{0.94}Sn_{0.06}Se$ single crystals are shown in Fig. 5.3 for various temperatures and two carrier densities /79L4/. The data demonstrate the pronounced Burstein-Moss shift between the two samples with different carrier densities. The authors analysed the influence of the free electrons on the absorption edge behaviour, i.e. the Burstein-Moss shift and the absorption of free carriers in the temperature range between 77 K and 450 K.

During the last decade the lead salts have proved to represent a useful material for infrared light emitting devices, particularly for wavelengths between 5 and 50 μm e.g. /77H, 79P/. The fundamental band gap of the lead chalcogenides being a function of temperature, pressure and magnetic field, the wavelength of the emitted photons can easily be tuned by an octave or more. On the other hand, emission data

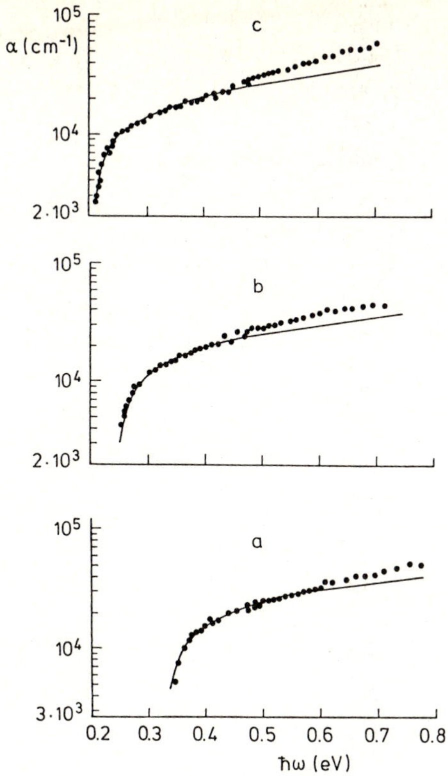

Fig. 5.2. $Pb_{1-x}Sn_xTe$. Dependence of the absorption coefficient α on the photon energy $\hbar\omega$ for (a) PbTe, (b) $Pb_{0.9}Sn_{0.1}Te$, and (c) $Pb_{0.8}Sn_{0.2}Te$. The solid lines represent calculated values /73D4/

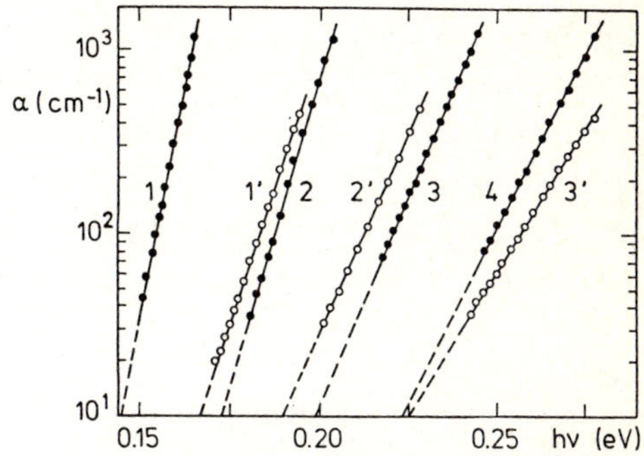

Fig. 5.3. $Pb_{0.94}Sn_{0.06}Se$. Absorption coefficient vs photon energy.
●: $n = 1.0\cdot10^{23}\ m^{-3}$, (1) T = 77 K, (2) T = 120 K, (3) T = 200 K, (4) T = 450 K;
o: $n = 1.8\cdot10^{24}\ m^{-3}$, (1') T = 77 K, (2') T = 180 K, (3') T = 450 K /79L4/

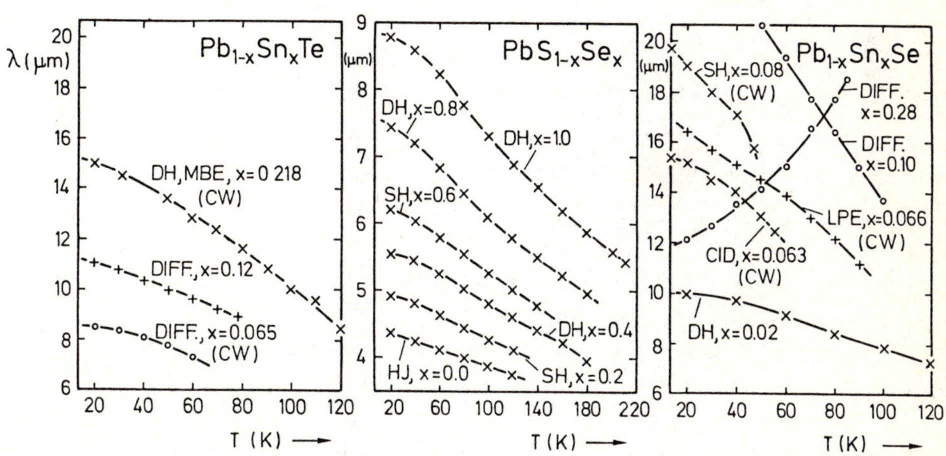

Fig. 5.4a-c. Tuning curves of actual diode lasers: a) $Pb_{1-x}Sn_xTe$ (x = 0.218 /73W/, x = 0.12 /73R/, x = 0.065 /77L3/; b) $PbS_{1-x}Se_x$ /79P/; c) $Pb_{1-x}Sn_xSe$ (x = 0.066 /77L3/, x = 0.10, 0.28 /69H/, x = 0.02, 0.063, 0.08 /79P/. The results are obtained from single hetero (SH) and double hetero (DH) structure diodes. The methods of fabrication are diffusion (DIFF), compositional interdiffusion (CID), and molecular beam epitaxy MBE /79P/

gave valuable information on the band-edge characteristics. Data of the temperature dependence of the emitted photon wavelength are shown in Figs. 5.4a-c for $Pb_{1-x}Sn_xTe$ /73R, 73W, 77L2/, $PbS_{1-x}Se_x$ /73S4, 74S, 78L5, 79P/, and $Pb_{1-x}Sn_xSe$ /69H, 77L3, 79P/. (For the latter see also Fig. 3.14).

MARTINEZ /73M2/ studied the pressure dependence of the laser wavelength in $Pb_{1-x}Sn_xSe$ diodes. In Fig. 5.5 the pressure dependent emission of a $Pb_{0.88}Sn_{0.12}Se$ laser diode is shown. This special composition shows at hydrostatic pressures near 7 kbar a band inversion thus causing a zero gap. The calculated values for the photon energy agree well at low pressures as shown in Fig. 5.5, (dotted lines) the discrepancy at high pressure is not understood. Further data on pressure dependent photoemission experiments are given, e.g., in /68B1/ for PbSe, and for $Pb_{1-x}Sn_xSe$ ($x = 0.085$) in /74Z1/.

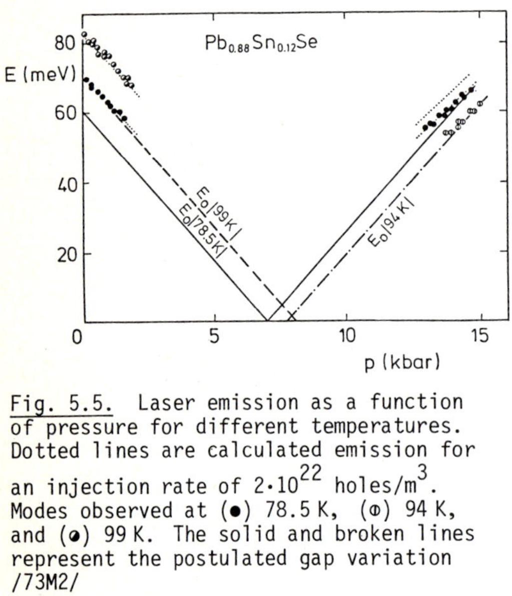

Fig. 5.5. Laser emission as a function of pressure for different temperatures. Dotted lines are calculated emission for an injection rate of $2\cdot10^{22}$ holes/m^3. Modes observed at (●) 78.5 K, (ⴲ) 94 K, and (◑) 99 K. The solid and broken lines represent the postulated gap variation /73M2/

Fig.5.6. $Pb_{1-x}Sn_xTe$. Magnetic field dependence of laser emission in three $Pb_{1-x}Sn_xTe$ diodes with $x = 0.19$, 0.22, and 0.28 /69C1/

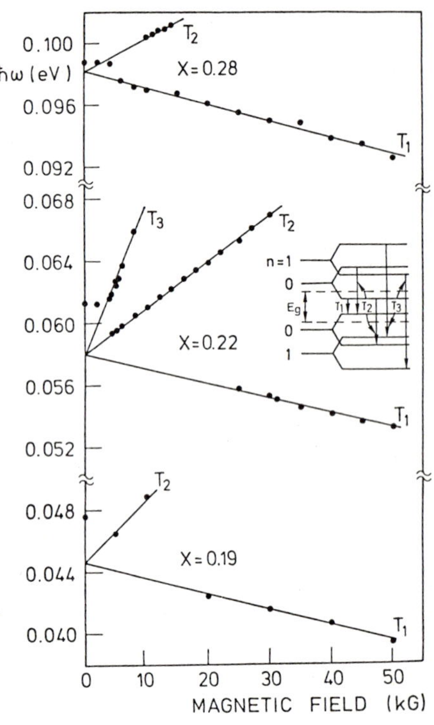

The magnetic field dependence of the emitted photon energy was investigated by CALAWA et al. /69C1/ in the $Pb_{1-x}Sn_xSe$ alloy. Typical results are shown in Fig. 5.6 for three different compositions. Due to the energy level splitting in magnetic fields, which is schematically shown in the inset of the figure, several emitted lines are observed. At low fields one generally observes the line T_2. As the field is increased the emission switches to T_1. However, if the diode current is increased the T_2 emission persists up to higher magnetic field values and in some cases a third line T_3 is observed. The solid T_1 line corresponds to the effective band gap. The T_2 and T_3 emission lines correspond to transitions between higher Landau levels as indicated in the inset. The emission energy at zero magnetic field is generally 2 - 4 meV above the extrapolated value due to band filling with free carriers. For $x < 0.15$ all laser lines increase with increasing magnetic field, however, for $x > 0.15$ the effective band gap decreases, the spin splitting being higher than the Landau level splitting.

5.3 Optical Properties at Energies Above the Fundamental Band Gap

The band-structure data presented in Chap. 3 are based mainly on the results of optical experiments. Unfortunately, almost all investigations were restricted to the binary compounds. The first extensive study of reflectivity of PbS, PbSe, PbTe, and SnTe in the energy range 0.5 -25 eV was performed by CARDONA and GREENAWAY /64C1/. Later reflectivity was studied by, e.g., /72C , 73K2 , 75H4 , 75M3/, electro-thermo reflectance of PbSe in the energy range 1.2 - 3.5 eV by /71N/, $Pb_{1-x}Sn_xTe$ 0.5 - 13 eV /72K1/ and 1.5 - 6 eV /77C/, electro-optical measurements of PbS, PbSe, and PbTe were carried out by /68A2 , 74G1/ at energies up to 6 eV. Photoemission experiments were reported from PbTe (5 - 11 eV) /65S/ and from PbS, PbSe, and PbTe in the range 14 - 26 eV /72C/. The spectra of photo-emitted electrons were measured on PbS, PbSe, and PbTe by CARDONA et al. /73C1/.

Typical reflectivity data are presented in Figs. 5.7a-c for the energy range 1 - 6.5 eV and in Figs. 5.8a,b for energies between 18 and 26 eV. The detection of

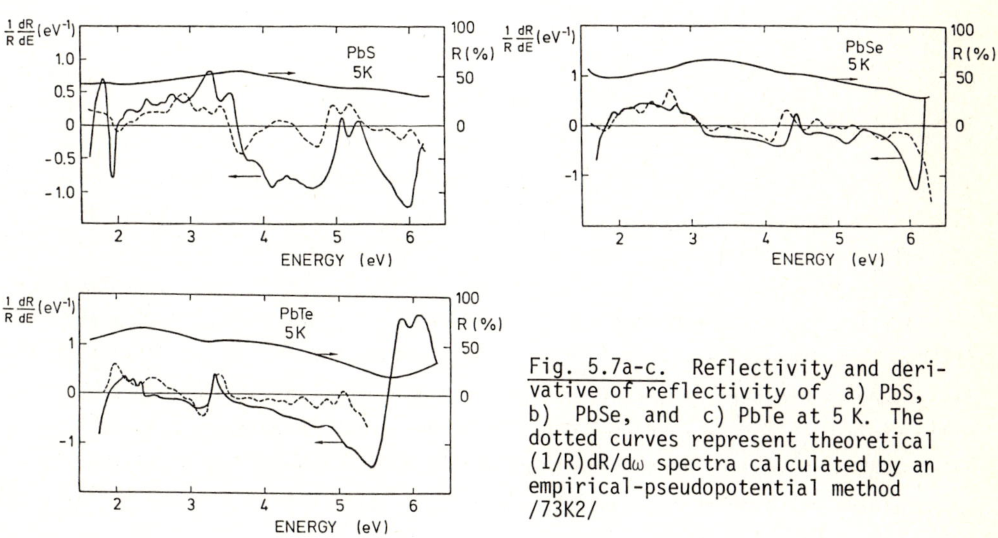

Fig. 5.7a-c. Reflectivity and derivative of reflectivity of a) PbS, b) PbSe, and c) PbTe at 5 K. The dotted curves represent theoretical (1/R)dR/dω spectra calculated by an empirical-pseudopotential method /73K2/

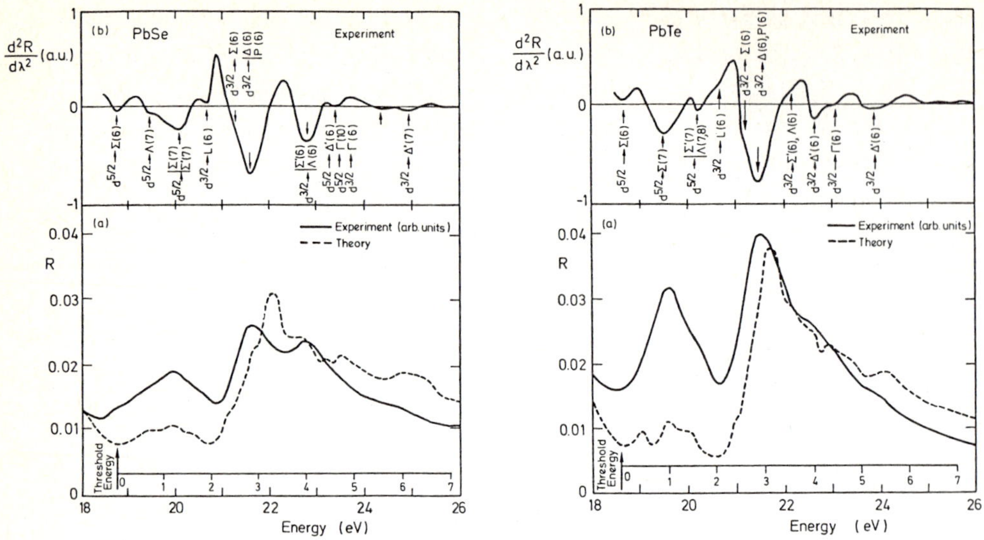

Fig. 5.8a,b. Reflectivity (a) and second derivative spectrum for measured reflectivity (b) for a) PbSe and b) PbTe. The theoretical reflectivity curve calculated by an empirical-pseudopotential method is indicated by the broken line /75M4/

weak structures was substantially improved by performing the first or second derivative of the reflectivity as shown in these figures.

GRANDKE et al. /78G1/ measured angle-resolved photoemission spectra of single-crystal PbS, PbSe, and PbTe. The emission was excited with 16.85 eV and 21.22 eV photons. The experimental results were compared with several band-structure calculations. The best overall agreement was found with calculations from first principle, i.e. augmented plane wave and orthogonal plane-wave calculations.

5.4 Dielectric Properties, Phonons

The dielectric properties of the family of IV-VI narrow-gap semiconductors have been studied because of the infrared device aspect as well as strong scientific interest. The latter arose since these compounds exhibit a structural lattice instability (see also Sect. 2.1) which is accompanied by some spectacular properties. The structural instability is caused by the critical ionicity of the binding and may induce a displacive phase transition from a NaCl-type to an arsen-like structure. The phase transition may be provoked by a variation of temperature, by the application of a magnetic field or pressure, and by a proper mixture of two different compounds.

COCHRAN's suggestions /64C3/ that PbTe might have a ferroelectric phase at low temperature was based on the observed anomalously large dielectric constant. The

Table 5.1. Transverse effective charge e_T^*, Phillips-ionicity f_i, optical dielectric constant ε_∞, average optical deformation potential Ξ, T = 300 K /73L2, 78K2/

	e_T^*/e	ε_∞	f_i	Ξ (eV)
PbS	4.8	17.2	0.77	16.2
PbSe	5.8	22.9	0.72	16.8
PbTe	6.5	32.8	0.63	14.8
SnTe	8.1	45		17.5

general tendency in the series of IV-VI narrow-gap compounds is shown in Fig. 2.6 and Table 5.1. With decreasing ionicity the temperature of the phase transition from the ferroelectric rhombohedral (if this phase does exist at all) to the cubic NaCl-like structure increases, similarly the lattice polarizability (e_T^{*2}) and the electronic polarizability ($\varepsilon_\infty - 1$) increase. LUCOVSKY and WHITE /73L2/ have shown that the proportionality of the high frequency dielectric constant to the square of the effective charge (see Fig. 2.6) can be explained by the resonance aspect of the chemical bonding in these compounds. Since 12 electrons are necessary for pair bonding at the octahedral sites of the atoms in the NaCl-like structure, they considered a resonantly bonded system with both ionic and metallic contributions. In all the narrow-gap IV-VI's the s levels are only weakly hybridized with the p levels. The crystal structure is predominantly a result of pure p bonding without s to p promotion. Thus in the rocksalt structure the six covalent bonds are half-filled. If a relative displacement of the sublattices is made (i.e. an optical phonon) the energies of the bonds formed on one side of an atom are lowered relative to those on the other side, thus sucking electrons from one side to the other and producing a large dipole moment and hence a large effective transverse charge. The strength of this coupling can be measured by an optical deformation potential, and if it is large enough there will be a phase transition to distorted structure.

LITTLEWOOD /82L/ recently made a classification of structures of superoctet $A^N B^{10-N}$ materials based on pseudopotential orbital radii. The pseudopotential model developed for the band structure allowed him to give a quantitative description of the optical properties and the crystal structure for the narrow-gap IV-VI binary compounds /79L1-3, 82L/. Other model calculations on lattice dynamics and structural phase were presented quite recently by POROD and VOGL /82P1/ and by MULA /82M2/.

In the following diagrams (Figs. 5.9 - 11) data of the phonon dispersion relations are presented for PbS, PbTe, and SnTe. These data were obtained from neutron scattering experiments. The solid lines in Figs. 5.9,10 represent calculated values from a model in which core-shell forces have been included in the nearest neighbour interaction /66C, 67E/.

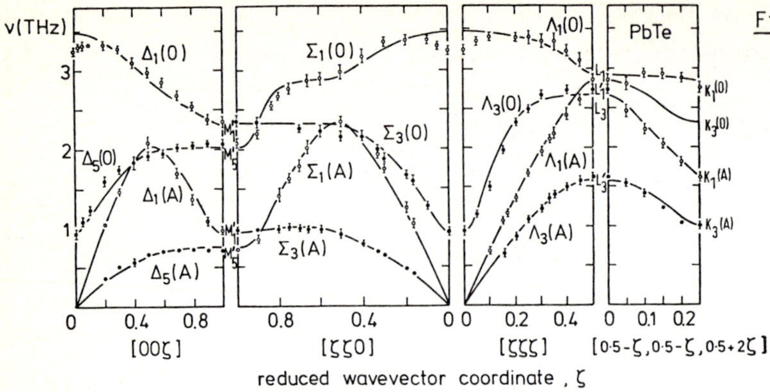

Fig. 5.9.

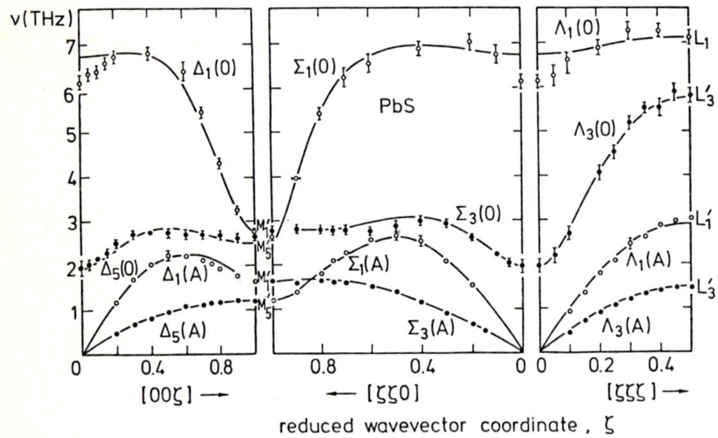

Fig. 5.10.

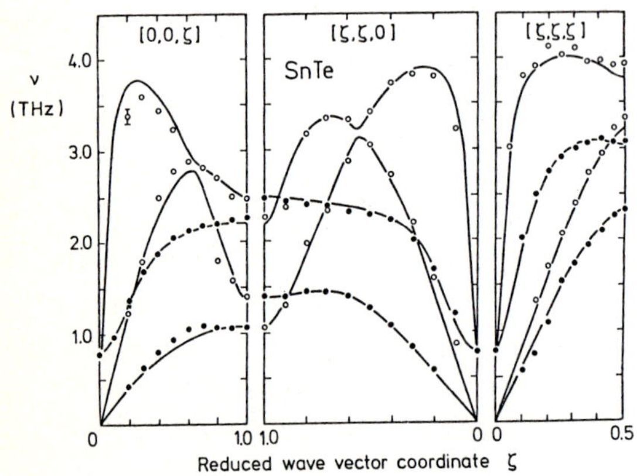

Fig. 5.11. SnTe. Phonon dispersion curve at 100 K:
○ longitudinal modes,
● transverse modes.
Solid lines represent calculated data of a shell model modified to include screening /69C2/

Fig. 5.9. PbS. Phonon dispersion curves at 296 K. The dots and circles are experimental data from inelastic neutron scattering. The solid lines represent computed data from a model in which coreshell forces have been included in the nearest-neighbour interaction /67E/

Fig. 5.10. PbTe. Phonon dispersion curves at 296 K. Dots and circles are experimental data from inelastic neutron scattering. Solid lines represent model calculations including core-shell forces in the nearest-neighbour interaction /66C/

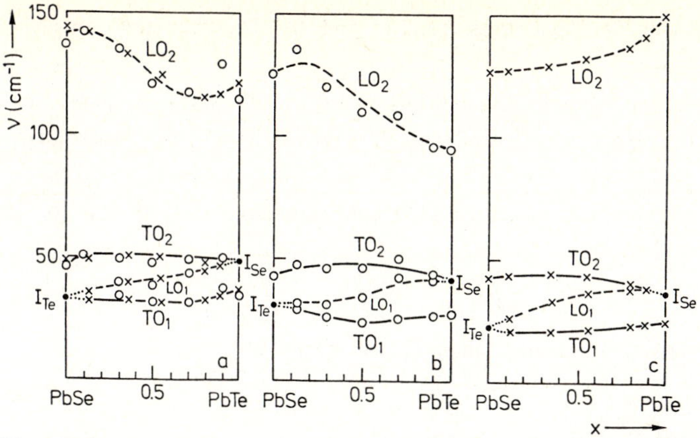

<u>Fig. 5.12a-c.</u> $PbSe_{1-x}Te_x$. Measured TO (——) and evaluated LO (---) mode frequencies at a) 300 K, b) 100 K, and c) 8 K. (×) represents data from thin films, (o) data from polycrystalline samples. Possible extrapolations of the LO_1 - TO_1 and LO_1 - TO_2 branches (·····) ending at the localized modes I_{Se} and I_{Te} /79F/ are shown

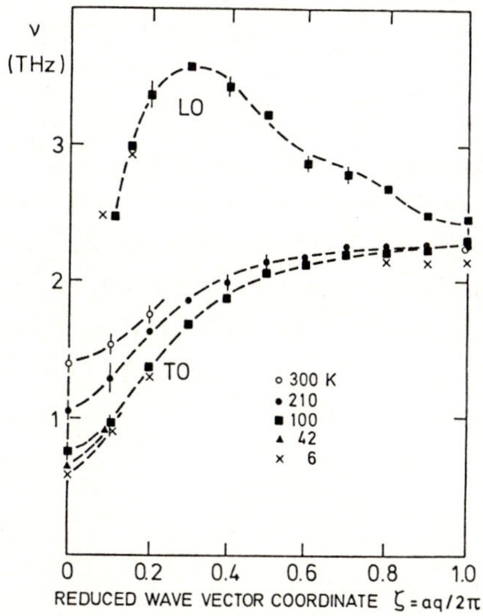

<u>Fig. 5.13.</u> SnTe. The dispersion relations for optical normal modes of vibration propagating along the <001> direction at various temperatures /66P/

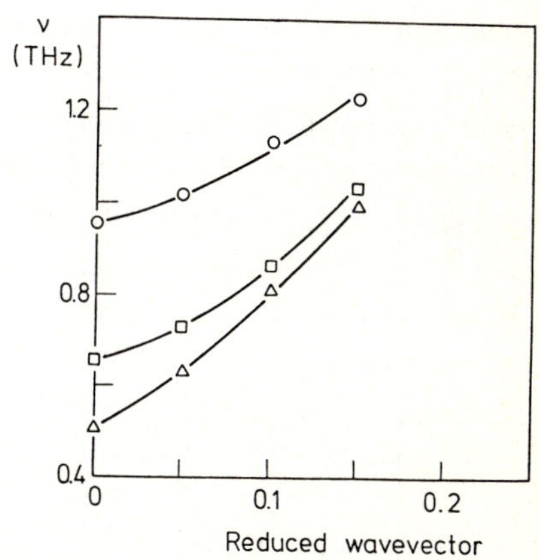

<u>Fig. 5.14.</u> $Pb_{0.80}Sn_{0.20}Te$. TO (<001>) phonon branch near zone-centre at (o) 239 K, (□) 78 K, and (△) 5 K /78D1/

The composition dependence of the LO and TO modes in $PbSe_{1-x}Te_x$ was analysed from reststrahlen spectra by FINKENRATH et al. /79F/. The experiments were carried out with polycrystalline bulk samples as well as with thin films. The results shown in Figs. 5.12a-c give evidence for a two-mode behaviour of this mixed crystal. Also the localized mode frequencies for Se impurities in PbTe and for Te impurities in PbSe are presented in Figs. 5.12a-c.

The variation of temperature reveals the softening of the TO modes as shown in Figs. 5.12 as well as in the dispersion relations of Figs. 5.13,14. The temperature dependence of the static dielectric constant and of the TO phonon energy was observed in many experimental investigations to obey approximately a Curie law in the para-electric range

$$\varepsilon_0^{-1} = \gamma_1(T - T_c) \quad , \tag{5.1}$$

or taking into consideration the Lyddane-Sachs-Teller relation

$$\omega_{TO}^2 = \gamma_2(T - T_c) \quad , \tag{5.2}$$

where $\gamma_{1,2}$ are constants and T_c the Curie temperature.

Equivalently, measuring the capacitance C of a p-n junction diode (Fig. 5.15) also a Curie law was found

$$(A/C)^{3/2} = \gamma_3(V_b - V)\,(T - T_c) \tag{5.3}$$

where A is the cross sectional area of the junction, γ_3 is a constant, V the applied static bias voltage and V_b is the intercept of the $C^{-3/2}$ vs V plot on the voltage axis /70B2 , 70B3 , 76J/.

Some experimental results of ω_{TO}^2 vs T are shown in Figs. 5.16 - 18, further data are published e.g. in /66P , 72A , 73D5/ from neutron scattering, and /72K3 , 76B3/ from infrared spectroscopy. As Fig. 5.16 shows, in PbTe the softening of the square of the energy is linear for temperatures above 20 K. Below this temperature ω_{TO}^2 decreases less rapidly than linearly and appears to become temperature independent as T approaches zero. This is similar to the (capacitance)$^{-3/2}$ vs T plot of Fig. 5.15 and to the $1/\varepsilon_0$ vs T plots of Fig. 5.19. An explanation of this feature was offered by NISHI et al. /80N2/. They explained the low-temperature discrepancies from the

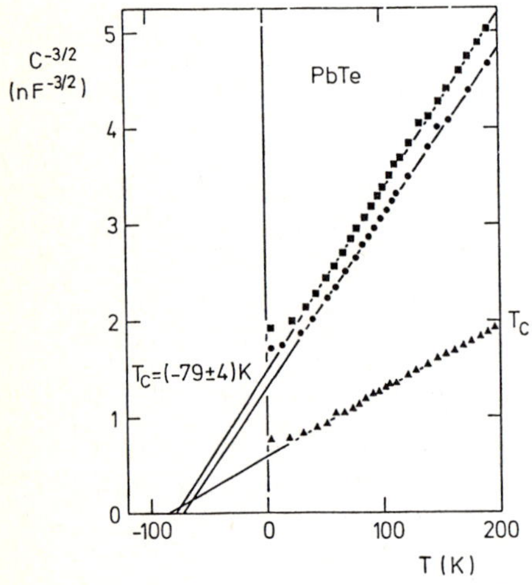

Fig. 5.15. PbTe. Plots of (capacitance)$^{-3/2}$ vs temperature for a fixed reverse bias of 1V for three PbTe diodes having different charge gradients. The implied Curie law behaviour of the dielectric constant yields a Curie temperature T_c of -79 K /70B2/

Curie law by anharmonic interaction of the lattice vibrations. Including fourth-order anharmonic terms in the Hamiltonian of lattice vibrations a good fit was obtained to the experimental data of $Pb_{1-x}Sn_xTe$ as shown in Fig. 5.19. The static dielectric constant was calculated from TO phonon energy using the Lyddane-Sachs-Teller relation.

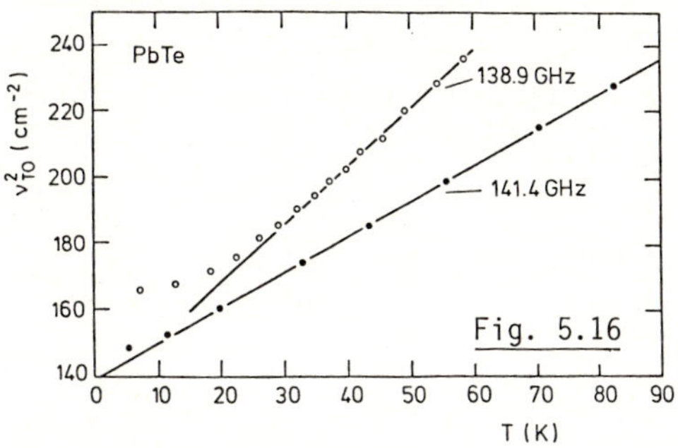

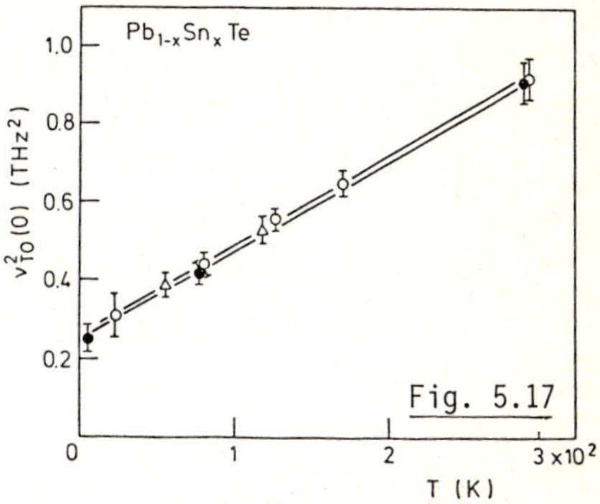

Fig. 5.16. PbTe. Plots of the square of the TO-mode energy ν^2_{TO} vs temperature as obtained for two samples from microwave magneto-plasma experiments. The data measured at 138.9 GHz correspond to an n-type sample, those at 141.4 GHz to a p-type sample /77F/

Fig. 5.17. $Pb_{1-x}Sn_xTe$. Soft-mode analysis for TO zone-centre phonons from neutron scattering with three crystals. ○,▲: x = 0.13; ●: x = 0.20 /78D1/

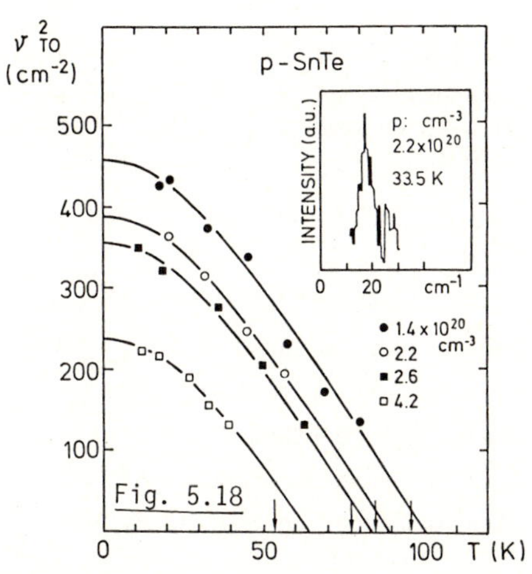

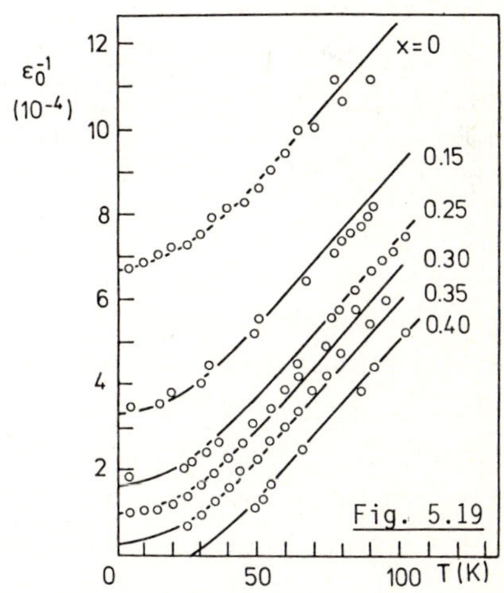

Fig. 5.18. SnTe. Plots of the square of the TO-mode energy vs temperature as obtained from Raman scattering. The experiments were performed with an argon-ion laser in back scattering geometry, a signal is shown in the inset. Data of samples with four different carrier densities are shown. The solid lines are calculated curves. The arrows indicate the transition temperature obtained from electrical resistivity measurements for the four carrier concentrations /78K2/

Fig. 5.19. $Pb_{1-x}Sn_xTe$. Inverse static dielectric constant versus temperature. Solid lines represent theoretical data considering anharmonic terms of lattice vibrations /80N2/

Many investigations are devoted to the static dielectric constant, see e.g.: $Pb_{1-x}Sn_xTe$ /66C , 66P , 68B2 , 76B3 , 76T2 , 76V3 , 77F1 , 77S4 , 78K2 , 79S3 , 80N2 , 81L/; PbSe /74B2/.

Some of the results are presented in Table 5.2. In Fig. 5.20 low temperature data of $1/\varepsilon_0$ are shown as a function of tin composition for $Pb_{1-x}Sn_xTe$. According to this investigation by NISHI et al. (in /78K2/) the static dielectric constant diverges at x = 0.36 at a temperature of 4.2 K, and at higher tin composition the ferroelectric phase will be established.

The temperature dependence of the energy width of the TO zone-centre phonon was investigated by DAUGHTON et al. /78D1/ and their results are shown in Fig. 5.21 for $Pb_{1-x}Sn_xTe$ (x = 0.13, 0.20).

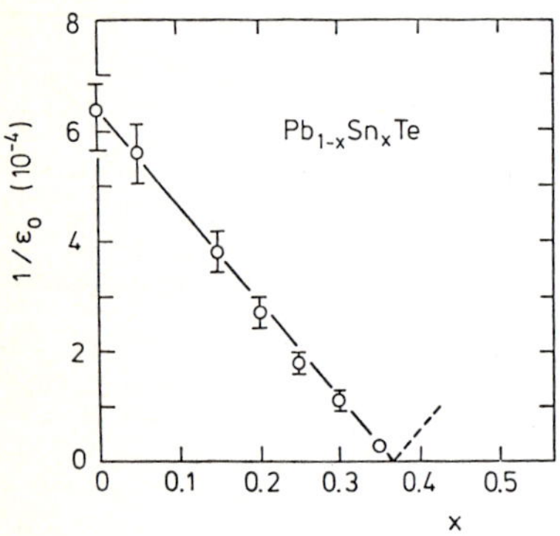

Fig. 5.20. $Pb_{1-x}Sn_xTe$. Inverse static dielectric constant as a function of tin composition at 4.2 K /78K2/

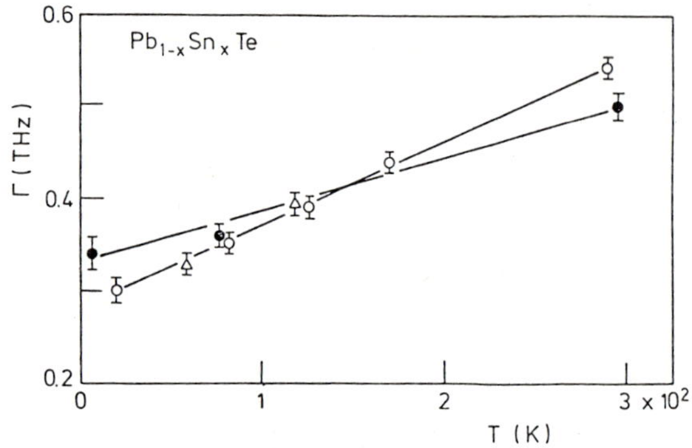

Fig. 5.21. $Pb_{1-x}Sn_xTe$. Energy width Γ of TO zone-centre phonon vs temperature T from neutron scattering with three crystals. ○,△: x = 0.13; ●: x = 0.20 /78D1/

The analysis of TO mode softening by COWLEY et al. /69C2/ first suggested that lattice stability depends upon free-carrier concentration. As shown in Fig. 2.8 the phase transition temperature decreases with increasing carrier density in $Pb_{1-x}Sn_xTe$. In addition, there is evidence that the static dielectric constant is also a function of carrier concentration as shown in Fig. 5.22 for several compositions of $Pb_{1-x}Sn_xTe$. This effect has been explained by interband coupling induced by TO phonons.

The theoretical model was introduced by KRISTOFFEL and KONSIN /68K3 , 73K3/. Later a number of changes to this model were made by several groups, e.g. /74K1 , 76N2 , 76N3 , 78D2/. KAWAMURA et al. /74K1 , 76K1 , 76V3 , 78K2 , 80N2/ particularly investigated the carrier dependence of the static dielectric constant in $Pb_{1-x}Sn_xTe$. They described the electron-TO-phonon coupling by the deformation potential and obtained the following relation for the effective renormalized TO phonon frequency

Table 5.2. Static dielectric constant

Material	T (K)	ε_0	Carrier Concentration ($10^{17}/cm^3$)	Comment	Ref.
PbTe	300	412		bulk crystal	66C
PbTe	300	384.5		epitaxial layer, infrared spectroscopy	76B3
PbTe	80	900	$n \approx 1 - 10$	magnetoplasma reflection, bulk crystal	80N2
PbTe	77	600	p = 2	from capacitance-voltage characteristic of Schottky barriers on epitaxial film	76J
PbTe	77	180	p = 7	from capacitance-voltage characteristic of Schottky barriers on epitaxial film	76J
PbTe	77	1590	n = 0.78	microwave transmission at 9 GHz, epitaxial layer	81L
PbTe	77	1980	p = 3.2	microwave transmission at 9 GHz, epitaxial layer	81L
PbTe	5	1333		epitaxial layer, infrared spectroscopy	76B3, 78B3
PbTe	4.2	3000	n = 4	magnetoplasma reflection, bulk crystal	77F1
PbTe	4.2	1500	n = 2.4	magnetoplasma reflection, bulk crystal	80N2
PbTe	4	1340	p	neutron scattering	72A
$Pb_{0.8}Sn_{0.2}Te$	300	450	intrinsic	microwave transmission at 9 GHz, epitaxial layer	81L
$Pb_{0.8}Sn_{0.2}Te$	300	230	n = 2	microwave transmission at 9 GHz, epitaxial layer	81L
$Pb_{0.8}Sn_{0.2}Te$	77	2010	p = 0.7	microwave transmission at 9 GHz, epitaxial layer	81L
$Pb_{0.8}Sn_{0.2}Te$	77	1850	n = 0.14	microwave transmission at 9 GHz, epitaxial layer	81L
$Pb_{0.8}Sn_{0.2}Te$	4.2	4000	n = 1.3	magnetoplasma reflection, bulk crystal	80N2

Table 5.2. (continued)

Material	T (K)	ε_0	Carrier Concentration ($10^{17}/cm^3$)	Comment	Ref.
$Pb_{0.75}Sn_{0.25}Te$	80	1750	$p \approx 1 - 10$	magnetoplasma reflection, bulk crystal	80N2
$Pb_{0.75}Sn_{0.25}Te$	4.2	5800	$p = 2.2$	magnetoplasma reflection, bulk crystal	80N2
PbSe	300	203	$n = 1$	infrared reflectivity, bulk crystal	74B2
PbS	77	181		calculated using Lyddane-Sachs-Teller relation	73D3
PbS	300	169		calculated using Lyddane-Sachs-Teller relation	73D3
PbS	373	160		calculated using Lyddane-Sachs-Teller relation	73D3
SnTe	300	1770		calculated from Lyddane-Sachs-Teller relation	68B2
SnTe	100	1200		calculated from Lyddane-Sachs-Teller relation	66P

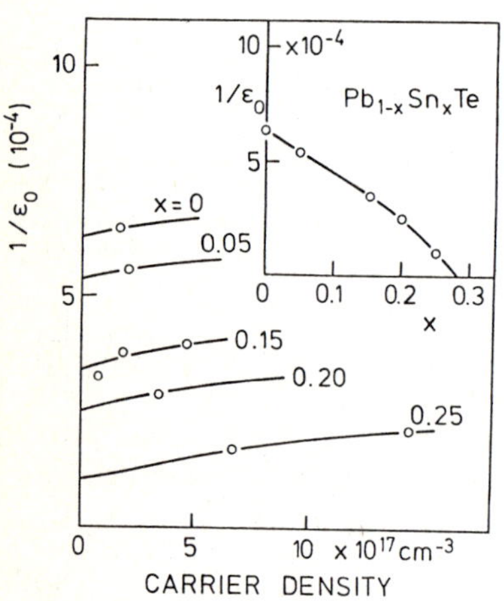

Fig. 5.22. $Pb_{1-x}Sn_xTe$. Inverse static dielectric constant as a function of carrier concentration at 1.6K /76K1/. The inset shows data extrapolated to zero carrier density

$$\tilde{\omega}_{TO}^2 = \omega_{TO}^2 - \frac{4\,\Xi_{cv}^2}{M a^2 W} \ln\left(\frac{W}{2E_F + E_0}\right) \quad , \tag{5.4}$$

where E_F is the Fermi energy measured from the band edge, E_0 the gap energy, W the sum of the bandwidths of conduction and valence bands, M the reduced mass of two different ions, a the lattice constant, and Ξ_{cv} the optical deformation potential.

The authors /74K2/ have obtained a good fit of the above relation to their experimental data as shown in Fig. 5.23. This figure shows the static dielectric constant as a function of the effective band gap $\tilde{E}_0 = E_0 + 2E_F$ at 1.6 K, corresponding to

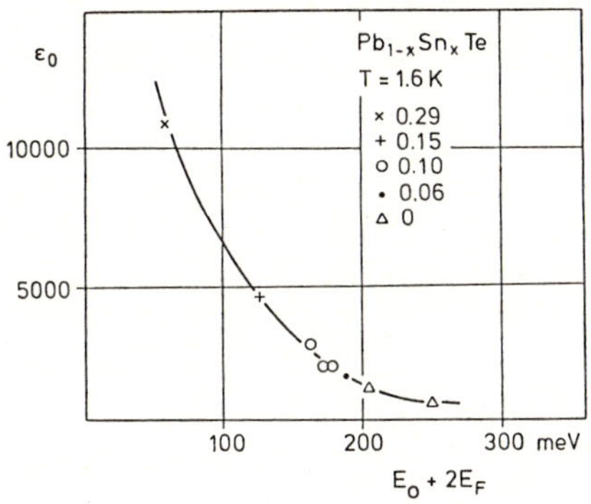

Fig. 5.23. $Pb_{1-x}Sn_xTe$. Static dielectric constant as a function of the sum of band gap and Fermi level energies. The data correspond to bulk samples with different x values, as indicated. The solid line is calculated considering the electron-TO-phonon coupling through interband optical deformation potential /74K2/

various carrier concentrations and x values. However, other authors, e.g. /80N2, 81L/ could not obtain such a good fit. In Fig. 5.24 data of various experimental studies are plotted versus effective band gap for the same mixed crystal system at 77 K.

The dependence of dielectric constant, i.e. TO phonon frequency on carrier concentration is still not understood satisfactorily. JANTSCH et al. /76J/

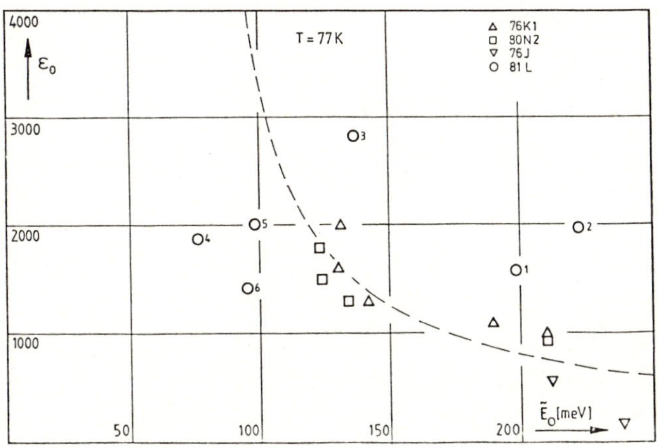

Fig. 5.24. $Pb_{1-x}Sn_xTe$. Static dielectric constant as a function of the effective band gap. The data are obtained from bulk material (Δ, □), and from epitaxial layers (∇, o) /81L/

studied the dependence of T_c and ε_0 on the concentration of vacancies in PbTe and on the growth conditions of the crystal - naturally, the number of vacancies being coupled with the growth conditions. They concluded that the dielectric behaviour of PbTe is sensitive to the degree of compensation between lead and tellurium vacancies.

In conclusion there are some hints that the above-mentioned theoretical models give a reasonable description of the experimental results of bulk material. However, due to the lattice mismatch between epitaxial layer and substrate the results from films may be strongly influenced by a mismatch induced strain. For instance, dropping the data of the static dielectric constant obtained from epitaxial layers in Fig. 5.24, the agreement between theory and the remaining experimental data from bulk material becomes acceptable.

6. Recombination of Carriers

6.1 General Remarks

The recombination process of free carriers plays a key role in limiting both the sensitivity and the rise time of a photodetector. On the other hand, the efficiency too of a light emitting diode is determined by the amount of non-radiative recombination transitions. Similarly, as in the narrow-gap alloys of $Hg_{1-x}Cd_xTe$, the direct band-band Auger process is important in the IV-VI narrow-gap compounds /80N1/. The Auger mechanism is expected to become more and more relevant with decreasing band gap. This is a consequence of the exponential dependence of the Auger lifetime on the gap energy.

In the following sections it will become evident that for the time being the analysis of the experimental data on carrier lifetime is by no means satisfactory. In the next section some representative experimental data on recombination in lead chalcogenides will be presented. After that models and calculated values for intrinsic recombination mechanisms will be introduced.

6.2 Experimental Data

Most of the recombination studies are devoted to the $Pb_{1-x}Sn_xTe$ alloy. Lifetime of carriers was determined from photoconductivity experiments /76B2 , 77L4 , 78D5 , 78H , 78L2 , 78S , 79K2 , 81W2/, from the decay of impact-ionized carriers /78S/, from photomagnetic effect /76B2 , 78H/, and in photoemission experiments /79K1 , 79Y1/. The results of the first extensive investigation of $Pb_{1-x}Sn_xTe$ ($0.17 \leqslant x \leqslant 0.22$) revealed an

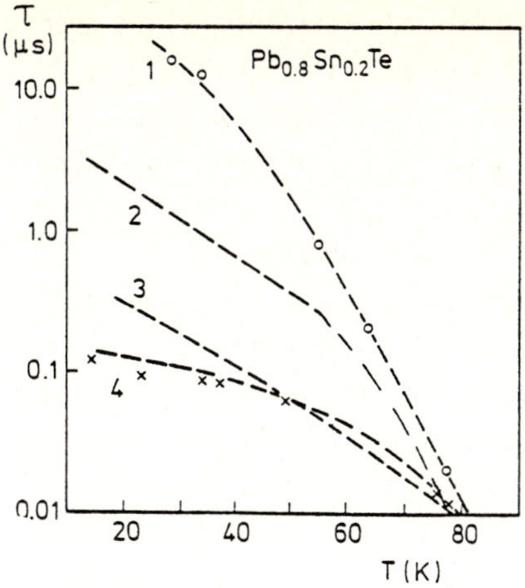

Fig. 6.1. $Pb_{0.8}Sn_{0.2}Te$. Photoconductive decay time vs temperature for four different samples /81W2/. Sample
1) $n = 1 \cdot 10^{16}/cm^3$ /81W2/;
2) $n = 5 \cdot 10^{16}/cm^3$ /78S/;
3) $n = 24 \cdot 10^{16}/cm^3$ /78S/;
4) $p = 10 \cdot 10^{16}/cm^3$ /81W2/

extremely large variation of the lifetime between different samples, e.g. at 77 K lifetimes between 10^{-12} and 10^{-8} s were found in different samples. A satisfactory interpretation of this behaviour was not given /76B2 , 78H/.

Low-temperature lifetime data of $Pb_{0.8}Sn_{0.2}Te$ measured by two groups are plotted in Fig. 6.1. Samples 1 - 4 have carrier concentrations 1; 5; 24; $10 \cdot 10^{16}$ cm^{-3} respectively, sample 4 is p-type, all the others are n-type. An interesting point is that in spite of the strong variation in carrier density the lifetime of all the samples

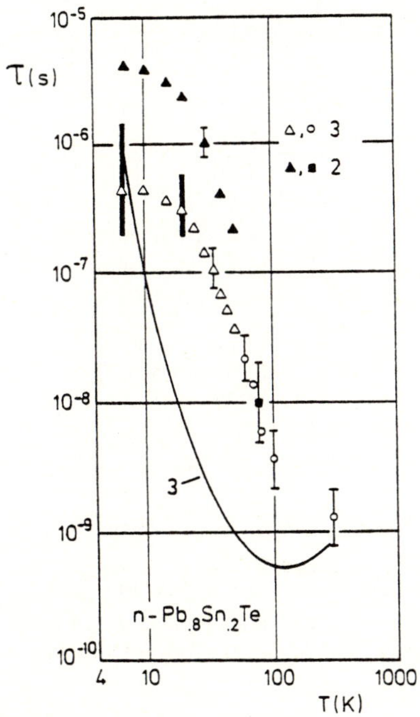

Fig. 6.2. $Pb_{0.8}Sn_{0.2}Te$. Photoconductive decay time vs temperature for the temperature range 6 - 300 K for samples 2 and 3 of Fig. 6.1. The solid line is calculated for Auger recombination for Sample 3 /78S/

is about 10^{-8} s near 80 K /81W2/. This result excludes any importance of the Auger mechanism at this temperature. Data of samples 2 and 3 are presented in Fig. 6.2 for the temperature range 6 to 300 K. The saturation observed below 20 K was found to be caused by minority carrier sweep-out /78S/. The solid line represents calculated data for dominant Auger recombination, which yielded much shorter lifetimes than the observed ones. This will be discussed in the following sections.

In the experiments performed with $Pb_{0.8}Sn_{0.2}Te$ mostly a linear transient carrier decay was observed, i.e. at ionization rates up to 10% the lifetime was not depending on excess carrier concentration. The situation was quite different in PbTe crystals /77L4 , 78L2 , 78S/. The transient decay of photoionized carriers, i.e. conductivity, was found to be strongly non-linear as shown in Fig. 6.3. Even in the case of low ionization rates of about 1% the instantaneous decay time varied from many ms to some μs. This behaviour is indicated by the hatched areas in Fig. 6.4, which cover the observed range of instantaneous lifetimes /78S/. The time-resolved analysis of the photo Hall effect by SCHLICHT et al. /78S/ revealed, that in the PbTe samples a minority carrier trapping takes place. Thus an intrinsic recombination, e.g. Auger or radiative transitions, can be ruled out at temperatures below

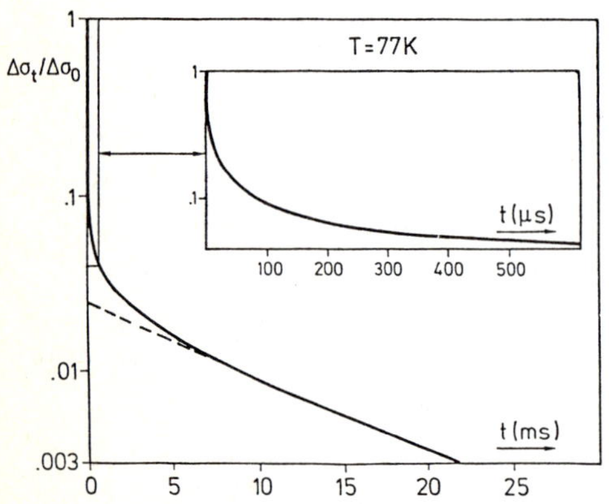

Fig. 6.3. PbTe, epitaxial layer. Change in conductivity due to a laser pulse (width ≈ 20 ns) versus time for a sample with $n(77\,K) = 6\cdot10^{16}/cm^3$ /78S/. $\Delta\sigma_t$ is the excess conductivity at time t, $\Delta\sigma_0$ is the peak value at t = 0

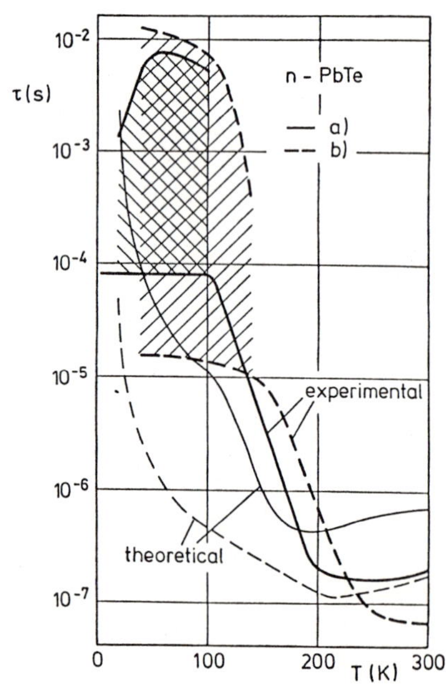

Fig. 6.4. PbTe. Photoconductive decay time vs temperature for two epitaxial layers a) $n(77\,K) = 10^{16}/cm^3$, b) $n = 6\cdot10^{16}/cm^3$. Calculated data for Auger recombination. The hatched areas indicate the range of observed instantaneous lifetime during the decay /78S/

140 K. LISCHKA and HUBER /78L2/ who observed similar recombination behaviour in PbTe samples, concluded that the temperature-dependent donor level found in PbTe /78L2 , 78S/ may act as a trapping or recombination centre.

Compared to the results of /78L2 , 78S/ a very short minority carrier lifetime was measured on a PbTe homojunction laser at low temperatures by KASEMSET and FONSTAD /79K1/. The data are shown in Fig. 6.5.

Likewise, short lifetimes were observed by YUNOVICH et al. /79Y1/ in $PbS_{1-x}Se_x$ films.

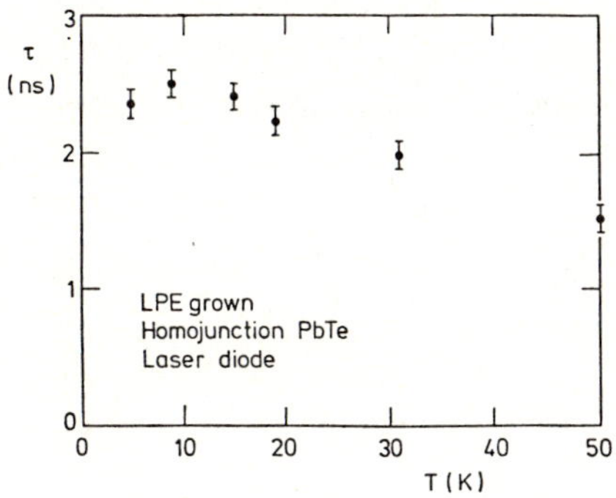

Fig. 6.5. PbTe. Minority carrier lifetime vs temperature measured on a homojunction laser /79K1/

6.3 Theoretical Investigations

The experimental data published so far show that in most of the investigated crystals the recombination is dominated by recombination centres. These are represented by lattice defects which may be caused either by impurities or by native defects such as interstitials or vacancies. Recombination transitions via energy levels of lattice defects are usually called extrinsic mechanisms, since they can be manipulated more or less by the preparation of the crystal. The other class of recombination transitions is called intrinsic mechanisms, since these transitions are determined by the physical properties of the crystal alone thus representing the limiting recombination mechanisms. The major intrinsic mechanisms are the Auger and the radiative recombination processes, single phonon and plasmon recombination are of minor importance. However, under special conditions they may dominate the recombination in narrow-gap semiconductors /75C , 78D3 , 80N1 , 80Z/. In the following sections we shall report on the theoretical aspects of the radiative and Auger recombination in the lead chalcogenides.

6.3.1 Radiative and Plasmon Recombination

The radiative recombination in lead chalcogenides was investigated by several groups recently (see for instance /79K1 , 79M , 80Z/). KASEMSET and FONSTAD calculated the rate of spontaneous emission following the general treatment of LASHER and STERN /64L/ by

$$r_{spon}(E)\,dE = \frac{\chi N e^2 E\,|M|^2\,\rho_{red}(E)\,f_u(1-f_\ell)\,dE}{\pi m_0^2 \hbar^2 c}\,, \qquad (6.1)$$

where $|M|^2$ is the momentum matrix element for the optical transition averaged over all possible directions of polarization, ρ_{red} is the reduced density of states, f_u and f_ℓ are the occupational probabilities of the upper and lower levels, respectively, χ is the susceptibility of free space, N is the index of refraction and m_0 is the free electron mass. Taking into consideration the band parameters by HEWES et al. /73H3/ and assuming degenerate statistics the simple relation for the lifetime in $Pb_{1-x}Sn_xTe$ was obtained

$$\tau_R = \frac{9.9\cdot 10^{-8}}{N E_0}\,s\,, \qquad (6.2)$$

where N is the index of refraction and E_0 the band gap energy in eV. The lifetime obtained by this relation is plotted versus tin fraction at 4.2 K in Fig. 6.6. For

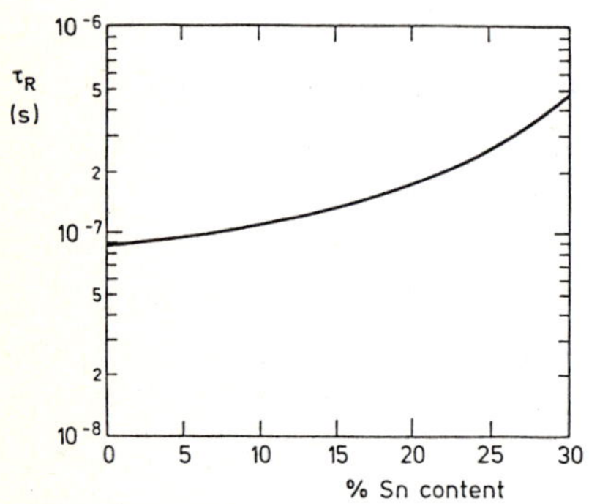

Fig. 6.6. $Pb_{1-x}Sn_xTe$. Calculated radiative lifetime at 4.2 K vs tin fraction using (6.2) /79K1/

instance the lifetime for PbTe is found to be 85 ns. The calculated values of Fig. 6.6 are considerably higher than the experimental ones measured on a PbTe homojunction laser by these authors (see Fig. 6.5). ZIEP et al. /80Z/, also using the model by LASHER and STERN /64L/ calculated the spontaneous lifetime for $Pb_{0.78}Sn_{0.22}Te$ and $Pb_{0.91}Sn_{0.09}Te$ as a function of temperature with the carrier concentration as parameter. Their calculated data are shown in Figs. 6.7,8. For comparison these authors have also presented calculated data for plasmon and Auger recombination. The latter mechanism will be discussed in the following section.

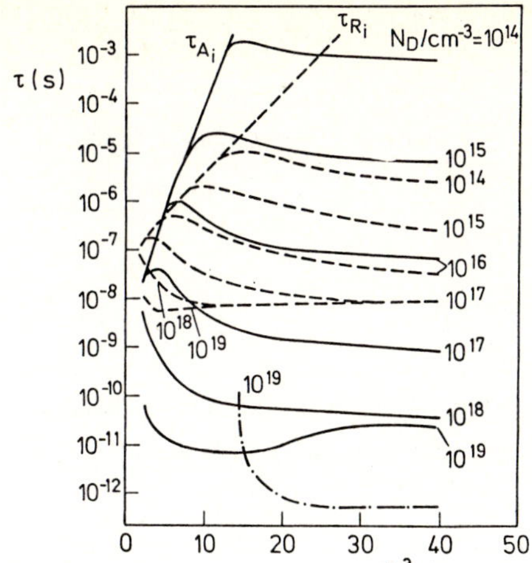

Fig. 6.7

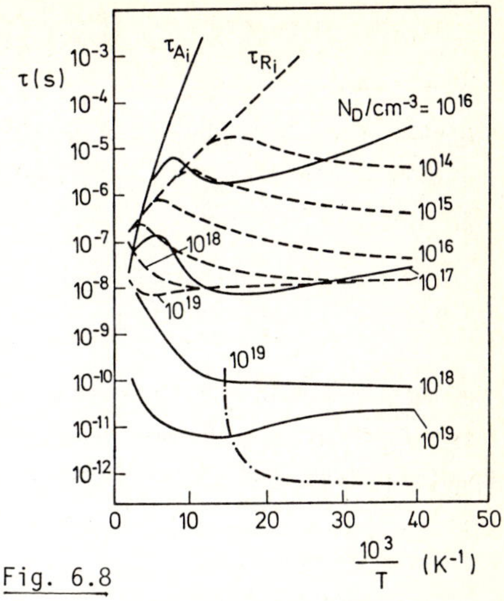

Fig. 6.8

Fig. 6.7. $Pb_{0.78}Sn_{0.22}Te$. Calculated lifetime vs temperature for Auger (——, τ_{Ai} for intrinsic material), for radiative (-----, τ_{Ri} for intrinsic material), and for plasmon recombination (-·-·-). Parameter is the donor concentration /80Z/

Fig. 6.8. $Pb_{0.91}Sn_{0.09}Se$. Calculated lifetime vs temperature for Auger (——, τ_{Ai} for intrinsic material), for radiative (-----, τ_{Ri} for intrinsic material), and for plasmon recombination (-·-·-). Parameter is the donor concentration /80Z/

In some experiments with $Pb_{0.8}Sn_{0.2}Te$ it was observed that the lifetime is significantly higher than expected from estimates of the intrinsic recombination mechanisms /78S, 81W2/. This may be caused by carrier trapping. Another mechanism, however, was discussed by WEISER et al. /81W2/. In order to explain a lifetime enhancement they suggested a possible recycling of photons. This is an effect in which a photon created by the recombination of an electron-hole pair is reabsorbed to create a new pair and the process is repeated several times until a photon can either escape or a non-radiative recombination takes place. Estimates made by these authors yielded an enhancement of the effective lifetime up to two orders of magnitude. This mechanism should be considered, since the data in question /78S, 81W2/ did not give evidence for any carrier trapping.

Plasmon recombination becomes very efficient at high carrier concentrations, i.e. if for the plasmon energy the condition holds

$$\hbar\omega_{pl} \gtrsim E_0 \quad . \tag{6.3}$$

Assuming a q independent undamped plasmon mode ZIEP et al. /80Z/ obtained for the small signal carrier lifetime the relation

$$\tau_{pl}^{-1} = \frac{e^2\pi^{1/2}E_0^{5/2}(z_p - 1)\exp\left[-\frac{1}{2}\beta^{-1}(z_p - 1)\right]\arcsin\sqrt{1 - P_\ell^2/P_t^2}}{\varepsilon_\infty P_t h^2\omega_{pl}\sqrt{1 - P_\ell^2/P_t^2}\,(k_BT)^{1/2}} I_1\left[\frac{1}{2}\beta^{-1}(z_p - 1)\right] \quad (6.4)$$

where P_ℓ, P_t are momentum matrix elements of the Kane-type two-band model, $z_p = \hbar\omega_{pl}/E_0$, $\beta = kT/E_0$, n_0 is the extrinsic carrier concentration, $I_1(x)$ the Bessel function, and

$$\hbar\omega_{pl} = \left[\frac{4\pi e^2 n_0(2P_t^2 + P_\ell^2)}{3\varepsilon_\infty E_0}\right]^{1/2} . \quad (6.5)$$

The numerical data made ZIEP et al. /80Z/ conclude, that at carrier concentrations above 10^{19} cm^{-3} the plasmon recombination dominates over radiative and Auger recombination (Figs. 6.7,8).

6.3.2 Auger Recombination

Auger recombination is based on the Coulomb interaction of the free carriers. The energy from an electron-hole recombination is released to a third particle, either to an electron or to a hole. The free carrier taking the recombination energy is pushed to an excited state in the conduction or valence band respectively. It was shown by BEATTIE and LANDSBERG /59B/ that the dominant temperature dependence for a non-degenerate and intrinsic semiconductor is given by

$$\tau_{Ai} \propto \left(\frac{E_0}{kT}\right)^{3/2} \cdot \exp\left\{\frac{E_0}{kT}\right\} \exp\left\{\frac{\mu}{1+\mu}\frac{E_0}{kT}\right\} \quad (6.6)$$

where is $\mu = m_c/m_v$ if $m_c < m_v$ and the lifetime is determined by electron-electron collisions, and $\mu = m_v/m_c$ if $m_c > m_v$ and the lifetime is determined by hole-hole collisions. The second exponential containing the factor μ results from momentum and energy conservation in the recombination process. Thus the Auger mechanism is expected to be strongly hampered by nearly symmetric conduction and valence bands in the lead chalcogenides. However, it was pointed out by EMTAGE /76E/ that due to the interaction between electrons (holes) of non-equivalent energy valleys this handicap may be compensated to some extent. Non-equivalent valleys are valleys elongated parallel to different diagonals in k-space. The Auger transition due to an e-e impact between two non-equivalent valleys is sketched in Fig. 6.9. EMTAGE obtained for this case instead of the second exponential of relation (6.6) an exponential

$$\tau_{Ai} \propto \exp\left(\frac{r E_0}{kT}\right) , \quad (6.7)$$

where $r = 1/K_m$ is the ratio of transverse-to-longitudinal effective masses. This ratio, therefore, plays the same part as μ in reducing markedly the temperature dependence, r being in the lead chalcogenides of the order of 0.1 (see Chap. 3).

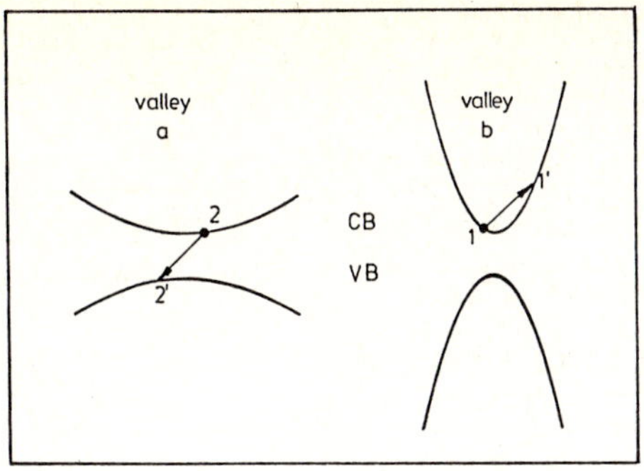

Fig. 6.9. Auger transition due to an e-e impact between two non-equivalent valleys /80N1/

Calculated data according to Emtage's model are shown in Figs. 6.2, 4, 7, 8. Two groups, LISCHKA and HUBER /77L4/, and HERRMANN /78H/, claim to have obtained an agreement between theoretical data and experimental values of some PbTe and PbSnTe samples. An agreement between experimental data and calculated Auger data was not obtained by SCHLICHT et al. /78S/ and WEISER et al. /81W2/. Instead, the observed lifetime was higher than that expected for Auger recombination. Also the dependence of lifetime on carrier density in the extrinsic range of conductivity was not observed (see Figs. 6.1, 7, 8). It was pointed out by NIMTZ /80N1/ that the screening of the Coulomb interaction might be not properly considered. Only the screening by valence electrons (ε_∞) and not a possible screening due to the highly polarized lattice ($\varepsilon_0 \approx 100 - 10\,000$) (see Chap. 5) was taken into consideration. An enhanced screening would result in a smaller Auger recombination rate.

Thus we conclude that in the lead chalcogenides the Auger effect is expected to be important; however, experimental evidence for this mechanism is not yet unambiguously presented.

References

40B P.W. Bridgman: Proc. Am. Acad. Arts Sci. *74*, 21 (1940)
48B P.W. Bridgman: Proc. Am. Acad. Arts Sci. *76*, 55 (1948)
50F H.Y. Fan: Phys. Rev. *78*, 808 (1950); *82*, 900 (1951)
51B S. Bhagavantam, T.S. Rao: Nature *168*, 42 (1951)
54B P.W. Bridgman: Proc. Am. Acad. Arts Sci. *83*, 151 (1954)
56B J. Bloem: Philips Res. Rep. *11*, 273 (1956)
56K E.O. Kane: J. Phys. Chem. Solids *1*, 249 (1956)
57S K. Shogenji, S. Uchiyama: J. Phys. Soc. Japan *12*, 1164 (1957)
58A1 R.S. Allgaier, W.W. Scanlon: Phys. Rev. *111*, 1029 (1958)
58A2 R.S. Allgaier: Phys. Rev. *112*, 828 (1958)
58H M. Hansen: *Constitution of Binary Alloys* (McGraw-Hill, New York 1958)
59B A.R. Beattie, P.T. Landsberg: Proc. Roy. Soc. *A 249*, 16 (1959)
59S1 W.W. Scanlon: J. Phys. Chem. Solids *8*, 423 (1959)

59S2 K. Shogenji: J. Phys. Soc. Japan *14*, 1360 (1959)
60A R.S. Allgaier: Phys. Rev. *119*, 554 (1960)
60B R.F. Brebrick, R.S. Allgaier: J. Chem. Phys. *32*, 1826 (1960)
61A R.S. Allgaier: J. Appl. Phys. *32*, 2185 (1961)
61C M.H. Cohen: Phys. Rev. *121*, 387 (1961)
61K H. Krebs, K. Grün, D. Kallen: Z. anorg. und allg. Chemie *312*, 307 (1961)
61P W. Paul: J. Appl. Phys. *32*, 2082 (1961)
61S F. Stern: J. Appl. Phys. *32*, 2166 (1961)
62C A.A. Chudinov: Sov. Phys. Solid State *4*, 553 (1962)
63B R.F. Brebrick: J. Phys. Chem. Sol. *24*, 27 (1963)
63C1 A.A. Chudinov: Sov. Phys. Solid State *5*, 1061 (1963)
63C2 A.A. Chudinov: Sov. Phys. Crystallogr. *8*, 374 (1963)
63H F. Herman, S. Skillman: *Atomic Structure Calculations* (Prentice-Hall, Engelwood Cliffs N.J. 1963)
63I K. Igaki, N. Ohashi: J. Phys. Soc. Japan *18* Suppl., 143 (1963)
63J L.E. Johnson, J.B. Conklin, G.W. Pratt: Phys. Rev. Lett. *11*, 538 (1963)
63S Y. Sato, M. Fujimoto, A. Kobayashi: Japan J. Appl. Phys. *2*, 683 (1963)
63Z W. Zawadzki: Phys. Stat. Sol. *3*, 1421 (1963)
64A R.S. Allgaier, B.B. Houston, R.F. Bis, J. Babiskin, P.G. Siebenmann: Proc. 7th Int. Conf. Phys. Semicond. (Dunod, Paris 1964) p. 659
64C1 M. Cardona, D.L. Greenaway: Phys. Rev. *133*, A1685 (1964)
64C2 K.F. Cuff, M.R. Ellett, C.D. Kuglin, L.R. Williams: Proc. 7th Int. Conf. Phys. Semicond. (Dunod, Paris 1964) p. 677
64C3 W. Cochran: Phys. Lett. *13*, 193 (1964)
64G G. Gilat, G. Dolling: Phys. Lett. *8*, 304 (1964)
64K A. Kobayashi, Y. Sato, M. Fujimoto: Proc. 7th Int. Conf. Phys. Semicond. (Dunod, Paris 1964) p. 1257
64L G. Lasher, F. Stern: Phys. Rev. *133*, A 553 (1964)
64M D.L. Mitchell, E.D. Palik, J.N. Zemel: Proc. 7th Int. Conf. Phys. Semicond. (Dunod, Paris 1964) p. 325
64N S.I. Novikova, N.Kh. Abrikosov: Sov. Phys. Solid State *5*, 1397 (1964)
64S D.R. Simpson: Econ. Geol. *59*, 150 (1964)
64W E.R. Washwell, K.F. Cuff: Symposium on "Radiative Recombination in Semiconductors", Paris (1964)
65B J.F. Butler, A.R. Calawa: in *Physics of Quantum Electronics*, ed. by P.L. Kelly, B. Lax, P.E. Tannenwald (McGraw-Hill, New York 1965), p. 458
65C J.B. Conklin, L.E. Johnson, G.W. Pratt: Phys. Rev. *137*, A 1282 (1965)
65D J.R. Dixon, H.R. Riedl: Phys. Rev. *138*, A 873 (1965)
65L D. Liberman, J.T. Waber, D.T. Cromer: Phys. Rev. *137*, A27 (1965)
65N P.M. Nikolic: Brit. J. Appl. Phys. *16*, 1075 (1965)
65R A.J. Rosenberg, F. Wald: J. Phys. Chem. Sol. *26*, 1079 (1965)
65S W.E. Spicer, G.J. Lapeyre: Phys. Rev. *139*, A 565 (1965)
65Z J.N. Zemel, J.D. Jensen, R.B. Schoolar: Phys. Rev. *140*, A 330 (1965)
66A1 R.S. Allgaier, B.B. Houston: J. Appl. Phys. *37*, 302 (1966)
66A2 S.V. Airapetyants, M.N. Vinogradova, I.N. Dubrovskaya, N.V. Kolomoets, I.M. Rudnick: Sov. Phys. Solid State *8*, 1069 (1966)
66B1 E.G. Bylander: Mater. Sci. Eng. *1*, 190 (1966)
66B2 J.R. Burke, B. Houston, H.T. Savage, J. Babiskin, P.G. Siebenmann: J. Phys. Soc. Japan *21*, 384 (1966)
66C W. Cochran, R.A. Cowley, G. Dolling, M.M. Elcombe: Proc. Royal Soc. *A293*, 433 (1966)
66D1 J.O. Dimmock, I. Melngailis, A.J. Strauss: Phys. Rev. Lett. *16*, 1193 (1966)
66D2 I.N. Dubrovskaya, Yu.I. Ravich: Sov. Phys. Solid State *8*, 1160 (1966)
66E L. Esaki, P.J. Stiles: Phys. Rev. Lett. *16*, 1108 (1966)
66F M. Fujimoto, Y. Sato: Japan J. Appl. Phys. *5*, 128 (1966)
66M1 I. Melngailis, A.R. Calawa: Appl. Phys. Lett. *9*, 304 (1966)
66M2 D.L. Mitchell, R.F. Wallis: Phys. Rev. *151*, 581 (1966)
66N S.I. Novikova, L.E. Shelimova: Sov. Phys. Solid State *7*, 2052 (1966)
66P G.S. Pawley, W. Cochran, R.A. Cowley, G. Dolling: Phys. Rev. Lett. *17*, 753 (1966)
66S D.N. Seidmann: Trans. Met. Soc. AIME *236*, 1361 (1966)

66T R.N. Tauber, A.A. Machonis, I.B. Cadoff: J. Appl. Phys. *37*, 4855 (1966)
67A1 A.A. Andreev, V.N. Radionov: Sov. Phys. Semicond. *1*, 145 (1967)
67A2 A.A. Andreev: Sov. Phys. Solid State *9*, 1232 (1967)
67B S. Bermon: Phys. Rev. *158*, 723 (1967)
67C A.J. Crocker: J. Phys. Chem. Sol. *28*, 1903 (1967)
67D I.N. Dubrovskaya, E.D. Nensberg, G.V. Nikitina, Yu.I. Ravich: Sov. Phys. Solid State *8*, 1789 (1967)
67E M.M. Elcombe: Proc. Royal Soc. *A300*, 210 (1967)
67L T.L. Loucks: *Augmented Plane Wave Method* (W.A. Benjamin, Menlo Park, California, 1967)
67M A.N. Mariano, K.L. Chopra: Appl. Phys. Lett. *10*, 282 (1967)
67P V. Prakash: Technical Report No. HP-13 from the Division of Engineering and Applied Physics, Harvard University (1967) (unpublished)
67S A.J. Strauss: Phys. Rev. *157*, 608 (1967)
67T R.N. Tauber, I.B. Cadoff: J. Appl. Phys. *38*, 3714 (1967)
67V Y.P. Varshni: Physica (Utrecht) *34*, 149 (1967)
67W R.C. Weast (ed.): Handbook of Chemistry and Physics, 48th ed. (The Chemical Rubber Co 1967)
68A1 A.A. Andreev: J. Physique, Suppl. *29*, C4 - 50 (1968)
68A2 D.E. Aspnes, M. Cardona: Phys. Rev. *173*, 714 (1968)
68B1 J.M. Besson, W. Paul, A.R. Calawa: Phys. Rev. *173*, 699 (1968)
68B2 E. Burstein, S. Perkowitz, M.H. Brodsky: J. Physique, Suppl. *29*, C4 - 78 (1968)
68C1 A.R. Calawa, T.C. Harman, M. Finn, P. Youtz: Trans. AIME *242*, 374 (1968)
68C2 I.A. Chernik, V.I. Kaidanov, M.I. Vinogradova, N.V. Kolomoets: Sov. Phys. Semicond. *2*, 645 (1968)
68D J.R. Dixon, R.F. Bis: Phys. Rev. *176*, 942 (1968)
68H1 F. Hermann, R.L. Kortum, I. Ortenburger, J.P. van Dyke: J. Physique, Suppl. *29*, C4 - 62 (1968)
68H2 B. Houston, R.E. Strakna, H.S. Belson: J. Appl. Phys. *39*, 3913 (1968)
68K1 H. Krebs: *Fundamentals of Inorganic Crystal Chemistry* (McGraw-Hill, London 1968)
68K2 S.S. Kabalkina, N.R. Serebryanaya, L.F. Vereshchagin: Sov. Phys. Solid State *10*, 574 (1968)
68K3 N. Kristoffel, P. Konsin: Phys. Stat. Sol. *28*, 731 (1968)
68M Yu.V. Maltsev, I.K. Smirnov, Yu.I. Ukhanov, A.N. Veis: J. Physique, Suppl. *29*, C4 - 99 (1968)
68R S. Rabii: Phys. Rev. *167*, 801 (1968); Errata *173*, 918 (1968)
68S P.J. Stiles: J. Physique, Suppl. *29*, C4 - 105 (1968)
68W I. Wakabayashi, H. Kobayashi, H. Nagasaki, S. Minomura: J. Phys. Soc. Japan *25*, 227 (1968)
69B1 J.R. Burke, H.R. Riedl: Phys. Rev. *184*, 830 (1969)
69B2 J.F. Butler, T.C. Harman: IEEE J. *QE-5*, 50 (1969)
69C1 A.R. Calawa, J.O. Dimmock, T.C. Harman, I. Melngailis: Phys. Rev. Lett. *23*, 7 (1969)
69C2 E.R. Cowley, J.K. Darby, G.S. Pawley: J. Phys. *C2*, 1916 (1969)
69D1 R. Dalven: Infrared Physics *9*, 141 (1969)
69D2 M.I. Dyakonov, A.L. Efros, D.L. Mitchell: Phys. Rev. *180*, 813 (1969)
69H T.C. Harman, A.R. Calawa, I. Melngailis, J.O. Dimmock: Appl. Phys. Lett. *14*, 333 (1969)
69K L.N. Kurbatov, A.D. Britov, I.S. Averyanov, V.E. Mashchenko, N.N. Mochalkin, A.I. Dirochka: Sov. Phys. Semicond. *2*, 1008 (1969)
69P N.J. Parada, G.W. Pratt: Phys. Rev. Lett. *22*, 180 (1969)
69R S. Rabii: Phys. Rev. *182*, 821 (1969)
69T Y.W. Tung, M.L. Cohen: Phys. Lett. *A29*, 236 (1969); Phys. Rev. *180*, 823 (1969)
70A G.A. Antcliffe, J.S. Wrobel: Appl. Phys. Lett. *17*, 290 (1970)
70B1 R.L. Bernick, L. Kleinman: Solid State Commun. *8*, 569 (1970)
70B2 R.T. Bate, D.L. Carter, J.S. Wrobel: Proc. 10th Int. Conf. Phys. Semicond. (Cambridge, Massachusetts 1970) p. 125
70B3 R.T. Bate, D.L. Carter, J.S. Wrobel: Phys. Rev. Lett. *25*, 159 (1970)
70B4 J.R. Burke, B. Houston, H.T. Savage: Phys. Rev. *B2*, 1977 (1970)

70G O.S. Gryaznov, Yu.I. Ravich: Sov. Phys. Semicond. *3*, 1092 (1970)
70K1 Ch. Keffer, T.M. Hayes, A. Bienenstock: Phys. Rev. *B2*, 1966 (1970)
70K2 L.N. Kurbatov, A.D. Britov, N.N. Mochalkin: Sov. Phys. Semicond. *4*, 95 (1970)
70L E.M. Logothetis, H. Holloway: Solid State Commun. *8*, 1937 (1970)
70M1 I. Melngailis, T.C. Harman: in "Semiconductors and Semimetals", Vol. 5, ed. by R.K. Willardson and A.C. Beer (Academic Press New York 1970) p. 111
70M2 O. Madelung: *Grundbegriffe der Halbleiterphysik* (Heidelberger Taschenbuch Nr. 71) (Springer, Berlin, Heidelberg, New York 1970)
70 O H. Overhof, U. Rössler: Phys. Stat. Sol. *37*, 691 (1970)
70R Yu.I. Ravich, B.A. Efimova, I.A. Smirnov: *Semiconducting Lead Chalcogenides* (Plenum Press, New York - London 1970)
70S J. Steininger: J. Appl. Phys. *41*, 2713 (1970)
71A A.A. Averkin, V.I. Kaidanov, R.B. Melnik: Sov. Phys. Semicond. *5*, 75 (1971)
71B J.R. Burke, J.D. Jensen, B. Houston: in "Physics of Semimetals and Narrow Gap Semiconductors" ed. by D.L. Carter and R.T. Bate (Pergamon, Oxford 1971) p. 393
71C1 M.L. Cohen and Y.W. Tsang: in "Physics of Semimetals and Narrow-Gap Semiconductors", ed. by D.L. Carter and R.T. Bate (Pergamon, Oxford 1971) p. 303
71C2 S.P. Chashchin, I.S. Averyanov, N.S. Baryshev: Sov. Phys. Semicond. *4*, 1538 (1971)
71D1 J.P. Donnelly, T.C. Harman, A.G. Fogt: Appl. Phys. Lett. *18*, 259 (1971)
71D2 R. Dalven: Phys. Rev. *B3*, 3359 (1971)
71D3 J.O. Dimmock: in "Physics of Semimetals and Narrow Gap Semiconductors", ed. by D.L. Carter and R.T. Bate (Pergamon, Oxford 1971) p. 363
71K I.V. Kucherenko, Yu.N. Korolev, A.P. Shotov: Sov. Phys. Semicond. *5*, 868 (1971)
71L1 A.I. Likhter, E.G. Pel, A.V. Rudnev: Sov. Phys. Semicond. *5*, 612 (1971)
71L2 G. Lippmann, P. Kästner, W. Wanninger: Phys. Stat. Sol. (a) *6*, K 159 (1971)
71M J. Melngailis, T.C. Harman, J.G. Mavroides, J.O. Dimmock: Phys. Rev. *B3*, 370 (1971)
71N T. Nishino, H. Ogawa, Y. Hamaha: J. Phys. Soc. Japan *30*, 1113 (1971)
71P N.J. Parada: Phys. Rev. *B3*, 2042 (1971)
71R Yu.I. Ravich, B.A. Efimova, V.I. Tamarchenko: Phys. Stat. Sol. (b) *43*, 11, 453 (1971)
71T1 Y.W. Tsang, M.L. Cohen: Phys. Rev. *B3*, 1254 (1971)
71T2 T.E. Thompson, P.R. Aron, B.S. Chandrasekhar, D.N. Langenberg: Phys. Rev. *B4*, 518 (1971)
72A H.A. Alperin, S.J. Pickart, J.J. Rhyne: Phys. Lett. *40A*, 295 (1972)
72C M. Cardona, C.M. Penchina, E.E. Koch, P.Y. Yu: Phys. Stat. Sol. (b) *53*, 327 (1972)
72D G. Dionne, J.C. Woolley: Phys. Rev. *B6*, 3898 (1972)
72K1 D.M. Korn, R. Braunstein: Phys. Rev. *B5*, 4837 (1972)
72K2 P. Kästner: Phys. Stat. Sol. (b) *53*, 753 (1972)
72K3 M.A. Kinch, D.D. Buss: Solid State Commun. *11*, 319 (1972)
72M1 A. Maitre, R. Le Toullec, M. Balkanski: Proc. 11th Int. Conf. Phys. Semicond. (PWN, Polish Scientific Publisher Warszawa 1972) p. 826
72M2 G. Martinez, M. Balkanski: Solid State Commun. *10*, 721 (1972)
72M3 J. Melngailis, T.C. Harman, W.C. Kernan: Phys. Rev. *B5*, 2250 (1972)
72M4 A. Mycielski, A. Aziza, M. Balkanski, M.Y. Moulin, J. Mycielski: Phys. Stat. Sol. (b) *52*, 187 (1972)
72N A.V. Novoselova, V.P. Zlomanov, S.G. Karbanov, O.V. Matvey, A.M. Gas'kov: Progr. in Solid State Chem. *7*, 85 (1972)
72S H. St-Onge, J.N. Walpole: Phys. Rev. *B6*, 2337 (1972)
73A M.S. Adler, C.R. Hewes, S.D. Senturia: Phys. Rev. *B7*, 5186 (1973)
73C1 M. Cardona, D.W. Langer, N.I. Shevchik, J. Tejeda: Phys. Stat. Sol. (b) *58*, 127 (1973)
73C2 I. Chambouleyron, J.M. Besson, M. Balkanski: J. Appl. Phys. *44*, 3222 (1973)
73D1 J.P. Donnelly, H. Holloway: Appl. Phys. Lett. *23*, 682 (1973)
73D2 I.A. Drabkin, B.A. Efimova, G.F. Zakharyugina, V.I. Kaidanov, R.B. Mel'nik, I.V. Nel'son: Sov. Phys. Semicond. *7*, 545 (1973)
73D3 R. Dalven: in "Solid State Physics", ed. by H. Ehrenreich, F. Seitz and D. Turnbull, Vol. 28 (Academic Press, New York 1973) p. 179

73D4 I.A. Drabkin, L.Ya. Morgovsky, I.V. Nelson, Yu.I. Ravich: Sov. Phys. Semicond. *6*, 1156 (1973)
73D5 G. Dolling, W. Buyers: J. Nonmetals *1*, 159 (1973)
73H1 T.C. Harman: J. Nonmetals *1*, 183 (1973)
73H2 C.R. Hewes, M.S. Adler, S.D. Senturia: J. Appl. Phys. *44*, 1327 (1973)
73H3 C.R. Hewes, M.S. Adler, S.D. Senturia: Phys. Rev. *B7*, 5195 (1973)
73K1 V.I. Kaidanov, R.B. Mel'nik, I.A. Chernik: Sov. Phys. Semicond. *7*, 522 (1973)
73K2 S.E. Kohn, P.Y. Yu, Y. Petroff, Y.R. Shen, Y. Tsang, M.L. Cohen: Phys. Rev. *B8*, 1477 (1973)
73K3 N. Kristoffel, P. Konsin: Ferroelectrics *6*, 3 (1973)
73K4 I.V. Kucherenko, M.S. Taktakishvili, A.P. Shotov: Sov. Phys. Semicond. *7*, 562 (1973)
73L1 A. Laugier: Rev. Phys. Appl. *8*, 259 (1973)
73L2 G. Lucovsky, R.M. White: Phys. Rev. *B8*, 660 (1973)
73L3 E.M. Logothetis, H. Holloway: J. Nonmetals *2*, 1 (1973)
73M1 F.R. McFeely, S. Kowalczyk, L. Ley, R.A. Pollack, D.A. Shirley: Phys. Rev. *B7*, 5228 (1973)
73M2 G. Martinez: Phys. Rev. *B8*, 4678, 4686, 4693 (1973)
73M3 A. Mooradian, A.J. Strauss, J.A. Rossi: IEEE *QE-9*, 347 (1973)
73M4 J. Melngailis, T.C. Harman, J.A. Kafalas: J. Nonmetals *1*, 329 (1973)
73P G.W. Pratt, Jr.: J. Nonmetals *1*, 103 (1973)
73R R.W. Ralston, I. Melngailis, A.R. Calawa, W.T. Linley: IEEE *QE-9*, 350 (1973)
73S1 A.J. Strauss, T.C. Harman: J. Electron. Mater. *2*, 71 (1973)
73S2 K. Seeger: *Semiconductor Physics* (Springer, Wien, New York 1973)
73S3 A.J. Strauss: J. Electron. Mat. *2*, 553 (1973)
73S4 K.J. Sleger, G.F. McLane, D.L. Mitchell: J. Nonmetals *1*, 223 (1973)
73S5 U. Schlichting, K.H. Gobrecht: J. Phys. Chem. Sol. *34*, 753 (1973)
73S6 H. St-Onge, F.J. Leonberger, J.N. Walpole: J. Nonmetals *2*, 49 (1973)
73V A.N. Veis, V.I. Kaidanov, N.A. Kostyleva, R.B. Mel'nik, Yu.I. Ukhanov: Sov. Phys. Semicond. *7*, 630 (1973)
73W J.N. Walpole, A.R. Calawa, R.W. Ralston, T.C. Harman, J.P. McVittie: Appl. Phys. Lett. *23*, 620 (1973)
74B1 E. Bangert, P. Kästner: Phys. Stat. Sol. (b) *61*, 503 (1974)
74B2 H. Burkhard, R. Geick, P. Kästner, K.-H. Unkelbach: Phys. Stat. Sol. (b) *63*, 89 (1974)
74G1 R. Glosser, J. Kinoshita, B. Rennex: Phys. Rev. Lett. *33*, 1027 (1974)
74G2 E.A. Gurieva, B.A. Efimova, Yu.I. Ravich: Sov. Phys. Semicond. *8*, 819 (1974)
74H1 T.C. Harman, I. Melngailis: Appl. Solid State Science *4*, 1 (1974)
74H2 H.T. Harper, B.A. Green, R.E. Leadon, J.A. Naber: IEEE *NS-21*, 30 (1974)
74H3 G.F. Hoff, J.D. Jensen: J. Appl. Phys. *45*, 3883 (1974)
74K1 H. Kawamura, S. Katayama, S. Takano, S. Hotta: Solid State Commun. *14*, 259 (1974)
74K2 H. Kawamura, S. Takano, S. Hotta, S. Nishi: Proc. 12th Int. Conf. Phys. Semicond., ed. by M.H. Pilkuhn (B.G. Teubner, Stuttgart 1974) p. 551
74K3 I.V. Kucherenko, M.S. Taktakishvili, A.P. Shotov: Sov. Phys. Semicond. *8*, 288 (1974)
74L1 G. Lucovsky, R.M. Martin, E. Burstein: in "Physics of IV-VI Compounds and Alloys" ed. by S. Rabii (Gordon and Breach, New York, London 1974)
74L2 A. Laugier, J. Cadoz, M. Faure, M. Moulin: J. Cryst. Growth *21*, 235 (1974)
74L3 G.V. Lashkarev, D.F. Migley, A.D. Shevchenko, K.D. Tovstyuk: Phys. Stat. Sol. (b) *63*, 663 (1974)
74 O M. Ocio: Phys. Rev. *B10*, 4274 (1974)
74P1 N. Piccioli, J.M. Besson, M. Balkanski: J. Phys. Chem. Sol. *35*, 971 (1974)
74P2 H. Preier, W. Riedel: Appl. Phys. Lett. *25*, 55 (1974)
74S K.J. Sleger, G.F. McLane, U. Strom, S.G. Bishop, D.L. Mitchell: J. Appl. Phys. *45*, 5069 (1974)
74T K.D. Tovstyuk, G.V. Lashkarev, V.B. Orletzkii, A.D. Shevchenko: Sov. Phys. Solid State *16*, 140 (1974)
74Z1 I.I. Zasavitskii, B.N. Matsonashvili, V.I. Pogodin, A.P. Shotov: Sov. Phys. Semicond. *8*, 467 (1974)
74Z2 W. Zawadzki: Adv. Phys. *23*, 435 (1974)

75A G.T. Alekseeva, B.A. Efimova, Yu.A. Logachev: Sov. Phys. Semicond. *9*, 83 (1975)
75B L.L. Boyer, B.M. Klein: Int. J. Quant. Chem. Symp. *9*, 511 (1975)
75C D.A. Cammack, A.V. Nurmikko, G.W. Pratt Jr., J.R. Lowney: J. Appl. Phys. *46*, 3965 (1975)
75D1 J.P. Donnelly, T.C. Harman: Solid State Electronics *18*, 1144 (1975)
75D2 I.A. Drabkin, Yu.Ya. Eliseeva, G.F. Zakharyugina, I.V. Nel'son, Yu.I. Ravich: Sov. Phys. Semicond. *8*, 1261 (1975)
75D3 R. Dalven: Phys. Rev. Lett. *35*, 66 (1975)
75G1 T.S. Gertovich, S.I. Grivena, V.G. Gutsulyak, V.B. Orletskii, K.D. Tovstyuk: Sov. Phys. Semicond. *9*, 668 (1975)
75G2 S.G. Gasan-Zade, V.B. Orletskii, E.A. Sal'kov, G.A. Shepel'skii: Sov. Phys. Semicond. *9*, 253 (1975)
75G3 T.S. Gudkin, I.A. Drabkin, V.I. Kaidanov, O.G. Sterlyadkina: Sov. Phys. Semicond. *8*, 1453 (1975)
75F D.M. Finlayson, I.A. Johnson: Phys. Stat. Sol. (b) *71*, 395 (1975)
75H1 J.S. Harris, J.T. Longo, E.R. Gertner, J.E. Clarke: J. Cryst. Growth *28*, 334 (1975)
75H2 L.A. Hemstreet, Jr.: Phys. Rev. *B11*, 2260 (1975)
75H3 L.A. Hemstreet, Jr.: Phys. Rev. *B12*, 1212 (1975)
75H4 G.H. Heckelmann, H.H. Landfermann, T.S. Wagner, U. Nielsen: Phys. Stat. Sol. (b) *69*, K 99 (1975)
75H5 H. Heinrich, W. Jantsch, J. Rozenbergs: Solid State Commun. *17*, 1145 (1975)
75I1 M. Iizumi, Y. Hamaguchi, K.F. Komatsubara, Y. Kato: J. Phys. Soc. Japan *38*, 443 (1975)
75I2 V.N. Ivanov, I.V. Kucherenko, V.N. Moiseenko, M.S. Taktakishvili, A.P. Shotov: Sov. Phys. Semicond. *9*, 453 (1975)
75K K.L.I. Kobayashi, Y. Kato, Y. Katayama, K.F. Komatsubara: Solid State Commun. *17*, 875 (1975)
75L1 G.V. Lashkarev, D.F. Miglei, K.D. Tovstyuk, A.D. Shevchenko: Sov. Phys. Semicond. *8*, 929 (1975)
75L2 G.V. Lashkarev, A.D. Shevchenko: Sov. Phys. Semicond. *8*, 1036 (1975)
75M1 G. Martinez, M. Schlüter, M.L. Cohen: Phys. Rev. *B11*, 651 (1975)
75M2 G. Martinez, M.L. Cohen, M. Schlüter: Phys. Rev. Lett. *35*, 1746 (1975)
75M3 G. Martinez, M. Schlüter, M.L. Cohen, R. Pinchaux, P. Thiry, D. Dagneaux, Y. Petroff: Solid State Commmun. *17*, 5 (1975)
75M4 G. Martinez, M. Schlüter, M.L. Cohen: Phys. Rev. *B11*, 660 (1975)
75N S. Narita, K. Suizu: Progr. Theor. Phys. Suppl. *57*, 187 (1975)
75 O V.P. Orletzkii, F.F. Sizov, G.V. Lashkarev, K.D. Tovstyuk: Sov. Phys. Semicond. *9*, 176 (1975)
75S1 M. Schlüter, G. Martinez, M.L. Cohen: Phys. Rev. *B11*, 3808 (1975)
75S2 M. Schlüter, G. Martinez, M.L. Cohen: Phys. Rev. *B12*, 650 (1975)
75V B.A. Volkov, I.V. Kucherenko, M.S. Taktakishvili, A.P. Shotov: Sov. Phys. Semicond. *8*, 1530 (1975)
75Z V.P. Zlomanov, A.I. Likhter, E.G. Pel', O.I. Tananaeva: Sov. Phys. Semicond. *9*, 919 (1975)
76A1 Yu.V. Andreev, K.I. Geiman, I.A. Drabkin, A.V. Matveenko, E.A. Mozhaev, B.Ya. Moizhes: Sov. Phys. Semicond. *9*, 1235 (1976)
76A2 F.A.H. Al-Watban, D.M. Finlayson: J. Phys. *C9*, 1939 (1976)
76B1 G. Bauer, H. Burkhard, H. Heinrich, A. Lopez-Otero: J. Appl. Phys. *47*, 1721 (1976)
76B2 P. Berndt, D. Genzow, K.H. Herrmann: Phys. Stat. Sol. (a) *38*, 497 (1976)
76B3 H. Burkhard, G. Bauer, A. Lopez-Otero: Solid State Commun.*18* , 773 (1976)
76B4 P.P. Bodiul, D.V. Gitsu, S.F. Migley: Phys. Stat. Sol. (b) *73*, K5 (1976)
76C G.P. Carver, B.B. Houston, J.R. Burke: Solid State Commun. *20*, 1089 (1976)
76D W.J. Daughton, E. Gürmen, C.W. Tompson: Proc. Conf. on Neutron Scattering, ed. by R.M. Moon, *Vol. 1* (Gatlinburg 1976) p. 145
76E P.R. Emtage: J. Appl. Phys. *47*, 2565 (1976)
76F D.K. Ferry: Phys. Lett. A *56*, 317 (1976)
76G1 T.S. Gertovich, S.I. Grineva, M.F. Olekseeva, V.B. Orletzkii, K.D. Tovstyuk, S.A. Khramtsova: Sov. Phys. Semicond. *10*, 473 (1976)

76G2 B.A. Green: J. Appl. Phys. *47*, 2243 (1976)
76J W. Jantsch, A. Lopez-Otero: Proc. 13th Int. Conf. Phys. Semicond., ed. by F.G. Fumi (Tipografia Marves, Roma 1976) p. 487
76K1 H. Kawamura, S. Nishikawa, S. Nishi: Proc. 13th Int. Conf. Phys. Semicond., ed. by F.G. Fumi (Tipografia Marves, Roma 1976) p. 310
76K2 K.L.I. Kobayashi, Y. Kato, Y. Katayama, K.F. Komatsubara: Phys. Rev. Lett. *37*, 772 (1976)
76K3 S. Katayama: Solid State Commun. *19*, 381 (1976)
76K4 I.V. Kucherenko, M.S. Taktakishvili, A.P. Shotov: Sov. Phys. Semicond. *10*, 807 (1976)
76L G.V. Lashkarev, A.I. Dmitriev, M.V. Radchenko, K.D. Tovstyuk: JETP Lett. *24*, 307 (1976)
76N1 S. Narita, Y. Takafuji: Solid State Commun. *20*, 357 (1976)
76N2 A. Natori: J. Phys. Soc. Japan *40*, 163 (1976)
76N3 A. Natori: J. Phys. Soc. Japan *41*, 782 (1976)
76N4 A. Nishiyama: J. Phys. Soc. Japan *40*, 471 (1976)
76P G.I. Peresada, E.G. Ponyatovskii, Zh.D. Sokolovskaya: Phys. Stat. Sol. (a) *35*, K177 (1976)
76S1 T. Seddon, S.C. Gupta, G.A. Saunders: Solid State Commun. *20*, 69 (1976)
76S2 F.F. Sizov, G.V. Lashkarev, M.V. Radchenko, V.B. Orletzkii: Sov. Phys. Semicond. *10*, 236 (1976)
76S3 F.F. Sizov, G.V. Lashkarev, M.V. Radchenko, V.B. Orletzkii, E.T. Grigorovich: Sov. Phys. Semicond. *10*, 1075 (1976)
76T1 V.P. Ten, A.E. Yunovich: Sov. Phys. Semicond. *10*, 513 (1976)
76T2 W.E. Tennant: Solid State Commun. *20*, 613 (1976)
76V1 H.R. Vydyanath: J. Appl. Phys. *47*, 4993 (1976)
76V2 H.R. Vydyanath: J. Appl. Phys. *47*, 5003 (1976)
76V3 V.L. Volkov, V.I. Litvinov, V.M. Baginskii, K.D. Tovstyuk: Solid State Commun. *20*, 807 (1976)
76V4 A.N. Veis, V.I. Kaidanov, Yu.I. Ravich, I. Ryabtseva, Yu.I. Ukhanov: Sov. Phys. Semicond. *10*, 62 (1976)
76V5 M.P. Volotskoi, V.I. Kaidanov: Sov. Phys. Semicond. *10*, 1124 (1976)
77A1 B.A. Akimov, R.S. Vadkhva, V.P. Zlomanov, L.I. Ryabova, S.M. Chudinov: Sov. Phys. Semicond. *11*, 637 (1977)
77A2 W.W. Anderson: IEEE *QE-13*, 532 (1977)
77A3 B.A. Akimov, A.I. Dmitriev, G.V. Lashkarev, V.B. Orletzkii, K.D. Tovstyuk, S.M. Chudinov: Sov. Phys. Solid State *19*, 231 (1977)
77B1 R.F. Brebrick: J. Electronic Mat. *6*, 659 (1977)
77B2 G.S. Bushmarina, B.F. Gruzinov, I.A. Drabkin, E.Ya. Lev, I.V. Nel'son: Sov. Phys. Semicond. *11*, 1098 (1977)
77B3 Yu.A. Bratashevskii, V.D. Prozorovskii, Yu.S. Kharionovskii: Sov. Phys. Semicond. *11*, 112 (1977)
77C G. Cappuccio, C. Corsi, A. Giansantelli, M. Grandolfo, C. Ranghiasci: Phys. Rev. *B15*, 3986 (1977)
77D1 I.A. Drozd, V.B. Alenberg: Sov. Phys. Semicond. *11*, 698 (1977)
77D2 A.I. Dmitriev, G.V. Lashkarev, S.M. Chudinov: Sov. J. Low Temp. Phys. *3*, 213, (1977)
77F1 G.M.T. Foley, D.N. Langenberg: Phys. Rev. *B15*, 4830 (1977)
77F2 G.M.T. Foley, D.N. Langenberg: Phys. Rev. *B15*, 4850 (1977)
77G S.C. Gupta, K.N.S. Rajwanshi, A.K. Sreedhar: Phys. Rev. *B15*, 2231 (1977)
77H J. Hesse: Japan J. Appl. Physics *16-1*, 257 (1977)
77I M. Inoue, Y. Kaku, H. Yagi, T. Tatsukawa: J. Phys. Soc. Japan *43*, 512 (1977)
77K1 M.M. Kondratenko, A.P. Lavrenchuk, V.N. Makogonenko, V.B. Orletzkii: Sov. Phys. Semicond. *11*, 58 (1977)
77K2 I.V. Kucherenko, Yu.A. Mityagin, L.K. Vodop'yanov, A.P. Shotov: Sov. Phys. Semicond. *11*, 282 (1977)
77K3 I.V. Kucherenko, V.N. Moiseenko, A.P. Shotov: Sov. Phys. Semicond. *11*, 90 (1977)
77L1 S.N. Lykov, Yu.I. Ravich, I.A. Chernik: Sov. Phys. Semicond. *11*, 1016 (1977)
77L2 W. Lo: IEEE *QE-13*, 591 (1977)
77L3 K.J. Linden, K.W. Lill, J.F. Buttler: IEEE *QE-13*, 720 (1977)
77L4 K. Lischka, W. Huber: J. Appl. Phys. *48*, 2632 (1977)

77L5 G.V. Lashkarev, A.I. Dmitriev, V.B. Orletzkii, M.V. Radchenko, K.D. Tovstyuk: Sov. Phys. Semicond. *11*, 202 (1977)
77M E.A. Mozhaev, K.I. Geiman, A.V. Matveenko, D.V. Mashovets, R.V. Parfen'ev: Sov. Phys. Semicond. *11*, 1403 (1977)
77P C. Pickering, W.R. Harding, N.G. Geeves, M.L. Young: J. Phys. *D10*, L73 (1977)
77R1 J.C. Ramage, F. Kuchar, R.A. Stradling, A. Lopez-Otero: J. Phys. *C10*, 5089 (1977)
77R2 J. Rozenbergs, H. Heinrich, W. Jantsch: Solid State Commun. *22*, 439 (1977)
77R3 J.C. Ramage, R.J. Nicholas, R.A. Stradling, F. Kuchar: J. Phys. *C10*, L611 (1977)
77S1 H. Sitter, K. Lischka, H. Heinrich: Phys. Rev. *B16*, 680 (1977)
77S2 R.B. Schoolar, J.D. Jensen, G.M. Black: Appl. Phys. Lett. *31*, 620 (1977)
77S3 F.F. Sizov, G.V. Lashkarev, E.K. Martynchuk: Sov. Phys. Semicond. *11*, 987 (1977)
77S4 S. Sugai, K. Murase, H. Kawamura: Solid State Commun. *23*, 127 (1977)
77T1 K. Takasaki, S. Tanaka: Phys. Stat. Sol. (a) *40*, 173 (1977)
77T2 K. Takita, Truong Dinh Thanh, S. Tanaka: Phys. Stat. Sol. (b) *80*, 149 (1977)
77T3 A. Toneva, A. Alazhazhian: Phys. Stat. Sol. (a) *44*, 621 (1977)
77V1 A.N. Veis, S.A. Nemov, V.A. Polovinkin, Yu.I. Ukhanov: Sov. Phys. Semicond. *11*, 588 (1977)
77V2 K.V. Vyatkin, I.V. Kucherenko, V.N. Moiseenko, A.P. Shotov: Sov. Phys. Semicond. *11*, 1074 (1977)
77V3 M.N. Vinogradova, E.A. Gurieva, S.V. Zarubo, L.V. Prokof'eva, V.I. Tamarchenko, T.T. Dedegkaev, T.B. Zhukova: Sov. Phys. Semicond. *11*, 902 (1977)
78A1 A. Aladgadgyan, A. Toneva: Phys. Stat. Sol. (b) *85*, K131 (1978)
78A2 B.A. Akimov, R.S. Vadkhva, S.M. Chudinov: Sov. Phys. Semicond. *12*, 1146 (1978)
78A3 C. Albers, Ch. Walkowiak, P. Schäfer, I. Rechenberg: Phys. Stat. Sol. (a) *47*, K31 (1978)
78B1 N.B. Brandt, O.N. Belousova, V.P. Zlomanov, Ya.G. Ponomarev, E.P. Skipetrov, V.I. Shtanov: Sov. Phys. JETP *47*, 340 (1978)
78B2 N.B. Brandt, Ya.G. Ponomarev, E.P. Skipetrov: Sov. Phys. Semicond. *12*, 1325 (1978)
78B3 H. Burkhard, G. Bauer, A. Lopez-Otero: Phys. Rev. *B18*, 2935 (1978)
78B4 J.R. Burke, G.P. Carver: Phys. Rev. *B17*, 2719 (1978)
78B5 V.M. Baginskii. A.I. Dmitriev: Solid State Commun. *26*, 671 (1978)
78D1 W.J. Daughton, C.W. Tompson, E. Gürmen: J. Phys. *C11*, 1573 (1978)
78D2 W.J. Daughton, B. De Facio: J. Phys. *C11*, 4307 (1978)
78D3 R. Dornhaus, G. Nimtz: Solid State Electronics *21*, 1471 (1978)
78D4 A.K. Das, B.R. Nag: J. Phys. Chem. Sol. *39*, 259 (1978)
78D5 R. Dornhaus, G. Nimtz, K.-H. Müller, M. Berger, T. Jakobus: Proc. of the III. Int. Conf. "Physics of Narrow Gap Semiconductors" (PWN - Polish Scientific Publishers, Warszawa 1978) p. 143
78G1 T. Grandke, L. Ley, M. Cardona: Phys. Rev. *B 18*, 3847 (1978)
78G2 D. Genzow, K.H. Herrmann, H. Kostial, I. Rechenberg: Phys. Stat. Sol. (b) *86*, K21 (1978)
78G3 D.M. Gureev, I.I. Zasavitsky, B.N. Matsonashvili, A.P. Shotov: Proc. of the III. Int. Conf. "Physics of Narrow Gap Semiconductors" (PWN - Polish Scientific Publishers, Warszawa 1978) p. 109; Sov. Phys. Semicond. *12*, 411 (1978)
78G4 R. Grisar, H. Burkhard, G. Bauer, W. Zawadzki: Proc. of the III. Int. Conf. "Physics of Narrow Gap Semiconductors" (PWN - Polish Scientific Publishers, Warszawa 1978) p. 115
78H K.H. Herrmann: Solid State Electronics *21*, 1487 (1978)
78J1 A. Jedrzejczak, D. Guillot, G. Martinez: Phys. Rev. *B 17*, 829 (1978)
78J2 J.D. Jensen, B. Houston, J.R. Burke: Phys. Rev. *B 18*, 5567 (1978)
78J3 W. Jantsch, J. Rozenbergs, H. Heinrich: Solid State Electronics *21*, 103 (1978)
78K1 I.V. Kucherenko, A.P. Shotov: Sov. Phys. Semicond. *12*, 1069 (1978)
78K2 H. Kawamura: Proc. of the III. Int. Conf. "Physics of Narrow Gap Semiconductors" (PWN - Polish Scientific Publishers, Warszawa 1978) p. 7
78L1 A. Lopez-Otero: Thin Solid Films *49*, 3 (1978)
78L2 K. Lischka, W. Huber: Solid State Electronis *21*, 1509 (1978)

78L3 M. Lanir, H. Levinstein: Infrared Phys. *18*, 259 (1978)
78L4 M. Lanir, A.H. Lockwood, H. Levinstein: Solid State Commun. *27*, 313 (1978)
78L5 W. Lo, D.E. Swets: Appl. Phys. Lett. *33*, 938 (1978)
78L6 G.V. Lashkarev, R.O. Kikodze, A.V. Brodovoi: Sov. Phys. Semicond. *12*, 633 (1978)
78M V.N. Moiseenko, I.V. Kucherenko, A.P. Shotov: Sov. Phys. Semicond. *12*, 1387 (1978)
78R S. Rabii: Phys. Rev. *B 18*, 1876 (1978)
78S B. Schlicht, R. Dornhaus, G. Nimtz, L.D. Haas, T. Jakobus: Solid State Electronics *21*, 1481 (1978)
78V1 A.N. Veis, V.I. Kaidanov, S.A. Nemov: Sov. Phys. Semicond. *12*, 943 (1978)
78V2 A.N. Veis, V.I. Kaidanov, R.F. Kuteinikov, S.A. Nemov, S.A. Rudenko, Yu.I. Ukhanov: Sov. Phys. Semicond. *12*, 161 (1978)
78V3 B.A. Volkov, I.V. Kucherenko, V.N. Moiseenko, A.P. Shotov: JETP Lett. *27*, 371 (1978)
78Y M.L. Young: J. Phys. *D11*, 1195 (1978)
79A1 G. Appold, R. Grisar, G. Bauer, H. Burkhard, R. Ebert, H. Pascher, H.G. Häfele: Proc. 14th Int. Conf. Phys. Semicond. (B.L.H. Wilson, Institute of Physics, Edinburgh 1979) p. 1101
79A2 B.A. Akimov, L.I. Ryabova, O.B. Yatsenko, S.M. Chudinov: Sov. Phys. Semicond. *13*, 441 (1979)
79B1 G. Bahir, T. Bernstein, R. Kalish: Appl. Phys. Lett. *34*, 486 (1979)
79B2 L.D. Borisova: Phys. Stat. Sol. (a) *53*, K19 (1979)
79B3 H. Burkhard, G. Bauer, W. Zawadzki: Phys. Rev. *B19*, 5149 (1979)
79C G.P. Carver, B. Houston, J.R. Burke: Solid State Commun. *30*, 461 (1979)
79F H. Finkenrath, G. Franz, N. Uhle: Phys. Stat. Sol. (b) *95*, 179 (1979)
79G1 Yu.I. Gorina, G.A. Kalyuzhnaya, K.V. Kiseleva, V.M. Sal'man, N.I. Strogankova: Sov. Phys. Semicond. *13*, 175 (1979)
79G2 B.F. Gruzinov, I.A. Drabkin, Yu.Ya. Eliseeva, E.Ya. Lev, I.V. Nel'son: Sov. Phys. Semicond. *13*, 767 (1979)
79G3 B.F. Gruzinov, I.A. Drabkin, G.F. Zakharyugina, A.V. Matveenko, I.V. Nel'son: Sov. Phys. Semicond. *13*, 190 (1979)
79G4 D.M. Gureev, I.I. Zasavitskii, B.N. Matsonashvili, A.P. Shotov: Sov. Phys. Semicond. *13*, 1245 (1979)
79G5 R. Grisar, in /79P/
79G6 A.D.C. Grassie, J.A. Agapito, P. Gonzalez: J. Phys. *C12*, L925 (1979)
79K1 D. Kasemset, C.G. Fonstad: IEEE *QE-15*, 1266 (1979)
79K2 K. Kobayashi, Y. Katayama, K. Narita, K. Komatsubara, M. Iizumi: Proc. 14th Int. Conf. Phys. Semicond. (B.L.H. Wilson, Institute of Physics, Edinburgh 1979) p. 441
79K3 I.V. Kutcherenko, V.N. Moiseenko, A.P. Shotov, B.A. Volkov, K.V. Vyatkin: Proc. 14th Int. Conf. Phys. Semicond. (B.L.H. Wilson, Institute of Physics, Edinburgh 1979) p. 449
79K4 T. Knittel, H.J. Goldsmid: J. Phys. *C 12*, 1891 (1979)
79L1 P.B. Littlewood, V. Heine: J. Phys. *C12*, 4431 (1979)
79L2 P.B. Littlewood: J. Phys. *C12*, 4441 (1979)
79L3 P.B. Littlewood: J. Phys. *C12*, 4459 (1979)
79L4 R.M. Luchytskii, V.S. Manzhara, P.M. Staryk: Phys. Stat. Sol. (b) *94*, 381 (1979)
79L5 G.V. Lashkarev, R.O. Kikodze, M.V. Radchenko, E.I. Slyn'ko, I.Z. Marchuk: Sov. Phys. Semicond. *13*, 902 (1979)
79L6 R.M. Luchitskii, P.M. Staryk: Sov. Phys. Semicond. *13*, 748 (1979)
79M1 J.S. Melvin, D.C. Hendry: J. Phys. *C12*, 3003 (1979)
79M2 K. Murase, S. Sugai, T. Higuchi, S. Takaoka, T. Fukunaga, H. Kawamura: Proc. 14th Int. Conf. Phys. Semicond. (B.L.H. Wilson, Institute of Physics, Edinburgh 1979) p. 437
79N G. Nimtz, B. Schlicht, E. Tyssen, R. Dornhaus, L.D. Haas: Solid State Commun. *32*, 669 (1979)
79P H. Preier: Appl. Phys. *20*, 189 (1979)
79S1 H. Schaber, R.E. Doezema: Solid State Commun. *31*, 197 (1979)

79S2 F.F. Sizov, E.K. Martynchuk, V.B. Orletzkii, Z.K. Vershigora: Sov. Phys. Semicond. *13*, 869 (1979)
79S3 K. Shimizu, Y. Nisida, S. Narita: J. Phys. Soc. Japan *46*, 1797 (1979)
79S4 V.A. Shenderovskii, P.N. Gorley: Phys. Stat. Sol. (a) *53*, K189 (1979)
79T1 H. Tanaka, A. Morita: J. Phys. Soc. Japan *46*, 523 (1979)
79T2 E. Tyssen: "Magnetotransport im Mischkristall $Pb_{1-x}Sn_xTe$", Thesis, University of Cologne (1979)
79T3 A. Toneva, R. Kamburova: Phys. Stat. Sol. (a) *56*, K113 (1979)
79T4 A. Toneva: Phys. Stat. Sol. (a) *52*, K197 (1979)
79V1 A.N. Veis, V.I. Kaidanov, S.A. Nemov, S.N. Emelin, A.Ya. Ksendzov, Yu.K. Shalabutov: Sov. Phys. Semicond. *13*, 106 (1979)
79V2 A.N. Veis, S.A. Nemov: Sov. Phys. Semicond. *13*, 969 (1979)
79V3 A.N. Veis, S.A. Nemov: Sov. Phys. Semicond. *13*, 1394 (1979)
79V4 A.N. Veis, V.I. Kaidanov, S.A. Nemov, L.S. Lashkareva, S.I. Semenov, Z.N. Soroko: Sov. Phys. Semicond. *13*, 975 (1979)
79V5 A.N. Veis, L.S. Lashkareva, S.A. Nemov: Sov. Phys. Semicond. *13*, 221 (1979)
79W K. Weiser, A. Klein, M. Ainhorn: Appl. Phys. Lett. *34*, 607 (1979)
79Y1 A.E. Yunovich, A.S. Averyushkin, I.A. Drozd, V.G. Ogneva: Sov. Phys. Semicond. *13*, 988 (1979)
79Y2 K. Yasuda, J. Shirafuji: Appl. Phys. Lett. *34*, 661 (1979)
79Z O. Ziep, D. Genzow: Phys. Stat. Sol. (b) *96*, 359 (1979)
80A1 R.S. Allgaier, J.B. Restorff, B. Houston, J.D. Jensen, A. Lopez-Otero: J. Appl. Phys. *51*, 2119 (1980)
80A2 D.G. Andrianov, S.A. Belokon', V.M. Lakeenkov, O.V. Pelevin, A.S. Savel'ev, V.I. Fistul', G.P. Tsiskarishvili: Sov. Phys. Semicond. *14*, 102 (1980)
80B1 L.I. Bytenskii, V.I. Kaidanov, R.B. Mel'nik, S.A. Nemov, Yu.I. Ravich: Sov. Phys. Semicond. *14*, 40 (1980)
80B2 G. Bauer: in "Lecture Notes in Physics" Vol. 133 (Springer, Berlin, Heidelberg, New York 1980) p. 427
80C I.A. Chernik: Sov. Phys. Semicond. *14*, 44 (1980)
80F D.M. Freik, B.F. Kostik, O.M. Voznyak, A.G. Avramenko, M.S. Zaverukha: Sov. Phys. Semicond. *14*, 946 (1980)
80G1 M. Grudzien, A. Rogalski: Acta Phys. Pol. *A58*, 765 (1980)
80G2 E.A. Gurieva, Yu.A. Nikulin, L.V. Prokof'eva, Yu.I. Ravich: Sov. Phys. Semicond. *14*, 1355 (1980)
80G3 K.H. Gresslehner, L. Palmetshofer: J. Appl. Phys. *51*, 4735 (1980)
80H H. Heinrich: in "Lecture Notes in Physics", Vol. 133 (Springer, Berlin, Heidelberg, New York 1980) p. 407
80K1 U. Kaufmann, J. Schneider: "Festkörperprobleme" *XX* (Vieweg, Braunschweig 1980) p. 87
80K2 S. Katayama, D.L. Mills: Phys. Rev. *B22*, 336 (1980)
80L1 P.B. Littlewood: J. Phys. *C13*, 4855 (1980); J. Phys. *C13*, 4875 (1980); J. Phys. *C13*, 4893 (1980)
80L2 K. Lischka, H. Preier: Phys. Stat. Sol. (b) *101*, K129 (1980)
80M H. Maier, J. Hesse: in *Crystals, Growth, Properties and Applications*, Vol. 4 (Springer, Berlin, Heidelberg, New York 1980) pp. 145 - 190
80N1 G. Nimtz: Physics Reports *63*, 265 (1980)
80N2 S. Nishi, H. Kawamura, K. Murase: Phys. Stat. Sol. (b) *97*, 581 (1980)
80P1 L.V. Prokof'eva, M.N. Vinogradova, S.V. Zarubo: Sov. Phys. Semicond. *14*, 1304 (1980)
80P2 L.V. Prokof'eva, E.A. Gurieva, M.N. Vinogradova, S.V. Zarubo, Yu.A. Nikulin: Sov. Phys. Semicond. *14*, 296 (1980)
80S F.F. Sizov, V.B. Orletzkii, M.V. Radchenko: Sov. Phys. Semicond. *14*, 1258 (1980)
80V A.N. Veis, V.I. Kaidanov, S.A. Nemov: Sov. Phys. Semicond. *14*, 628 (1980)
80Z O. Ziep, D. Genzow, M. Mocker, K.H. Herrmann: Phys. Stat. Sol. (b) *99*, 129 (1980)
81A1 T.P. Abryutina, K.I. Geiman, B.G. Girich, D.M. Gureev, I.I. Zasavitskii, A.V. Matveenko, B.N. Matsonashvili, M.I. Nikolaev, O.V. Pelevin, A.P. Shotov: Sov. Phys. Semicond. *15*, 543 (1981)

81A2 T.D. Aitikeeva, A.I. Lebedev, A.E. Yunovich, K. Herrmann, A.W. Jalyschko, P. Schäfer: Phys. Stat. Sol. (a) *67*, 171 (1981)
81A3 A.A. Averkin, G.S. Bushmarina, I.A. Drabkin, Yu.Z. Sanfirov: Sov. Phys. Semicond. *15*, 117 (1981)
81A4 B.A. Akimov, N.B. Brandt, A.A. Zhukov, C.I. Ryabova, D.R. Khokhlov: Sov. Phys. Semicond. *15*, 1294 (1981)
81A5 E.A. Akopyan, G.A. Galandarov, A.Sh. Mekhtiev, F.E. Faradzhev: Sov. Phys. Semicond. *15*, 1167 (1981)
81B T.V. Bocharova, A.N. Veis, Z.M. Dashevskii, V.A. Kotel'nikov, R.Yu. Krupitskaya: Sov. Phys. Semicond. *15*, 103 (1981)
81C S.P. Chashchin, I.P. Guzhova, N.S. Baryshev, Yu.S. Kharionovskii: Sov. Phys. Semicond. *15*, 324 (1981)
81D A.I. Dmitriev, G.V. Lashkarev, M.V. Radchenko: Sov. Phys. Semicond. *15*, 585 (1981)
81G1 B.L. Gelmont, T.R. Globus, A.V. Matveenko: Solid State Commun. *38*, 931 (1981)
81G2 B.F. Gruzinov, I.A. Drabkin, E.A. Zakomornaya: Sov. Phys. Semicond. *15*, 190 (1981)
81I T. Ichiguchi, K. Murase: "Physics in High Magnetic Fields", in: Solid State Sci. Vol. 24, ed. by M. Cardona and P. Fulde (Springer, Berlin, Heidelberg, New York 1981) p. 249
81K1 I.V. Kucherenko, A.E. Svistov, A.P. Shotov: Sov. Phys. Semicond. *15*, 510 (1981)
81K2 W. Kaszuba, A. Rogalski: Acta Phys. Pol. *A59*, 397 (1981)
81K3 I.V. Kucherenko, A.E. Svistov, A.P. Shotov: Sov. Phys. Semicond. *15*, 1226 (1981)
81L H. Lehmann, G. Nimtz, L.D. Haas, T. Jakobus: Appl. Phys. *25*, 291 (1981)
81M A.J. Miller, G.A. Saunders, Y.K. Yogurtcu: J. Phys. *C14*, 1569 (1981)
81R J.B. Restorff, R.S. Allgaier, B. Houston: J. Appl. Phys. *52*, 6185 (1981)
81S1 T. Suski, J. Karpinski, K.L.I. Kobayashi, K.F. Komatsubara: J. Phys. Chem. Sol. *42*, 479 (1981)
81S2 K. Sieranski, J. Szatkowski: Phys. Stat. Sol. (b) *107*, K87 (1981)
81V1 V.S. Vinogradov, I.D. Voronova, T.Sh. Ragimova, A.P. Shotov: Sov. Phys. Semicond. *15*, 207 (1981)
81V2 A.N. Veis, S.A. Nemov: Sov. Phys. Semicond. *15*, 715 (1981)
81W1 K. Weiser: Phys. Rev. *B23*, 2741 (1981)
81W2 K. Weiser, E. Ribak, A. Klein, M. Ainhorn: Infrared Physics *21*, 149 (1981)
81Z A. Zemel, D. Eger, H. Shtrikman, N. Tamari: J. Electronic Mat. *10*, 301 (1981)
82A R.S. Allgaier, J.B. Restorff, B. Houston: J. Appl. Phys. *53*, 3110 (1982)
82F Z. Feit, A. Zemel, D. Eger, I. Sternberg: "Lecture Notes in Physics" Vol. 152 (Springer, Berlin, Heidelberg, New York 1982) p. 69
82L P.B. Littlewood: "Lecture Notes in Physics" Vol. 152 (Springer, Berlin, Heidelberg, New York 1982) p. 238
82M1 K. Murase, S. Nishi: "Lecture Notes in Physics" Vol. 152 (Springer, Berlin, Heidelberg, New York 1982) p. 261
82M2 G. Mula: "Lecture Notes in Physics" Vol. 152 (Springer, Berlin, Heidelberg, New York 1982) p. 252
82P1 W. Porod, P. Vogl: "Lecture Notes in Physics" Vol. 152 (Springer, Berlin, Heidelberg, New York 1982) p. 247
82P2 L. Palmetshofer, K.H. Gresslehner, L. Ratschbacher, A. Lopez-Otero: "Lecture Notes in Physics" Vol. 152 (Springer, Berlin, Heidelberg, New York 1982) p. 391
82P3 H. Pascher: "Lecture Notes in Physics" Vol. 152 (Springer, Berlin, Heidelberg, New York 1982) p. 202
82T R. Triboulet, G. Didier, A. Lasbley, F. Morales, B. Toulouse, C.M. Pelletier, R. Granger, S. Rolland: "Lecture Notes in Physics" Vol. 152 (Springer, Berlin, Heidelberg, New York 1982) p. 54
82Z I.I. Zasavitsky, B.N. Matsonashvili, G.V. Flusov: "Lecture Notes in Physics" Vol. 152 (Springer, Berlin, Heidelberg, New York 1982) p. 449
83L Landolt-Börnstein: *Zahlenwerte und Funktionen* ed. by K.H. Hellwege and O. Madelung, Neue Serie Vol. III 17f (Springer, Berlin, Heidelberg, New York 1983)

The Properties and Applications of the $Hg_{1-x}Cd_xTe$ Alloy System

By R. Dornhaus and G. Nimtz

1. Introduction

This article is concerned with experimental and theoretical studies on the $Hg_{1-x}Cd_xTe$ alloy system. We have attempted to present most of the important published data. However, there were frequent cases where an author cited the references as "to be published" or "private communication" and subsequently missed every opportunity actually to publish such work. Thus we were unable to get all the details of the cited data or get information on their background.

Throughout this article we have endeavoured to use SI units consistently (and where convenient appropriate accepted working units such as "eV" and "cm^{-1}"). Although "m^{-3}" for particle density etc. may be most unpopular with established workers in the field we count on their indulgence and cooperation in eliminating the confusing mixture of units which is still found in the physical literature.

The earliest investigations on the mixed crystal $Hg_{1-x}Cd_xTe$ were aimed at the development of infrared detectors /1/, especially for wavelengths around 10 μm controllable by the composition. This is the region of the second atmospheric window and thus of great interest for communication. It covers also the wavelength of the maximum of thermal radiation at room temperature and could be useful for measuring temperature gradients in the environment or in medical applications. Last not least it straddles the wavelengths of CO_2-lasers. Photodetectors made of this mixed crystal have been working now for more than one decade and have proved themselves the most useful ones for the 10 μm region. They had to compete with other narrow band gap materials, above all with the lead salts. However, crystals of $Hg_{1-x}Cd_xTe$ available nowadays have carrier densities about two orders of magnitude smaller than the lead salts so that their correspondingly larger detectivities more than outweigh the greater difficulty in preparation.

With increasing quality and size of the crystals the HgCdTe alloys attracted the interest of solid state physics. The historical development is similar to that of the classical semiconductors which were used as detectors (with a cat's whisker in the "crystal set" of early radio, and in tens of thousands in the mixer cartridges of wartime microwave radar) long before their basic properties were understood. Today, besides the infrared application, interest centres on phenomena connected with the variation of effective mass and effective g-factor of the conduction band electrons coupled with the variation of the band gap. This shows an approximately linear dependence on the mole fraction x of the components from the positive value for CdTe to the negative one for HgTe. In between there is a composition with zero band gap where according to Kane's theory $m^* \to 0$ and $g^* \to \infty$ at the conduction band edge. Crystals close to this composition are, therefore, ideal materials for studying various magnetic quantum and also spin-dependent scattering effects assumed to become more and more pronounced with increasing g-factor which itself represents the spin-orbit interaction. Spin-dependent carrier transport was first observed in magnetic material but is now to be investigated in non-magnetic semiconductors which have a simpler band structure than the magnetic ones. The problem of energy levels of lattice defects in compound semiconductors such as vacancies and interstitials is also of great interest. As preliminary investigations have shown such defects can generate energy states which interfere with valence band or conduction band states. Various groups are investigating this problem with $Hg_{1-x}Cd_xTe$ alloys, by studying its dependence on the band gap.

In recent years the electronic properties of binary or pseudo-binary alloys have also attracted considerable theoretical interest. The problems of random systems, of which substitutional alloys represent the simplest example, have proved formidable. Many attempts to attack them have been made using a variety of different models. A comparison between experimental and theoretical results may permit a test of different competing theoretical approximations. This may require a reexamination of experimental results, which for lack of a suitable theory have been analysed on the assumption that the alloys were normal crystals. When considering for example the phonon spectrum, alloy scattering or bowing parameters, it has become obvious that the peculiarities of random alloys are important and have to be accounted for in order to obtain a satisfactory theoretical description.

There have been reviews on the $Hg_{1-x}Cd_xTe$ alloy system which have been published during the last few years by LONG and SCHMIT /2/, HARMAN /3/, and HARMAN and MELNGAILIS /4/. They have the emphasis on the photo detector properties rather than transport properties.

The present article is aiming at a fairly complete review of the current state of the art of this alloy system, useful as a basis for future research. It is in general restricted to compositions with $0 < x < 1$, data of the pure compounds are only given or discussed where those of the mixed crystal are not available.

2. The Crystal

In this chapter the basic physical properties, methods of preparation and imperfections of the $Hg_{1-x}Cd_xTe$ alloy are presented. Some of these properties, for example the elastic or thermal ones, are known only for the pure compounds. However, these data for the pure compounds do not differ markedly so that they represent a good approximation to the values for intermediate compositions, which are not expected to lie outside those for x = 0 and x = 1.

In the first section the lattice, elastic, thermal, and related properties are given. In the second section the main features of crystal preparation, and in the last section the present knowledge of lattice imperfections are discussed.

2.1 Basic Properties

The $Hg_{1-x}Cd_xTe$ alloy system is formed by II-VI compounds which are isomorphous with zincblende. In the zincblende structure each Te-ion has four nearest neighbours, which in the alloy may be either Hg or Cd. The five possible basic units around a Te-ion site are shown in Fig. 1a. The Brillouin zone of the zincblende lattice is shown in Fig. 1b with the identification of the main symmetry points and lines used throughout this article. It is usually assumed that in the alloy the anions are distributed randomly with a mean density in accordance with the mole fraction x. On the other hand in mixed crystals a trend for clustering, that is to say non-random distribution, has been often observed. In $Hg_{1-x}Cd_xTe$ there is so far no

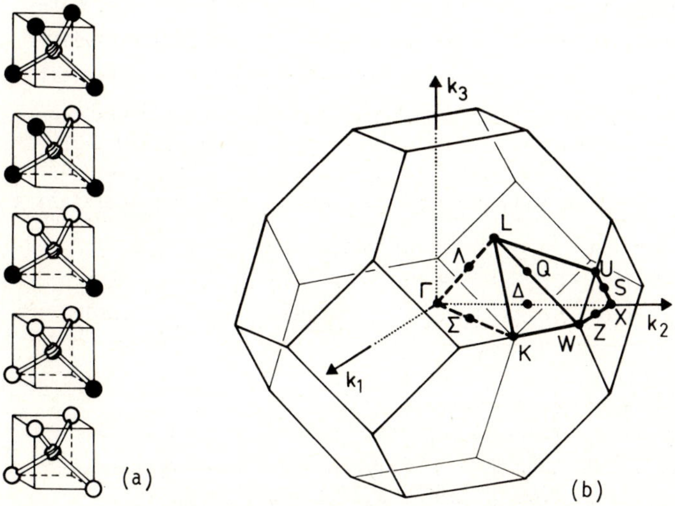

Fig. 1. (a) Basic units of nearest-neighbour ions around a Te-ion site. ○ Hg, ● Cd, ⊘ Te. (b) Brillouin zone of the zincblende structure

definite experimental proof for such an ordering within a distance of some lattice units. The previously observed splitting of the CdTe reststrahl, discussed in Sect. 5.2, may be a pointer to such a clustering.

The lattice parameters of the compounds HgTe and CdTe are very close to each other, but it was observed by various authors /5,6,7/ that their variation with x is not a linear one (Fig. 2). This is characteristic for many physical properties of the $Hg_{1-x}Cd_xTe$ alloy as will be seen in later chapters. Figure 2 includes a plot of the measured density versus composition.

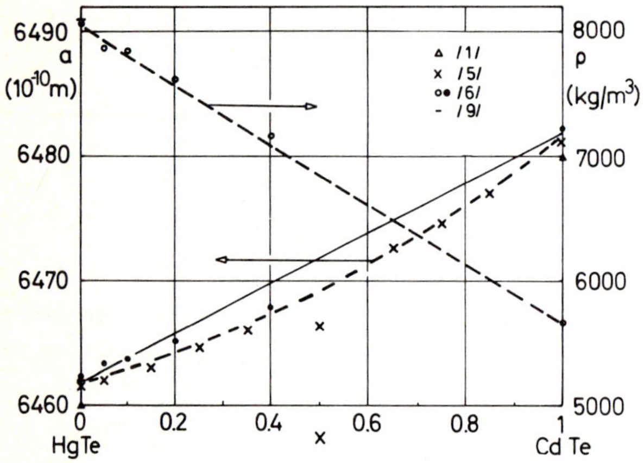

Fig. 2. Lattice parameter a and density ρ for various compositions /1,5,6,9/

In accordance with the zincblende symmetry ($\bar{4}3$ m; T_d) the elastic behaviour is determined by three elastic constants. Published data, available on the pure compounds only, describe approximately the elastic properties of the alloy. As seen in Table 1, the values for the two compounds are within the range of experimental error which judging from the scatter of values obtained by different authors seems

Table 1. Elastic stiffness C, deformation potential D, and piezoelectric constant e

Compound	Ref.	Temp. [K]	C_{11} [10^{10} N/m^2]	C_{12}	C_{44}	Du	Du' [eV]	Dd^c-Dd^v	e_{14} [C/m^2]
HgTe	/8/	300	5.08	3.58	2.05				
	/9/	290	5.48	3.81	2.05				
	/9/	4.2	5.92	4.14	2.19				
CdTe	/12/	300	5.35	3.68	1.99	1.77	4.18	-4.5	
	/16/	77	6.15	4.30	1.96				0.0335

to be approximately ± 10 %. Sample preparation was reported by ALPER and SAUNDERS /9/ to have an influence of the same order of magnitude. The data of Table 1 are obtained at various temperatures. In the temperature range between 1.4 K and 300 K the elastic constants of HgTe increase with temperature by approximately 8 % /9/.

The linear thermal expansion coefficient α of HgTe was investigated by various authors /9-11/. The results show pronounced deviations in the dependence on temperature. For the temperature range between 77 K and 300 K $\alpha \approx 4 \cdot 10^{-6}\ K^{-1}$ appears to be a good approximation.

Preliminary results of the variation of the thermal conductivity with composition have been reported by CHASMAR et al. /13/: they observed a minimum at $x \approx 0.5$. Data on the thermal conductivity and specific heat for $Hg_{1-x}Cd_xTe$, HgTe and CdTe are presented in /14/ and on Debye-Waller factors in /15/.

Crystals with zincblende structure are piezoelectric with one independent component of the piezoelectric tensor. There is no published work on piezoelectric and acoustoelectric effects in the alloy system. The piezoelectric constant e_{14} for CdTe /16/ is listed in Table 1. In the alloy the piezoelectric constant which depends on the ionic charge could vary markedly with composition.

The temperature dependence of the elastic constants of HgTe has been measured previously by COTTAM and SAUNDERS /370/ between 2 and 320 K. The data were calculated from ultrasound velocities measured by an ultrasonic pulse superposition method. The elastic constants of as-grown and annealed crystals are presented in Fig. 3. For comparison, experimental data of other authors are included. In the same figure a fit to the phenomenological model for the elastic constant temperature dependence by LAKKAD /371/ is shown /370/.

Experimental data of the linear thermal expansion coefficient obtained for various compositions were recently published /373,374/ (Table 2). The new data have revealed that the linear thermal expansion coefficient becomes negative at temperatures below 64 K. The experimental values are presented in Fig. 4 for x = 0, 0.201, 0.303, 1. The measurements of /373,374/ were carried out in a three-terminal capacitance dilatometer.

The heat capacity of HgTe was measured by COLLINS et al. /373/ in the temperature range between 1.5 and 20 K; their data are given in Table 2. Low-temperature values of the Debye temperature for HgTe and CdTe are plotted in Fig. 5 In Table 3, low-temperature values of Debye temperature θ_D, adiabatic bulk modulus B, density ρ, Grüneisen parameter γ, Phillips ionicity factor f_i, and reduced shear modulus c'/c_0 are presented for HgTe and CdTe.

An experimental and theoretical analysis of the thermal conductivity was carried out by ALIEV et al. /377/ for the composition range $0 \leqq x \leqq 0.3$ at temperatures between 65 and 300 K. The investigation was extended to 2 K for samples with x = 0.1 and 0.14 /378/. According to their results the thermal conductivi-

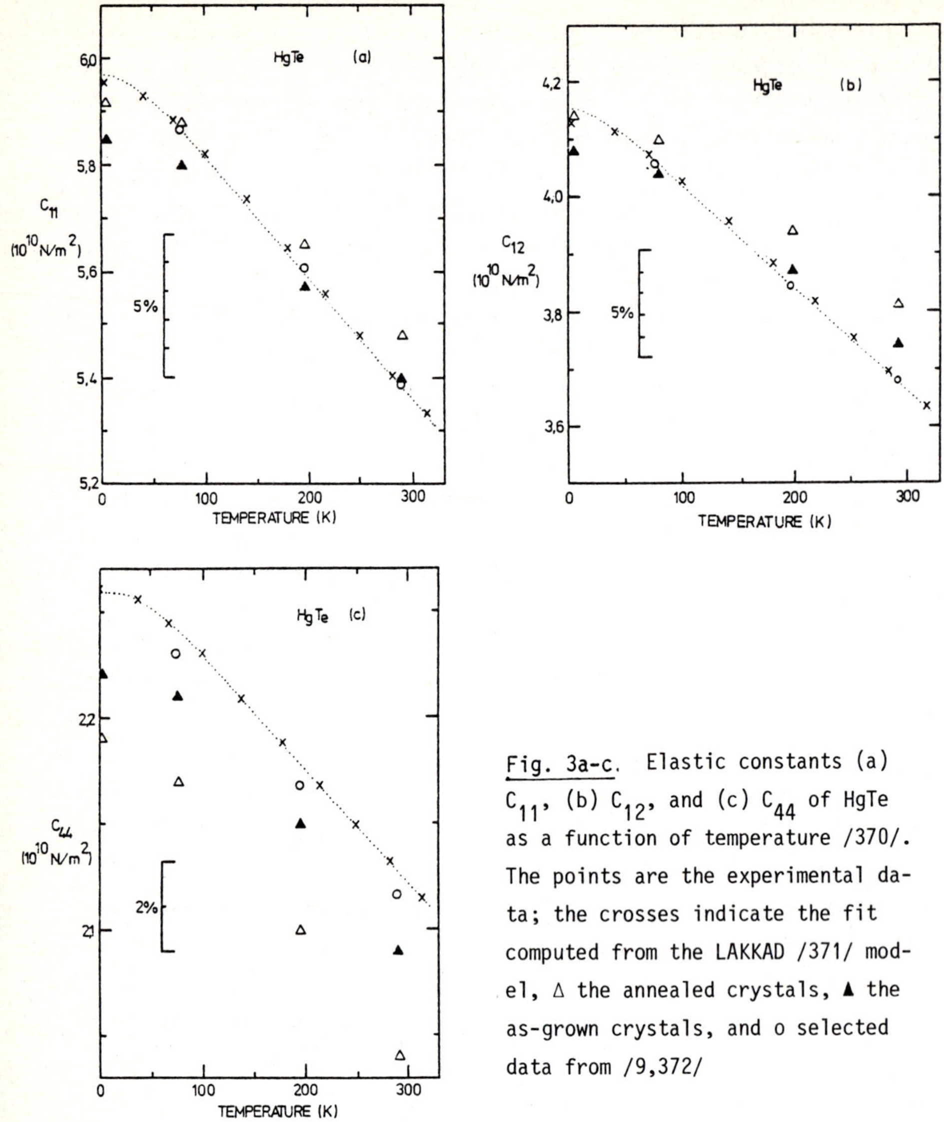

Fig. 3a-c. Elastic constants (a) C_{11}, (b) C_{12}, and (c) C_{44} of HgTe as a function of temperature /370/. The points are the experimental data; the crosses indicate the fit computed from the LAKKAD /371/ model, Δ the annealed crystals, ▲ the as-grown crystals, and o selected data from /9,372/

Table 2. Experimental values of the linear expansion coefficient α and of heat capacity C_p for HgTe. Values in brackets are interpolated /373/

T [K]	α [$10^{-8}K^{-1}$]	C_p [$J\ mol^{-1}K^{-1}$]	T [K]	α [$10^{-8}K^{-1}$]	C_p [$J\ mol^{-1}K^{-1}$]
2	-0.28	0.0116	4	-11.65	(0.28)
2.5	-0.69	0.0270	5	-34.5	0.83
3	-2.20	0.063	6	-69	(1.70)

Table 2 (cont.)

T [K]	α [$10^{-8}K^{-1}$]	C_p [J $mol^{-1}K^{-1}$]	T [K]	α [$10^{-8}K^{-1}$]	C_p [J $mol^{-1}K^{-1}$]
7	-110	(2.77)	26	-378	16.8
8	-152	3.95	28	-364	
10	-229	6.40	30	-348	
12	-287	8.3	57.5	-33	
14	-333	10.3	65	32	
16	-360	11.3	75	122	
18	-379	12.8	85	190	
20	-387	14.0	150	(380)	
22	-390	15.1	200	(450)	
24	-386	16.0	283	475	

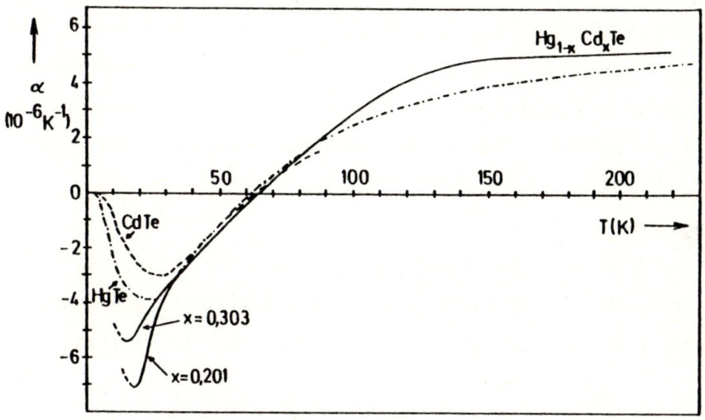

Fig. 4. Plots of the linear thermal expansion coefficient α vs T for various compositions of $Hg_{1-x}Cd_xTe$. x = 0.201, 0.303 /374/, x = 0 /373/, x = 1 /375/

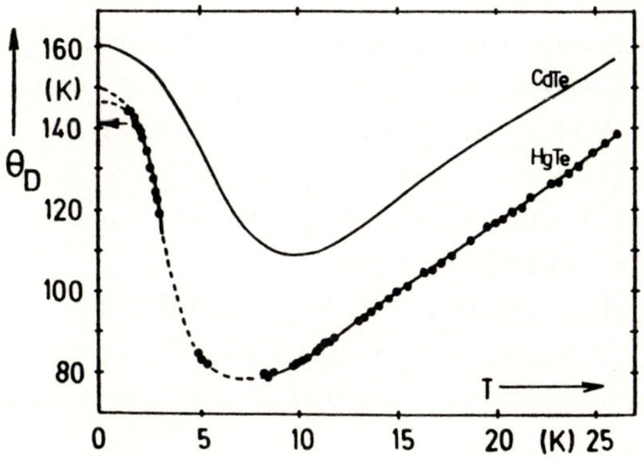

Fig. 5. Debye temperature θ_D vs temperature T for HgTe /373/ and for CdTe /376/. The values are calculated from individual measurements of the heat capacity. The arrow denotes $\theta_D(T \to 0)$ = 143 K calculated by /373/ from the data from /370/

Table 3. Values of Debye temperature θ_D, adiabatic bulk modulus B_s at 4 K, density ρ at 4 K, Grüneisen parameter γ_0, Phillips ionicity factor f_i, and reduced shear modulus c'/c for HgTe and CdTe. The subscript 0 and the superscripts th and el denote values for T → 0; the latter are deduced from thermal and elastic data, respectively

	θ_{D0}^{th} [K]	θ_{D0}^{el} [K]	B_s [G Pa]	ρ [g cm^{-3}]	γ_0^{th}	f_i	c'/c_0	Ref.
HgTe	147.5 ± 2	142.8 ± 2	47.6	8.10	-1.2 ± 0.2	0.65	0.24	/373/
CdTe	160 ± 2	160.8 ± 0.5	44.5	5.87	-0.85	0.717	0.23	/375/

ty is dominated by the scattering of phonons on point defects at temperatures $T < \theta_D$. For $T > \theta_D$, e.g., 300 K, in addition to the interaction with the defects both the three-phonon umklapp and the normal processes have to be considered. An extensive investigation of the lattice thermal conductivity of the binary compound HgTe was carried out by WHITSETT and NELSON /379/ in the temperature range between 1.7 and 150 K. According to their analysis, at the lowest temperatures the thermal conductivity is limited not by boundary scattering of phonons, but by hole-phonon scattering. The relaxation times for various phonon-scattering mechanisms were deduced from a fit of the experimental data using the Callaway phenomenological model.

Calculations of the interatomic force parameters and the ionicity have been performed for mercury chalcogenides by KUMAZAKI /380/. The calculations are based on a valence-force-field approach. The obtained data are compared with various II-VI and I-VII compounds.

2.2 Phase Diagram and Crystal Growth

In view of two surveys /2,3/ published on this topic during the last few years the main problems and methods will be reported only briefly. Most of the problems in crystal growth of the HgTe-CdTe pseudobinary system arise from the marked difference between the liquidus and solidus curves. A (T,x) phase diagram is shown in Fig. 6 /17/. The results of the early investigations on the phase diagram /3,6,18/ have shown a large variation between results for both liquidus and solidus lines obtained by different investigations as indicated by the shaded areas in Fig. 6. Determining the (P,T) phase diagram and the segregation coefficients SCHMIT and SPEERSCHNEIDER /17/ found that the discrepancy is caused by the dependence of the phase transition on the Hg pressure. This pressure dependence is clearly demonstrated

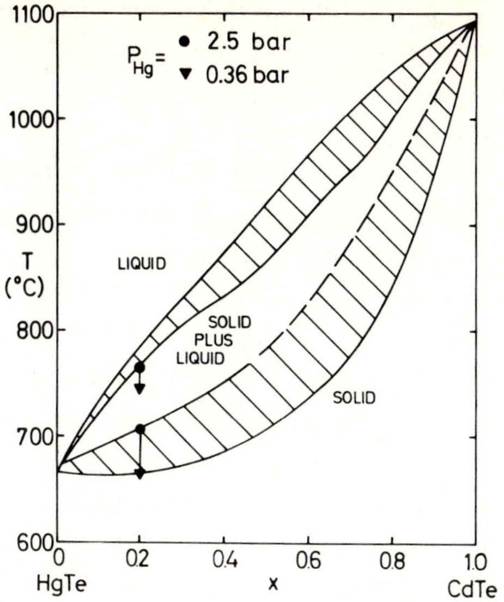

Fig. 6. (T,x) Phase diagram for $Hg_{1-x}Cd_xTe$ /17/. The two pairs of experimental points at x = 0.2 demonstrate the influence of mercury pressure on the phase transition

for a composition with x = 0.2 in Fig. 6, the phase transitions are shifted to lower temperatures when the Hg pressure is decreased form 2.5 bar to 0.36 bar /17/.

Besides the segregation of CdTe with respect to HgTe there are also problems connected with the segregation of any excess Te in the melt during crystal growth /19/. High quality single crystals of the alloy system have been prepared from ingots either with the three elements or with the two compounds. Growth of crystals by the Bridgman method /2,3,6,19,20/ and by zone melting techniques /21,22/ has been reported. The material as grown, at least in the composition range near x = 0.2, is p type with 10^{22} - 10^{23} holes/m^3. In a post-crystal-growth annealing process in a Hg atmosphere the crystal can become n type with electron densities as low as 10^{20} electrons/m^3. The mechanism assumed to cause the type inversion is discussed in the following section.

A large number of investigations have been carried out with the preparation of hetero-structures by solid-state diffusion between samples of bulk CdTe and bulk HgTe /23,24/ and by epitaxial growth /25-31/. The epitaxial growth proceeds via evaporation from a HgTe source and the diffusion into a single crystal CdTe substrate. As a result of the interdiffusion of the two compounds on the CdTe substrate, a film with CdTe is grown. The CdTe content decreases with increasing thickness of the condensed film. Thus the films have a graded band gap structure following the variation of composition with thickness. Such structures have been applied as photovoltaic detectors as discussed in Sect. 6.3.

TUFTE and STELZER /30/ have shown that the growth rate and the surface composition of epitaxial layers can be controlled by the use of excess Hg pressure. Recently this effect was studied more extensively by BAILLY et al. /32/ and by SVOB et al. /33/. The experimental results have shown that with increasing Hg pressure the interdiffusion process between the two compounds is reduced. It is assumed that with increasing Hg pressure the number of vacancies decreases by which the diffusion proceeds /33/. The interdiffusion process was studied also by means of the KIRKENDALL effect /34/.

Recently increasing interest has arisen in the growth of epitaxial layers of the mixed crystal system $Hg_{1-x}Cd_xTe$. Reviews on various methods are contained in the Proceedings of the First US-workshop on the Physics and Chemistry of Mercury Cadmium Telluride (Minneapolis, Minnesota, October 28-30, 1981), which are published in the "Journal of Vacuum Science and Technology 21, 117-263 (1982)". See also the Proceedings of "The 1983 U.S. Workshop on the Physics and Chemistry of Mercury Cadmium Telluride" (Dallas, Texas, February 8-10, 1983), which will be published in the "Journal of Vacuum Science and Technology (1983)"

Liquid phase epitaxial (LPE) growth was recently successfully achieved by various authors /381-385/. Growth was done on CdTe substrates either from solutions of Hg, from HgTe or from Te solutions. For growth from Hg solution, the Hg pressure is typically 8 atm for growth temperatures of about 500 °C, which prohibits open-tube growth /385/. However, the Hg pressure can be lowered to about 0.1 atm by lowering the growth temperature below 240 °C. For growth from a Te-rich solution, the Hg pressure can be kept below 0.1 atm for growth temperatures below 500 °C. This low pressure makes open-tube slider growth possible.

The LPE layer properties are comparable to bulk-grown n-type crystals. For a layer composition with x = 0.40 the variation in composition was measured to be $\Delta x \leq 0.02$, the layer dimensions being about 1×1 cm^2. An electron density of $4 \cdot 10^{21}$ m^3 was found in the annealed layer. The as-grown layers are p-type but can be converted to n-type conduction under Hg-rich anneals.

Epitaxial layers obtained by the evaporation-diffusion method under isothermal growth were investigated by VANYUKOV et al. /386,387/. These papers particularly are concerned with the uniformity of the composition of the layers, the kinetics of the growth process, and the transport properties of the layers. Measurements of the homogeneity of single crystals of $Hg_{.8}Cd_{.2}Te$ grown by a zone-melting technique are reported by YUDIN et al. /388a/.

Molecular beam epitaxy (MBE) was studied by FAURIE and MILLION /388b-d/. The epilayers were grown on CdTe (111) substrates between 100 and 120 °C. The results obtained by this technique are promising. After an anneal procedure n-type material ($x \sim 0.2$) was obtained with carrier densities $n(77K) \sim 10^{21}$ m^{-3} and mobilities up to nearly 20 m^2/Vs /388c/.

The growth of HgCdTe layers by metal organic vapor phase epitaxy was reported by MULLIN and IRVINE /388e-g/. Further data on epitaxial layers are presented in Sect. 7.7.

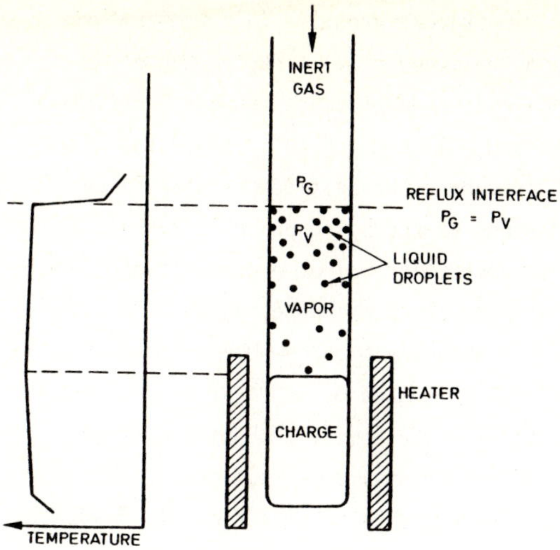

Fig. 7. Principle of the high-pressure reflux technique. (P_G) pressure of inert gas; (P_V) pressure of refluxing vapor /389/

A high-pressure reflux technique (HPR) was developed by STEININGER /389/ for the growth of $Hg_{1-x}Cd_xTe$ crystals with x values of up to 0.6. The principle of the HPR technique is sketched in Fig. 7. The crystal growth crucible consists of a high-temperature charge chamber and a cooled reflux tube. The charge contains an excess of the volatile species for refluxing. The ampoule is pressurized (up to 80 atm) and heated while a steep temperature gradient is maintained across the reflux tube. Upon heating, the volatile species evaporate and enter the reflux tube. Because of the temperature gradient in the tube, the vapor becomes supersaturated and condenses in fine droplets which fall back into the charge to be evaporated again. The stream of vapor molecules removes the non-condensable inert gas from the high-temperature zone. Consequently the charge is maintained in an atmosphere of volatile vapor which prevents its decomposition. At the reflux interface, pure vapor and inert gas are in pressure equilibrium. By controlling the pressure of the inert gas, one can therefore control the vapor pressure of the volatile constituent as well as the stoichiometry of the charge. The combination of inert-gas pressure and steep temperature gradient acts like a flexible membrane by trapping the escaping vapor /389/. The best ingots had longitudinal composition profiles with $x = 0.210 \pm 0.015$ over most of their length, being about 6 cm and having a 2.5 cm diameter. Low-temperature ($T \leqq 50$ K) mobility and carrier densities of the annealed crystals are reported to be $\mu \cong 20\ m^2/V\ s$ and $n \cong 2 \cdot 10^{20}\ m^{-3}$. A high-pressure reflux technique has been used also for the determination of extensive phase diagramm data for the Hg-Cd-Te system /390/.

Crystal growth by a modified bridgman technique was reported by FIORITO et al. /391/. In order to improve the compositional uniformity of the ingots the authors perform the growth with a special temperature profile. This temperature profile is to establish three regions along the ampoule: 1) a region where the charge will be entirely melted, 2) a region with a temperature between the liquidus and solidus line and consequently having a liquid and solid two-phase mixture, and 3) the freezing region. By this procedure the melt composition is controlled by the solid phase in region 2. As long as the solid phase is present in region 2 the crystal grows with a constant composition value. A compositional variation from $x = 0.20$ to $x = 0.25$ was obtained along a charge length of about 15 cm. Crystal growth by a bridgman technique and by a high-speed quenching technique are investigated and compared by DITTMAR /392/ and by SCHUBERT et al. /393/.

Various methods have been used to characterize the composition of a crystal and the homogeneity of the crystal and to detect crystal imperfections. Electroreflectance measurements in an electrolyte were reported by VANIER et al. /394/ and by TYAGAI et al. /395/. The technique is said to be sensitive enough to determine changes of composition $\Delta x \cong 0.002$ for samples with $0.2 < x < 0.3$. The spatial resolution is reported to be about $0.15 \cdot 10^{-3}$ m. Both investigations were carried out taking the optical spectra in the region of the E_1 transition. The variation of the E_1 structure with composition is about 1.2 eV between $x = 0$ and $x = 1$ (see Sects. 3.3,4,6).

Various other methods were applied and tested recently in order to characterize HgCdTe crystals. Chemical etching and chemical microanalysis were investigated to determine the composition of successive layers in a crystal /396/. An electron microprobe X-ray analysis is described by DITTMAR /397/, and an X-ray topographic assessment of HgCdTe crystals is given by BYE /398/. The short-range order of HgCdTe alloys is studied by means of X-ray diffuse scattering by BUBLIK and ZAITSEV /399/.

Crystal imperfections and surface conditions of photovoltaic detectors were investigated by an electron-beam-induced-current technique by LANIR et al. /400/. The current was induced by scanning the pn-junction device with a focused monoenergetic electron beam of a scanning electron microscope. The secondary electron emission induces a current which has proven to be very sensitive to material imperfections in the range of depth between 0.5 and 3 μm. Accordingly the electron-beam-induced-current technique can probe deeper regions of the material than the secondary-electron-emission technique.

2.3 Imperfections

Many important physical properties in semiconductors or semimetals such as carrier mobility and carrier density at low temperatures are controlled by imperfections of the crystal. There are three types of imperfection of major interest: dislocations, native point-defects, and foreign atom impurities.

2.3.1 Dislocations and Native Point Defects; Radiation Defects

The different types of dislocations in the zincblende structure have been studied by HOLT /35/. One of these, the so-called 60^0 dislocation was investigated in CdTe. BUCH and AHLQUIST /36/ have shown that this dislocation determines the plastic deformation and acts also as donor or electron trap depending on whether the core of the dislocation is formed by cations or anions in the CdTe compound. So far there is no work available on dislocations in the mixed crystal but there is abundant evidence for native point-defects in the form of vacancies and interstitials. It was soon observed that depending on preparation the samples are n or p type. An excess of both types of cations in the crystal yields n-type material, whereas an excess of anion atoms yields p-type material. Thus it is established that free carrier density and carrier type in the extrinsic regime can be controlled by appropriate departures from stoichiometry /2,4,6,17,20,37/. Usually n- or p-type material of $Hg_{1-x}Cd_xTe$ with $x \approx 0.2$ is produced by an annealing process in Hg-rich or Te-rich vapour. It is assumed that three types of defects cause this behaviour, namely Hg and Te vacancies and Hg interstitials. Evidence for Hg vacancies was obtained in semiconducting as well as in semimetallic compositions /38-40/. The Hg vacancy corresponds to an acceptor state, whose energy depends on the mole fraction x, i.e. the band gap as is shown in **Fig. 8** /39/. An interesting feature of this acceptor state is that it is resonant with the conduction band for $x < 0.16$. This acceptor state was studied theoretically by MAUGER and FRIEDEL /39/ and by BASTARD and NOZIERES /41/. ELLIOTT et al. /40/ deduced from thermal freeze-out of carriers and from photoluminescence measurements an acceptor level of about 20 meV

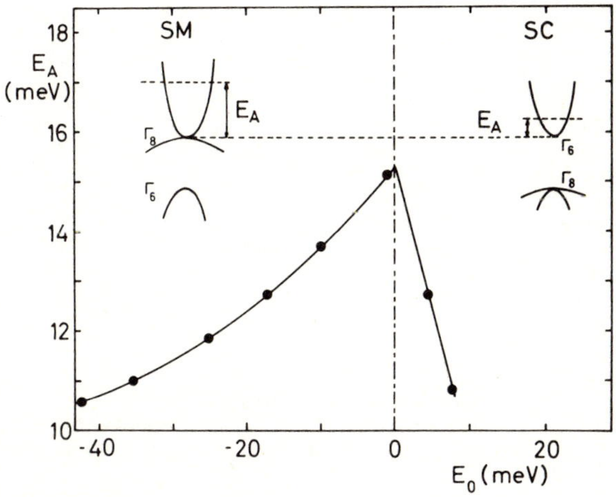

Fig. 8. Acceptor level energy E_A as defined in the insert versus band gap energy /39/

above the valence band. This energy value was found in semiconducting p-type material with $0.2 \leq x \leq 0.5$. It is sensible to assume this acceptor state also to be related to a Hg vacancy. Pressure dependent measurements of galvanomagnetic transport properties with p-type samples of $x \approx 0.15$ yielded two different acceptor levels /42/. In annealed samples ELLIOTT et al. /42/ could explain the experimental results by an acceptor state 9 meV above the valence band, whereas in as-grown material an acceptor state band of approximately 20 meV above the valence band seems to govern the transport properties.

The infrared emission of hot carriers observed recently /43/ may provide further support to the assumption that a Hg vacancy generates an acceptor state. Comparing the reflectance and the emission spectra which are presented in Figs. 106,127, it is obvious that in the emission spectrum at the CdTe reststrahl frequency the emission drops to zero but not at the HgTe reststrahl. As described in Sect. 6.4 the p-type layer was obtained by the evaporation of Hg-atoms at the surface of n-type $Hg_{.8}Cd_{.2}Te$. Thus the HgTe sublattice is strongly disturbed and consequently the HgTe reststrahl affected. It is speculated that a high density of Hg vacancies causes the transparency of the surface layer at the HgTe reststrahl frequency /43/.

Hg-rich material was found to be n type, which may be caused either by Te vacancies or by Hg interstitials /2,4,6,17,20,37/. From the analysis of various transport effects with n-type $Hg_{.8}Cd_{.2}Te$ DORNHAUS et al. /44/ concluded that the conduction band electrons might be generated by Te vacancies rather than by Hg interstitials. In the same samples a resonant electronic state was observed in Shubnikov-de Haas experiments and also in the far infrared transmission /45/. The experimental data can be described by an electronic state which is resonant with the conduction band as sketched in the insert of Fig. 9. The energy level of this state is approximately 8 meV above the conduction band at 4.2 K. The absorption coefficient for the transition from conduction band to the resonant level was calculated by analogy with the deuteron photoionization cross section and is compared with experimental data evaluated from transmission measurements in Fig. 9. There are some

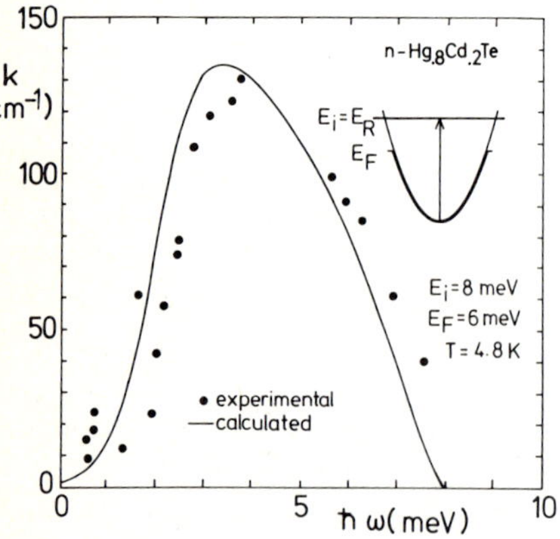

Fig. 9. Absorption coefficient k as deduced from far infrared transmission experiments and as calculated according to the model sketched in the insert /45/

theoretical investigations about the states of vacancies in compounds /46-48/ but so far none related to the II-VI compounds which are forming the $Hg_{1-x}Cd_xTe$ alloy.

Damage due to proton /4/ and electron irradiation /37,49,50/ can also produce a type conversion. With both proton and electron radiation, the number of electrons was increased and this effect was applied to create n-type layers on p-type bulk material (Chap. 6.3). The defects caused by the radiation are proposed to be either Te vacancies or Hg interstitials /37/. The statistical behaviour of the electron irradiation induced states can be described by two defect levels, as was shown by LEADON and MALLON /50/, one situated in the band gap near the valence band and a second one above the conduction band. Thus the latter is resonant to the conduction band similarly to the state observed by DORNHAUS et al. /45/ which was mentioned above. However, further investigations, both theoretical and experimental, are necessary to give a definite answer on this important question on defects in the $Hg_{1-x}Cd_xTe$ alloy.

Various experimental studies have confirmed the earlier published data on the ionization energy of acceptors particularly for the Hg vacancy in both semimetallic and semiconducting compositions /401-405a/. For the composition range x = 0.2 - 0.6 an acceptor ionization energy of about 14 meV was found if the crystals were heat treated after the quenching procedure /401/. In as-grown crystals, i.e. quenched material, however, much lower acceptor ionization energies were observed of about 5 meV /401,405a/. There are some arguments that the energy of the acceptor level does depend on the number of acceptors /401/. There are also some arguments that a clustering of Hg vacancies takes place, forming e.g. pairs or triplets of vacancies with acceptor levels shifted compared to a single Hg vacancy. Further indications for an acceptor ionization energy of about 14 meV were found in photoluminescence experiments (see Sect. 5.4). The photoluminescence experiments also revealed evidence for donor binding energies of about 1 meV (x = 0.32) and 4.5 meV (x = 0.48), which are consistent with values calculated in the hydrogenic model (Sect. 2.3.2) /406,407/. According to investigations by VYDYANATH /417/ the native acceptor defects are doubly ionized.

SWARTS et al. /405c/ published a theoretical study of bulk vacancies in HgCdTe alloys. Using the empirical tight binding method they calculated the band structure. Based on this band structure data they calculated the states associated with bulk vacancies. This was done with the Slater-Koster Green's function method. They found that the anion vacancy levels are always resonances far up in the conduction band, about 1.5 eV above the conduction band edge. On the other hand the states associated with cation vacancies can produce bound states, especially in Cd-rich alloys. These results were qualitatively confirmed by a generalized effective mass calculation by WANG et al. /405d/. The theoretical analysis for both cation and anion vacancies are in agreement with the experimental observations: The Hg vacancy was found to produce shallow acceptor levels in the band gap /39/, whereas in undoped n-type material no evidence for bound donor levels was found performing magneto-transport experiments /405e/. Such material behaves as a dilute metal and the free electrons are assumed to be produced by Te vacancies, their levels being resonant with the conduction band.

Experimental results on crystal damage due to ion irradiation have been reported by MARGALIT et al. /404/. Similar to radiation effects due to electron, gamma and neutron irradiation /408/, MARGALIT et al. /404/ observed that an n-type conductivity was always produced. Thus they concluded that in their ion implantation experiments an ion damage-induced conductivity was obtained rather than substitution. RYSSEL et al. /409/, however, found evidence for a substitutional doping taking place after annealing. The annealing was performed for 15 to 16 min in purified nitrogen at temperatures up to 350 °C. As shown in Table 4 for ion implantation of Ag, Cu, P, and As, a p-type conductivity was produced, while n-type conductivity was obtained with Al, B, In, Hg, Zn and F.

Ion	N [m^{-2}]	E [keV]	T_I [°C]	Ion	N [m^{-2}]	E [keV]	T_I [°C]
n-type conductivity				p-type conductivity			
Al	10^{18}	150	25,300	Ag	10^{17}	200	25
	$5 \cdot 10^{18}$	150	25		10^{18}	200	25
	$5 \cdot 10^{19}$	150	25	Cu	10^{17}	150	25
	10^{20}	150	250,350		10^{18}	150	25
B	$5 \cdot 10^{18}$	120	25	P	10^{18}	200	25
	10^{20}	100	250	As	10^{18}	120	25
In	10^{19}	150	25				
Hg	10^{17}	30	25				
	$5 \cdot 10^{17}$	30	25				
	10^{18}	100	25				
Zn	10^{18}	150	25				
	$5 \cdot 10^{18}$	150	25				
F	10^{18}	200	25				

Table 4. Compilation of ion implantation doping in $Hg_{.8}Cd_{.2}Te$. N denotes the ion dose, E the energy of ions and T_I the temperature at which the ion implantation was performed /409/

2.3.2 Foreign Atoms

For shallow donor and acceptor states in semiconductors the hydrogenic model presents a good approximation /51/. According to this the ionization energy of an impurity E and the Bohr radius a_B is

$$E = 13.6\, m_0^*/(m\varepsilon_0) \qquad \text{(in eV)}$$
$$a_B = 5.29\cdot 10^{-11}\, m\varepsilon_0/m_0^* \qquad \text{(in m)} \qquad (1)$$

in a medium with the band edge effective mass m_0^* and the dielectric constant ε_0. For a composition with $x \approx 0.2$ where m_0^* is small the energy value of a donor state is expected to be only of the order of 1 meV and even at very small impurity densities an overlap of the wavefunctions is expected. Thus a thermal freeze-out of carriers should not occur and the identification of shallow donor states should be rather hard. In semimetallic $Hg_{1-x}Cd_xTe$ the acceptor and donor levels fall into the regions of allowed states of the conduction and valence bands, thus there are acceptor and donor resonance. GEL'MONT and D'YAKONOV /52/ have investigated theoretically this problem. Their study shows that in zero gap materials with a much larger hole than electron effective mass discrete states exist only for the acceptor levels not for the donor levels.

The properties of foreign atoms have been investigated in a number of experiments /3,20,53,54/. However, reliable data are available only for Cu, which acts as acceptor. Its energy level is approximately 1 meV above the Γ_8 band edge in the semimetallic region of the alloy system /41,54/. Donor levels related to foreign atoms have not been identified, which may be caused by the above-mentioned small ionization energy expected in the alloy for not too large x values. It can be concluded that in the most widely investigated material the extrinsic transport properties are dominated rather by native point-defects than by foreign impurities. There have been more efforts to obtain and investigate pure materials with a view to technical application and obviously it has been easier to reduce the number of chemical impurities than that of defects.

The doping properties of a number of impurities have been investigated experimentally /409-417/ and theoretically /418-420/. The impurities were introduced into $Hg_{1-x}Cd_xTe$ either during the crystal growth, or by diffusion or ion implantation. The latter process is also discussed at the end of Sect. 2.3.1 and Table 4. The results obtained for the compositions $x = 0,2$, 0.4 are presented in Table 5. According to JOHNSON and SCHMITT /410/ it can be concluded:

1) The electrical activity of impurities is as expected from the substitution of various groups on the respective metal and Te sublattices.
2) Ag, Cu, Au and Li comprise the fast diffusing acceptors; Ga and In are fast diffusing donors.

Table 5. Doping properties of impurities in $Hg_{1-x}Cd_xTe$

Acceptors	Donors	Ref.
Ag	Ga	/410/
Cu	In	
Au	Al	
Li	Si	
P	Br	
As		
P	B	/413/
	I	/416/

Table 6. Diffusion coefficient D and activation energy Q for Au and Cu impurities. [$D = D_0 \exp(-Q/kT)$] /412/

Element	Molar composition x	T [K]	D [m^2/s]	Q [eV]	D_0 [m^2/s]
	0.3	473	$1.1 \cdot 10^{-11}$		
Cu	0.3	573	$4.4 \cdot 10^{-11}$	0.28	$3.3 \cdot 10^{-9}$
	0.3	673	$8.2 \cdot 10^{-11}$		
	0.3	773	$1.4 \cdot 10^{-10}$		
	0.2	473	$1.1 \cdot 10^{-13}$		
	0.2	573	$5.2 \cdot 10^{-13}$		
Au	0.2	673	$1.7 \cdot 10^{-12}$	0.4	$5.6 \cdot 10^{-10}$
	0.2	773	$3.6 \cdot 10^{-12}$		

3) Al and Si are donors; high temperatures are necessary to obtain diffusion.
4) Sn is inactive when diffused at ≦ 300 °C.
5) P and As are slowly diffusing acceptors; Br is a slowly diffusing donor.

These conclusions seem to be representative for most of the performed investigations. Data of the diffusion coefficient for Cu and Au were estimated by BECLA and MELANUIK /412/ and are given in Table 6.

In a series of papers VYDYANATH et al. /414-417/ published data on the defect structures of undoped and Cu-, In-, I-, P-doped $Hg_{1-x}Cd_xTe$ crystals (x = 0.2, 0.4). The following results were obtained from the analysis of Hall and mobility data:

Native acceptor defects are doubly ionized. Native donor defects appear to be negligible in concentration, implying that the p- to n-type conversion in these alloys is mainly due to residual foreign donor impurities.

Copper appears to be incorporated on metal lattice sites acting as a single acceptor with little compensation. Results on the heavily copper-doped samples indicate that the quench from the equilibrium temperatures was imperfect, resulting in a large fraction of the copper precipitating during the quench /414/.

Indium-doped crystals were n-type under moderate to high partial pressure of Hg, whereas they were p-type at very low Hg pressures. The electron concentration was much lower than the indium concentration in the crystals and was found to increase with increasing Hg pressures at the equilibration temperatures and with increase in the total indium present in the crystals. It is concluded that most of the indium is incorporated as In_2Te_3 dissolved in $Hg_{.8}Cd_{.2}Te$ with only a small fraction of indium acting as single donors occupying Hg lattice sites /414/.

Iodine-doped single crystals were n-type after anneals at high Hg pressures, whereas they turn p-type after anneals at low Hg pressures; the electron concentration increases with increase in Hg pressure. The results are explained on the basis that the crystals are saturated with $(Hg, Cd)I_2$, with the iodine being present as donors occupying tellurium lattice sites ($I^{\bullet}_{Te}$) and a fraction being present as $(I_{Te}V_{Hg})'$ species formed from the iodine on tellurium lattice sites ($I^{\bullet}_{Te}$) which pair with the doubly ionized native acceptor defects (V''_{Hg}). The solubility of the $I^{\bullet}_{Te}$ species increases with an increase in Hg pressure, whereas that of the $(I_{Te}V_{Hg})'$ species increases with a decrease in Hg /416/.

Phosphorus-doped $Hg_{.8}Cd_{.2}Te$ was found to be p-type with the hole concentration being much less than the total amount of phosphorus present in the crystals. The hole concentration was found to increase with an increase in partial pressure of Hg in contrast to the behaviour observed in undoped crystals. Also, the hole concentration obtained in the doped samples at low Hg pressures was less than that in undoped crystals. It was concluded that phosphorus behaves amphoterically in $Hg_{.8}Cd_{.2}Te$ acting as a single acceptor occupying interstitial and tellurium lattice sites at high Hg pressures; at intermediate Hg pressures, the majority of the phosphorus appears to be present as electrically neutral pairs formed from the association of the interstitial and substitutional phosphorus species $(P_iP_{Hg})^{\times}$. At low Hg pressures, a large fraction of the phosphorus appears to be present as electrically active associates $(P_{Hg}V_{Hg})'$ and $(P_{Hg}V_{Hg})^{\bullet}$ /417/.

The following symbols were used above for lattice sites occupied interstitially P'_i (acceptor), substitutionally $P^{\bullet}_{Hg}$ (donor), substitutionally P'_{Te} (acceptor), as neutral pairs $(P_iP_{Hg})^{\times}$, etc...

2.3.3 Long-Term Instability of n-Type $Hg_{1-x}Cd_xTe$, Diffusion Effects at Room Temperature, Effects of Annealing on the Electrical Properties

Recently, drastic long-term changes of the extrinsic carrier density and of the recombination properties were observed in $Hg_{.8}Cd_{.2}Te$ crystals by NIMTZ et al. /421/. It was found that some rather pure n-type and high-mobility material changes to p-type after a five-year storage under room-temperature conditions as shown in Fig. 10. On the other hand, in all the stored crystals a dramatic drop of the carrier lifetime was measured if compared with the data measured immediately after fabrication. It was confirmed by the authors that both effects, the change in the Hall effect and of the carrier lifetime, are bulk effects and not due to the surface. Similar drastic changes of the Hall effect with storage time were observed by SCHMIT et al. /422/ in Cominco samples. In Honeywell samples, these authors found only a slight change in the Hall data. Data on the carrier lifetime which are much more sensitive to the aging process according to the investigations by NIMTZ et al. /421/ and SCHLICHT et al. /405a/ were not given by SCHMIT et al. /422/. It is shown in Fig. 10. that undoped n-type material can easily be converted to p-type conduction by a heat treatment, the conversion being dependent on temperature and exposure time. On the other hand, it is well known that Hg is the most volatile component in the alloy. A significant quantity of Hg is always lost under free evaporation conditions from unpassivated surfaces at room temperature /423,424/. This fact and the results of annealing experiments by SCHMIT and STELZER /425a/ made NIMTZ et al. /421/ conclude

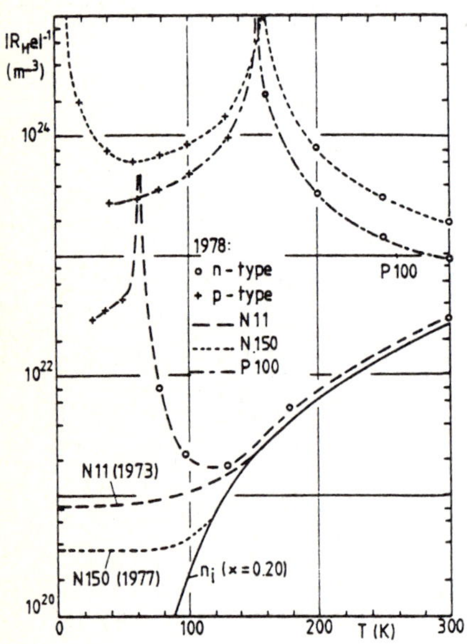

Fig. 10. Hall data of $Hg_{.8}Cd_{.2}Te$ crystals. The dashed curves without symbols represent data obtained a short time after fabrication; the symbols correspond to data measured after a five-year storage under room-temperature conditions (N 11) or heat treatment (N 150) at T = 350 °C for 8 h. Data of an as-grown p-type crystal (P 100) are included /405a/

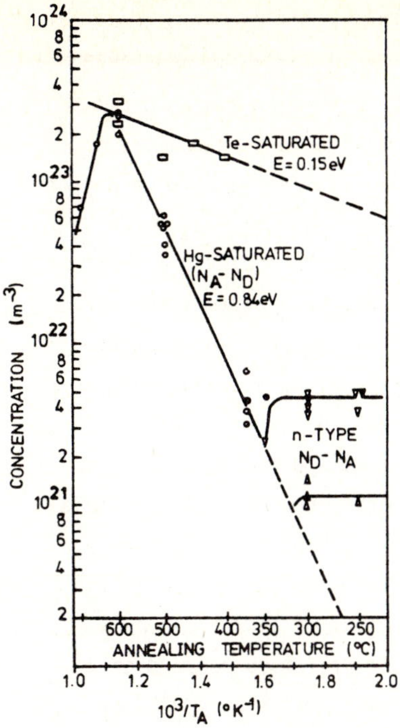

Fig. 11. Concentration as a function of annealing temperature for $Hg_{.6}Cd_{.4}Te$. The squares represent hole concentration for samples annealed with excess Te. All other symbols represent data on samples annealed with excess Hg; triangles for electron, circles for hole, and diamonds for acceptor concentrations /425a/

that fairly pure n-type material of the HgCdTe alloy is not thermodynamically stable. Annealing data of SCHMIT and STELZER /425a/ are shown in Fig. 11 for $Hg_{.6}Cd_{.4}Te$. (Quite recently an extensive study on effects of annealing on the electrical properties for compositions 0.17-0.31 was published by JONES et al. /425b/). Low-density n-type material is obtained only by an annealing procedure under Hg-saturated conditions. The horizontal branches are assumed to be due to residual donors. For both the Hg- and Te-saturation conditions, empirical expressions for the dependence of the acceptors on annealing temperature were found /425a/,

$$p(\text{Hg saturated}) = 1.46\cdot10^{28}\ \text{m}^{-3}\ \exp(-0.84/kT_A) \quad (66)$$

$$p(\text{Te saturated}) = 1.90\cdot10^{24}\ \text{m}^{-3}\ \exp(-0.15/kT_A)\ . \quad (67)$$

One may extrapolate the relation for the Te-saturated condition to 300 K which corresponds to room-temperature storage conditions and obtains about 10^{21} holes/m^3. NIMTZ et al. /421/ concluded that pure n-type material may not have a long-term stability except approaching this acceptor density.

In Fig. 10 Hall data of heat-treated n-type material at the Cd-saturated condition are presented. These experiments have confirmed that long-term changes observed at room temperature within a period of some years can be provoked within some hours if the samples are kept at higher temperatures. It was also shown that this p-type conductivity can be converted back to n-type in an annealing treatment under Hg-excess conditions /405,426/. Hall data of an as-grown p-type sample are also plotted in Fig. 10.. It was found that the carrier density can be reversibly manipulated by proper annealing or heat treatment as expected from the diagram of Fig. 11.

The following model was proposed by NIMTZ et al. /421/ in order to explain the observed instability in the $Hg_{1-x}Cd_xTe$ alloy. It has been established by numerous experimental and theoretical investigations that Hg vacancies cause the p-type conductivity in as-grown material. A Hg vacancy has an acceptor level near the valence band edge in the energy gap (Sect. 2.3.1). In contrast, it was found that Hg evaporates very easily if compared with Cd and Te. Accordingly, the surface of $Hg_{1-x}Cd_xTe$ crystals usually shows a marked depletion of Hg. This depletion causes a gradient of Hg vacancies and thus a diffusion current of Hg vacancies into the bulk sample. With the increasing number of Hg vacancies the sample becomes eventually p-type. A diffusion coefficient of approximately $D(300\ K) = 4 \cdot 10^{-17}\ m^2/s$ was deduced from the long-term Hall-type conversion data /421/ which was recently confirmed in the composition analysis by NITZ et al. /424/. One has to be aware that any Hg loss at the surface of a $Hg_{1-x}Cd_xTe$ crystal is expected to be strongly dependent on the surface condition and on the crystallographic orientation. MORGEN et al. /427/ carried out similar experiments by ultraviolet photoemission spectroscopy and by X-ray-induced photoemission on cleaved p-type material. They found, however, a considerably higher surface stability then the data reported above for chemically or mechanically treated surfaces. More details on the Hg diffusion are presented in Sect. 7.1.

Recently MICKLETHWAITE and REDDEN /426a/ suggested a modified model in order to explain the long-term instability of n-type $Hg_{1-x}Cd_xTe$ at room temperature. In addition to surface exchange with the ambient they expected a significant participation of Hg interstitials. They assumed the investigated samples to be inhomogeneous originally: a core material with about 10^{22}-10^{23} holes/m^3 is due to Hg vacancies, whereas n-type layers near the surface with $0.5 - 8 \cdot 10^{20}$ electrons/m^3 (T = 77 K) are due to Hg interstitials. The long-term Hall-type changes follow according to this model from a cross diffusion of Hg in the presence of interstitial-vacancy concentration gradients and/or surface exchange with the ambient.

The experimental data obtained from studying the surface of the mixed crystal /423,424/ and the annealing data /425a,b/ seem to favour the model by NIMTZ et al. /421/.

In order to avoid the loss of significant quantities Hg by vaporization, the alloy surface has to be passivated. More details and references on passivation and on the composition analysis of $Hg_{1-x}Cd_xTe$ surfaces are presented in Chap. 7.

3. Band Structure

In this chapter we present and discuss band structure calculations and experimental results on band edge characteristics of the mixed II-VI compound $Hg_{1-x}Cd_xTe$. After briefly considering the two components HgTe and CdTe in the first section we deal with three different band structure calculations based on the empirical pseudopotential method, the Korringa-Kohn-Rostocker (KKR)-method and a tight binding approximation.

In the next section we turn to the semimetal-semiconductor-transition in this pseudo binary alloy system, which arises from the fact that HgTe has an inverted band order with a negative fundamental gap $\Gamma_6 - \Gamma_8$ whereas CdTe has standard II-VI bands. In the third section band edge characteristics will be discussed in terms of the Kane model. The next two sections deal with the temperature dependence of the band gap and pressure effects. The last section is concerned with the influence of disorder induced by the statistical distribution of HgTe and CdTe in this mixed crystal.

After earlier photoemission studies on the density of valence states in CdTe /428/, the experimental results of SHEVCHIK et al. /429/ gave a wide overview on these data for II-VI compounds including HgTe and CdTe. Although the upper parts of the valence bands have been found to have shapes in qualitative agreement with theory, they are 1 to 2 eV wider than predicted. The shapes of the d-bands of the cation have been resolved, which permitted an accurate determina-

Table 7. Binding energies (in eV) of the outermost d-levels. The energies and the corresponding chemical shifts Δ^{TVB} are relative to the top of the valence band. W gives the "workfunction" used to obtain Δ^{V} relative to the vacuum level /429/

Compound	Cation $d_{5/2}$	Cation $d_{3/2}$	Anion		W	$\Delta^{TVB}_{cat.}$	$\Delta^{V}_{cat.}$	$\Delta^{TVB}_{an.}$	$\Delta^{V}_{an.}$
CdTe	10.09	10.72	39.75	41.25	6.2	0.66	-1.44	0.73	-0.58
HgTe	7.70	9.55	39.71	41.1	5.9	-2.3	-3.70	0.79	-0.21
Element									
Cd	10.62	11.58			4.1				
Hg	6.14				4.5				
Te	40.5	41.95			4.9				

tion of their positions relative to the top of the valence band and their spin-orbit splitting. Some data are listed in Table 7 /429/. For a discussion of recent results of ultraviolet photoemission spectroscopy on $Hg_{1-x}Cd_xTe$ (x = .2, .31, .39, 1.0) /430a-d/ and their interpretation /430e-h/, see Sect. 3.6.

3.1 Band Structure Calculations

Before discussing the band structure of the $Hg_{1-x}Cd_xTe$ alloy, recent calculations of the electronic band structure of the two components CdTe and HgTe will be mentioned without attempting to be exhaustive. Empirical pseudopotential calculations including spin-orbit-coupling have been carried out by CHADI et al. /55/ using slightly modified form factors for CdTe /56/ and HgTe /57/. The results are shown in Figs.12,13. KKR calculations on HgTe are also available /58/, details and comparison will be given later.

As already stated by OVERHOF /59/, BALKANSKI /60/ and others, theoretical investigations on mixed crystals have suffered from the lack of a theory scheme, in which starting from first principles the band edges of a mixed crystal can be described with sufficient accuracy. As a consequence of random mixing of Hg and Cd in the cation-sublattice one should expect an effect of disorder on the energy bands of the crystal. But most of the experimental results seem to call for a model where the electronic states do not deviate from those of a perfect crystal and may be described using the concept of energy bands. We will discuss this aspect in a later section in more detail.

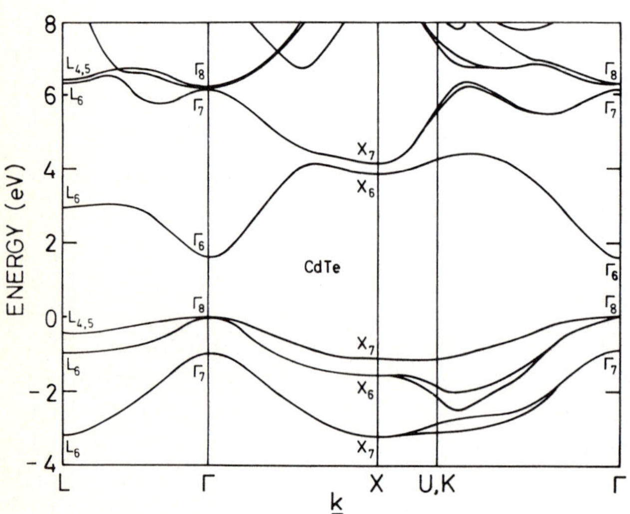

Fig.12. Electronic band structure of CdTe in the principal symmetry directions /55/

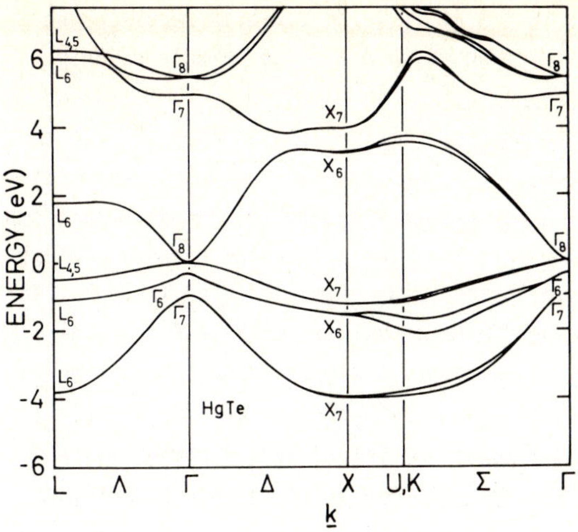

Fig.13. Electronic band structure of HgTe in the principal symmetry directions /55/

3.1.1 KKR Model Calculations

OVERHOF /59/ has calculated the energy bands with the Green's function method including relativistic effects (KKR Method) /61/. In this method the many-particle-problem is reduced to a one-particle problem by a suitable choice of a potential. The corresponding Schrödinger equation

$$\{ -\Delta + V(\underline{r}) \} \psi_{\underline{k}}(\underline{r}) = E(\underline{k})\psi_{\underline{k}}(\underline{r}) \tag{2a}$$

is replaced by an equivalent integral equation

$$\psi(\underline{r}) = \int G(\underline{r},\underline{r}')\, V(\underline{r}')\, \psi(\underline{r}')d\underline{r}' \tag{2b}$$

using a structural Green function

$$G(\varkappa\, ,\underline{k};\, \underline{r}-\underline{r}'') = \sum_{\underline{\ell}} \exp(i\varkappa\,|\underline{r}-\underline{r}'' + \underline{\ell}|)\, \exp(i\underline{k}\underline{\ell})/|\underline{r}-\underline{r}'' + \underline{\ell}| \qquad \varkappa^{2} = E \tag{3}$$

and inserting a muffin-tin potential.

$$V(\underline{r}) = \sum_{\underline{\ell}} v_a(\underline{r} - \underline{\ell})$$

Within non-overlapping spheres atomic potentials of HERMAN and SKILLMAN /62/ for Hg, Te and Cd were taken and superimposed along the nearest neighbour direction.

The constant potential V_0 in the region outside the spheres was taken as a fitting parameter to adjust the band gap $\Gamma_6 - \Gamma_8$. The solution of (2) was obtained by a variational procedure which gives a determinant depending on $\underline{k}$ and $E = \varkappa^2$. The zeros of this determinant give the relation $E(\underline{k})$.

The merits of this method lie in the fact that it is possible to separate the contribution of potential and crystal structure which has to be calculated only once ("structure constants").

In Overhof's model the main effect of admixing CdTe to HgTe is the shift of the Γ_6 band (metal s electron) to higher energies. Since the upper valence bands originate mainly from tellurium p electrons, they are expected to be almost unchanged by substituting Hg by Cd according to the mole fraction x. For the p and d partial waves the constant potential V_0 was fixed to the value - 10.6 eV (see Table 8). The shift of the s bands was obtained by using different values for V_0 for the s partial waves. To check the validity of the described model, a standard KKR calculation for CdTe was compared with model results, taking an s shift of 3.26 eV and thus fitting them to the fundamental band gap of CdTe (Fig.14). A fairly good agreement in the calculated band structures with deviations smaller than 0.2 eV in the vicinity of the Γ point was obtained.

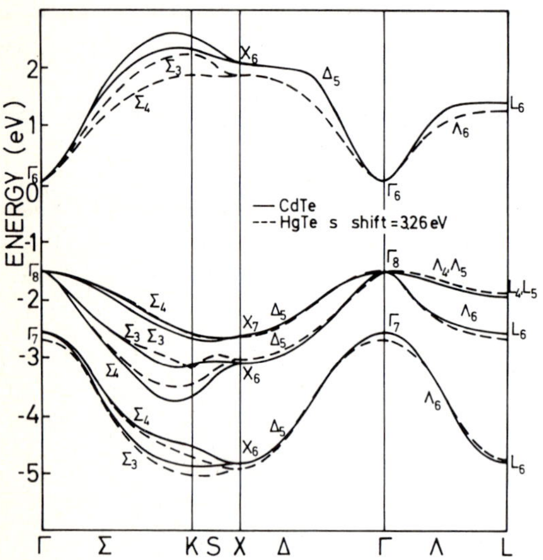

Fig.14. Calculated CdTe energy bands compared to HgTe bands obtained with an 3.26 eV s shift /59/

Table 8. Constants used in the KKR calculations /59/

	Lattice constant	Atomic radius metal	Te	Constant potential (eV)	s shift (eV)
	a_B x *)				
HgTe	12.17	2.58	2.69	- 10.6	3.26
CdTe	12.24	2.53	2.77	- 14.65	-

*) a_B = 0.53 Å Bohr radius.

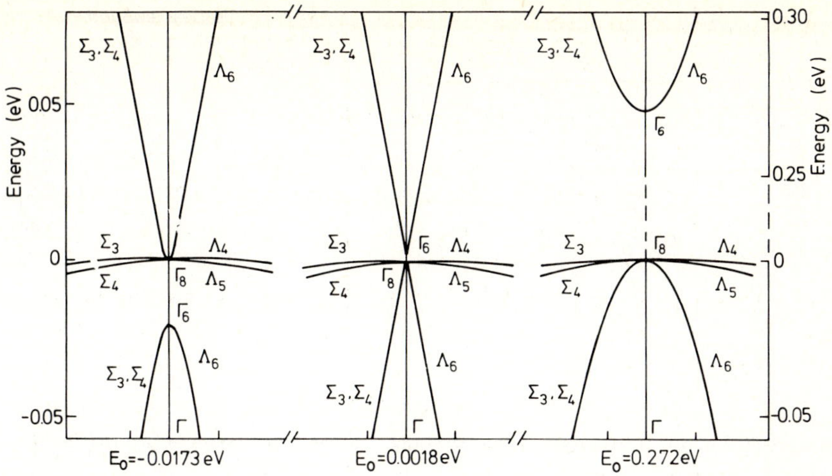

Fig.15. Electronic band structure around the Γ point of $Hg_{1-x}Cd_xTe$ with different values of E_0 /59/

From this model it is not possible to derive the band gap E_0 as a function of the CdTe admixture, so the calculated values E_0 for different s shifts were compared with experimental results to determine the corresponding composition. The energy band structure near the Γ point obtained in this way is shown in Fig. 15. A comparison of mass parameters with results of pseudopotential calculations in terms of the Kane model will be given in Sect. 3.3.

3.1.2 Pseudopotential Calculations

There are two pseudopotential band structure calculations available. KATSUKI and KUNIMUNE /63/ and CHADI and COHEN /64/. KATSUKI et al. use the angular momentum dependent (non-local) ionic model potential of ANIMALU and HEINE /65,66/:

$$v^M = v^{AH} + v^{SO}$$

with a crystal potential term

$$v^{AH} = \begin{cases} -A_2 - (A_0 - A_2)\,P_0 - (A_1 - A_2)\,P_1 & r \leq R_M \quad \text{(see Fig. 16)} \\ -Z/r & r > R_M \end{cases} \qquad (4a)$$

and a spin orbit term

$$v^{SO} = \begin{cases} (\lambda_1 - \lambda_2)\,\underset{\sim}{s}\,\underset{\sim}{\ell}\,P_0 + \lambda_2\,\underset{\sim}{s}\,\underset{\sim}{\ell}\,P_1 & r \leq R_M \\ 0 & r > R_M \end{cases} \qquad (4b)$$

where the A_i denote the effective square-well depth, P_ℓ is a projection operator, which picks out the ℓ-th orbital angular momentum component of any wave function,

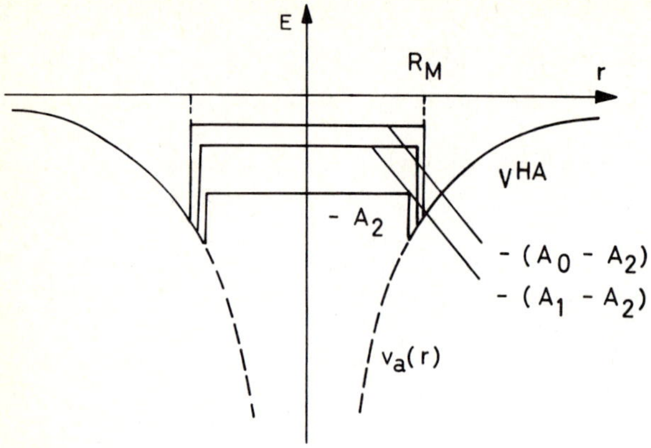

Fig. 16. Model pseudopotential of HEINE-ABARENKOW

and where the strengths of spin-orbit interactions λ_i are evaluated from the free ion term values of ANIMALU and HEINE /65,66/ (see Table 9). The valence electrons are taken into account by screening the ionic potential with the $\underline{k}$-dependent dielectric constant

$$\varepsilon_p(\underline{k}) = 1 + \left(\frac{\hbar\omega_p}{E_g^p}\right)^2 \left(1 - \frac{E_g^p}{4E_F}\right) \left[1 + \frac{E_F}{E_g^p}\left(\frac{k}{k_F}\right)^2 \left(1 - \frac{E_g^p}{4E_F}\right)^{1/2}\right]^{-2} \tag{5}$$

derived by PENN /67/. Here ω_p, E_F, k_F and E_g^p denote the plasma frequency, the free electron Fermi energy and momentum and an adjustable energy gap E_g^p, respectively.

Table 9. Constants used in the Animalu-Heine-Penn model potential /63/

	Effective square-well depth A_ℓ (eV)			Radius R_M	Z	Strength of spin-orbit interaction λ_ℓ	
	A_0	A_1	A_2	a_B x		λ_1(eV)	λ_2(eV)
Cd	23.95	26.67	30.21	2.6	2	0.54	0.41
Hg	26.40	30.21	23.13	2.6	2	1.36	1.22
Te	86.00	90.35	76.20	2.0	6	1.63	0.27

Some corrections concerning correlation and orthogonality were added to the ionic potential, orthogonality and Hubbard exchange correction added to the dielectric constant /65/. E_g^p, the only free parameter in this Animalu-Heine-Penn-Model (AHP), is adjusted so as to get the calculated band structure to agree with experimental optical transition energies for the pure compounds.

Values

$$E_g^p/E_F = 0.654 \quad \text{for CdTe}$$
$$ 0.241 \quad \text{for HgTe}$$

are obtained. The band structure of the alloy system $Hg_{1-x}Cd_xTe$ was calculated in the approximation of the virtual crystal model (Sect. 3.6) at high symmetry points Γ, X, L, K and on symmetry lines Δ, Λ, Σ. The results are shown in Figs. 17,18, 19 and in Table 10.

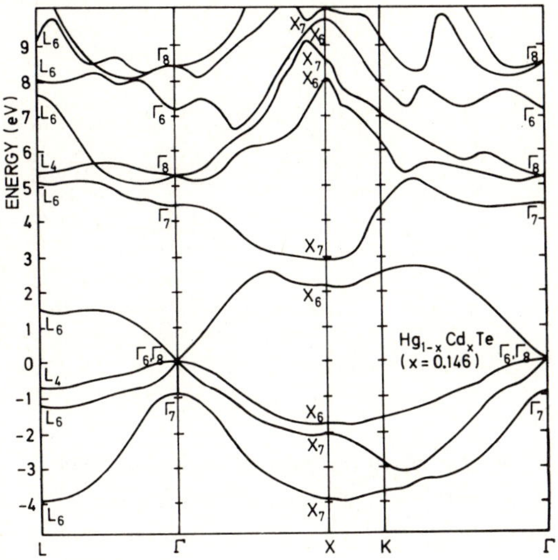

Fig. 17. Electronic band structure of $Hg_{.854}Cd_{.146}Te$ in the principal symmetry directions /63/

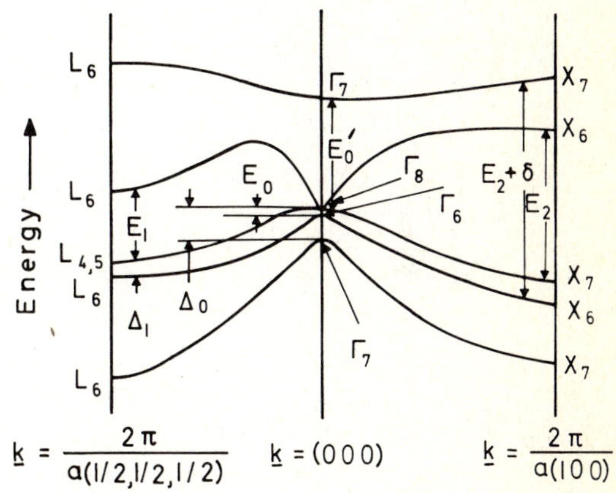

Fig. 18. Schematic band structure of HgTe with different observed interband transitions /63/

As shown, all energy levels seem to depend almost linearly on the composition x (we shall discuss this result in Sect. 3.6 in more detail). The fundamental gap $E_0 = E(\Gamma_6) - E(\Gamma_8)$ increases linearly too, from negative to positive as x increases. The semimetal-semiconductor transition is expected from the calculation to occur at x = 0.146 at 77 K.

The pseudopotential Hamiltonian for an electron in the crystal consists of the kinetic-energy term plus a weak potential which depends on position

$$H = -(\hbar^2/2m)\,\nabla^2 + V(\underline{r}). \qquad (6)$$

The potential V can be expanded in reciprocal lattice vectors $\underline{G}$ and can be expressed as the product of a structure factor $S(\underline{G})$ and a pseudopotential form factor V_G.

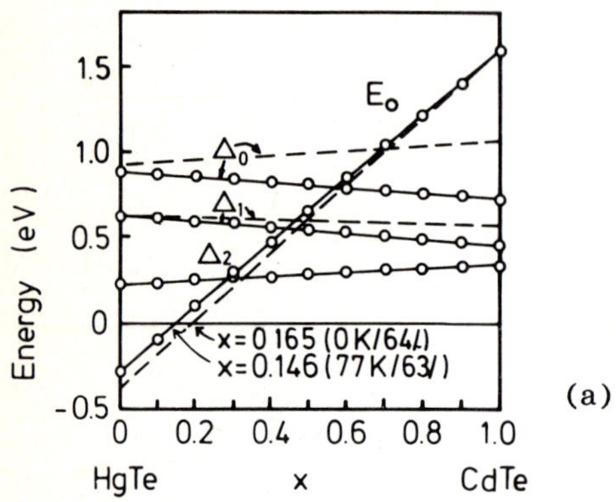

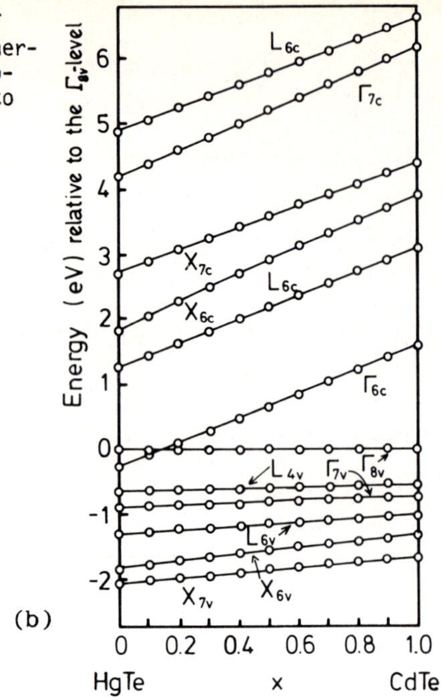

Fig. 19. (a) Calculated composition dependence of the energy gap E_0 and the spin-orbit splitting energies for $Hg_{1-x}Cd_xTe$ /63,64/. (b) Calculated composition dependence of the energy levels relative to the Γ_{8v} level of $Hg_{1-x}Cd_xTe$ /63/

Table 10. Calculated energies relative to the Γ_{8v} state (in eV) /63/

	HgTe	$Hg_{.854}Cd_{.146}Te$	CdTe
Γ_7	-0.893	-0.866	-0.728
Γ_8	0	0	0
Γ_6	-0.280	0	1.591
Γ_7	4.179	4.490	6.171
Γ_8	5.040	5.267	6.461
L_6	-1.298	-1.247	-1.008
L_4	-0.672	-0.649	-0.546
L_6	1.252	1.534	3.091
L_6	4.866	5.144	6.606
L_4	5.194	5.438	6.695
X_7	-2.056	-1.989	-1.658
X_6	-1.826	-1.733	-1.320
X_6	1.830	2.165	3.900
X_7	2.705	2.973	4.410
K_3	-2.884	-2.794	-2.409
K_3	-1.646	-1.575	-1.258
K_3	2.283	2.586	4.173
K_3	4.170	4.376	5.527

It is convenient to break up the potential into a symmetric and an antisymmetric part

$$V(\underline{r}) = \sum_{|G| \leq G_0} (S^S(\underline{G})\, V_G^S + iS^A(\underline{G})\, V_G^A)\, e^{-i\underline{G}\underline{r}} . \qquad (7)$$

Cubic semiconductors of the zincblende type have the fcc structure with two atoms per unit cell. Taking the origin of coordinates so as to give

$\underline{r}_1 = a(1/8, 1/8, 1/8) = \underline{\tau}$ and $\underline{r}_2 = -\underline{\tau}$ gives

$S^S(\underline{G}) = \cos \underline{G}\,\underline{\tau}$ and $S^A(\underline{G}) = \sin \underline{G}\,\underline{\tau}$.

If V_1 and V_2 denote the pseudopotentials due to single atoms in the lattice and Ω is the volume of the unit cell, V_G^S and V_G^A can be expressed as

$$V_G^S = \tfrac{1}{2}(V_1(\underline{G}) + V_2(\underline{G})) \quad V_G^A = \tfrac{1}{2}(V_1(\underline{G}) - V_2(\underline{G})) \quad V_i(\underline{G}) = \frac{2}{\Omega}\int V_i(\underline{r})e^{-i\underline{G}\underline{r}}d^3\underline{r}. \quad i=1,2$$

The first five of the reciprocal lattice vectors have squared magnitudes of 0, 3, 4, 8 and 11. Only these are considered to have a non-zero potential. The symmetric structure factor is zero for $G^2 = 4$, and the antisymmetric for $G^2 = 0$ and $G^2 = 8$. Keeping $V_0^5 = 0$, since it merely adds a constant to all energy levels, only three symmetric and three antisymmetric form factors remain for one compound.

In their empirical pseudopotential calculation CHADI and COHEN /64/ have used form factors obtained by fitting theoretical optical reflectance spectra to the experimental data /55/. Spin-orbit interactions were included in their calculation using the WEISZ scheme /68/ as modified by BLOOM and BERGSTRESSER /69/, parameters of the calculation are given in **Table 11.** The two spin-orbit parameters λ_1 and λ_2 were constrained to have the same ratio as the splittings in the free atoms, leaving only one adjustable parameter. The pseudopotentials of the alloys were taken to be the average of the HgTe and CdTe pseudopotentials including the spin-orbit part of the Hamiltonian. The form factors were scaled for the small changes in the lattice constant, which was assumed to vary linearly with x (Sect. 3.6).

The calculated variation of the energy gap E_0 and the spin-orbit splittings Δ_0 and Δ_1 are shown and compared with those from /63/ in **Fig. 19a.**

Table 11. Form factors used in the empirical pseudopotential calculation /64/ (in eV)

	$V_S(3)$	$V_S(8)$	$V_S(11)$	$V_A(3)$	$V_A(4)$	$V_A(11)$	$V_A(12)$	Spin-orbit parameters
CdTe	-3.18	-0.57	0.56	2.05	0.93	0.07	0	0.015
HgTe	-3.57	-0.48	0.68	1.36	0.57	0.27	0.26	0.035

3.1.3 Tight-Binding Model

KISIEL and LEE /70/ have calculated the energy levels at the Γ point and along the (111) direction Λ to L at the zone boundary on the basis of a tight-binding method including relativistic effects. According to HERMAN et al. /71/, the four-component Dirac Hamiltonian can be reduced by applying two successive canonical transformations of the Foldy-Wouthuysen type using the fine structure constant α as expansion parameter to a form, where negative and positive energy states are decoupled to order α^2. For the positive energy states one gets

$$H = - \nabla^2 + V(\underline{r}) - i(\frac{\alpha^2}{4}) \underline{\sigma} \, |\nabla V(\underline{r}) x \nabla| - (\frac{\alpha^2}{4}) \nabla^4 + (\frac{\alpha^2}{8}) \nabla^2 V(\underline{r}). \tag{8}$$

So the Hamiltonian taken is of the form

$$H = H_1 + H_{so} + H_{mv} + H_D \quad , \tag{8'}$$

where terms H_{so}, H_{mv} and H_D represent the spin-orbit, mass-velocity and Darwin interactions respectively. Here H_1 is the usual non-relativistic part, containing a potential energy term based on the virtual crystal approximation

$$xV_{Cd} + (1-x) V_{Hg} \quad . \tag{9}$$

Since relativistic effects arise mainly in the potential near the nucleus of an atom /72/, their interpretation in terms of diagonal matrix elements of the tight-binding wave function should be acceptable. In the alloy these terms appear weighted with the concentration of Cd and Hg. The spatial part of the tight-binding wave function is of the form

$$\psi = \sum_{\underline{R}_j, \underline{\tau}_i, \ell} a_{\ell, \underline{\tau}_i} \exp i\underline{k}(\underline{r} - \underline{R}_i + \underline{\tau}_i) \, \Phi_\ell \, (\underline{r} - \underline{R}_j + \underline{\tau}_i) \tag{10}$$

where $\underline{R}_j$ runs over the lattice vectors, $\underline{\tau}_i$ over the atoms in the unit cell, Φ_ℓ over the set of radial and angular wave functions and the coefficients $a_{\ell,\underline{\tau}_i}$ are obtained by solving the resulting secular equation of the tight-binding method.

In their calculations with main emphasis on the relativistic shifts of the upper valence and the lowest conduction band in the $Hg_{1-x}Cd_xTe$ alloy, KISIEL and LEE limit the set of Φ_ℓ to the 5s and 5p wave functions of Te and Cd and the 6s and 6p functions of Hg, combining these with the spin functions for $s_z = \pm \frac{1}{2}$ in order to obtain basis functions for representations of the double group. Using these functions the matrix elements of H are evaluated following the Koster/Slater formalism and additionally considering the mass-velocity, Darwin and spin-orbit terms. In calculating the kinetic and potential energy terms, values based on wave functions and potentials of HERMAN et al. /62/ are adjusted to reproduce the experimentally determined energy gaps at Γ and L. Contributions of relativistic terms in the Hamiltonian are based on values for CdTe and atomic Hg of /71/.

The mass-velocity term gives an additional contribution to s functions, while for the p function this contribution is very small. The Darwin term applies only to the s functions. The spin-orbit term is the only term depending upon the crystal symmetry and thus lifts the degeneracy.

Values of parameters (in eV) used are given in **Table 12.** The effect of relativistic terms on the order of energy levels at Γ is shown in **Fig. 20,** the curvature of the conduction band along Λ in **Fig. 21.**

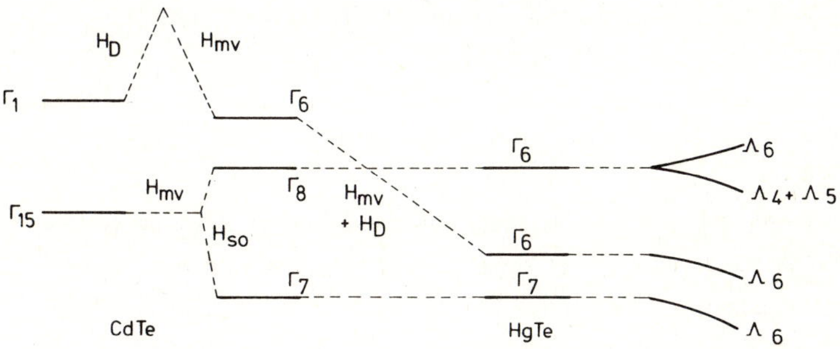

Fig. 20. Effect of relativistic terms on the order of energy levels at Γ /70/

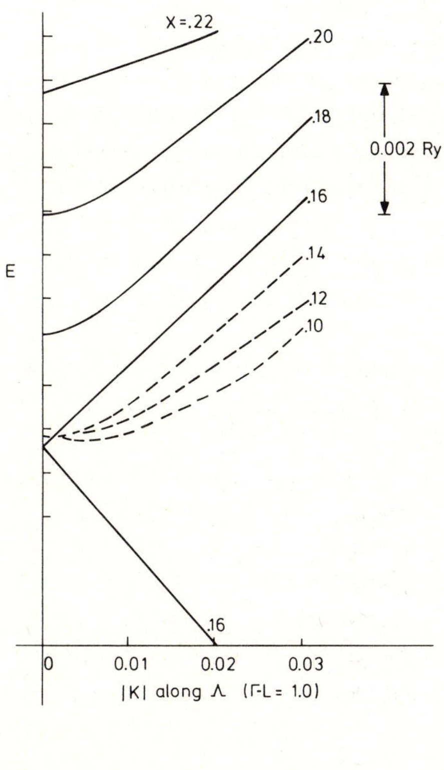

Fig. 21. Energy bands from Γ along Λ showing the change in curvature as the energy gap E_0 disappears /70/

Table 12. Values of parameters (in eV) used for the relativistic interactions. The spin-orbit splitting due to Cd and Hg mainly affects higher conduction bands and is omitted /70/

	spin-orbit	mass-velocity (s)	mass-velocity (p)	Darwin (s)
Cd	-	-1.878	0.680	1.097
Hg	-	-5.499	0.816	3.013
Te	0.599	-5.153	0.694	2.883

CASULA and KISIEL /431/ have used the method of linear combination of localized orbitals to calculate the band structure of HgCdTe and especially the bowing parameters for the E_0, E_1, and $E_1+\Delta_1$ gaps in the virtual-crystal approximation.

3.2 The Semimetal-Semiconductor Transition (SM-SC Transition)

The key feature in discussing the changes in the band structure of $Hg_{1-x}Cd_xTe$ alloys as a function of the mole fraction of CdTe is the inverted order of the Γ_6, Γ_8 bands in HgTe as compared to the normal II-VI compounds with ZnS structure.

As was shown in the last section this inverted order arises largely from relativistic effects lowering the Γ_6 level until it falls below Γ_8, which as a result of degeneracy becomes both the top of the valence band and the bottom of the conduction band. On substituting Cd for the heavy Hg atom the energy spectrum of the alloys becomes smoothly restructed from the HgTe spectrum of the CdTe spectrum (Figs.15,20,21). This means that the $Hg_{1-x}Cd_xTe$ alloys exhibit a transition from a semimetal behaviour in HgTe /73,74,75,76,77/ to a semiconducting behaviour in CdTe [1]. The band gap E_0, which, similar to the Groves and Paul model of white tin, is negative in HgTe, changes almost linearly with x from $\sim$- 0.3 eV to 1.6 eV at 0 K (Fig. 19a) thus leading to a gapless state $E_0 = 0$ at $x \approx 0.16$ at liquid helium temperatures.

A large number of measurements supporting this band crossover have been reported /38,53,73,78-81/ and will be discussed in the following sections.

---

[1] The absence of a symmetry centre in the zincblende lattice leads to a certain shift of the maximum of the heavy - hole band away from the point k = 0 and as a result the valence band overlaps the conduction band /59,63,64/. The overlap, however, is small and does not exceed 10^{-3} eV.

3.3 The Band Edge Characteristics

In discussing the band edge characteristics of the alloy system $Hg_{1-x}Cd_xTe$ one must bear in mind that in the most interesting range of concentrations x one deals with a small gap or zero-gap material. A small-gap semiconductor is a material, in which, following ZAWADZKI /82/, the value of electron energy E, as counted from the bottom of the conduction band, can become comparable to the energy gap E_0. This in turn means a small mass and a small density of states, so that even a relatively small number of electrons populates the conduction band up to quite high energies above the band edge. The proximity of the valence band is strongly felt and the $E(\underline{k})$ curve becomes strongly non-parabolic.

3.3.1 The Three Level Model of a Small-Gap Semiconductor

In order to discuss this behaviour including the spin-properties of conduction electrons, one has to start with a Hamiltonian of the form

$$H = \underline{P}^2/2m + V_0(\underline{r}) + (\hbar/4m^2c^2)\,(\underline{\sigma}\times\nabla V_0)\underline{P} + \mu_B\,\underline{B}\,\underline{\sigma} \tag{11}$$

where $\underline{P} = \underline{p} + e\underline{A}$, $\underline{A}$ is the vector potential of the magnetic field $\underline{B}$, μ_B is the Bohr magneton, $\underline{\sigma}$ denotes the Pauli spin operator and $V_0(\underline{r})$ the periodic potential of the lattice. Following the Kane model to solve the problem one has to consider a finite number of close-lying levels treating them exactly and leaving out all other levels in the first approximation /83,84/. To describe the properties of conduction electrons, according to ZAWADZKI /85/ one may restrict the theoretical description to a three-level model with the Γ_6 level separated by the energy gap E_0 from the two-fold degenerate Γ_8 level which in turn is split off by the spin-orbit energy Δ from the Γ_7 level (Fig. 22). In this model the heavy-hole band is completely flat, the finite curvature of this band results from the interaction with other bands. Modifications of this simple model will be discussed later.

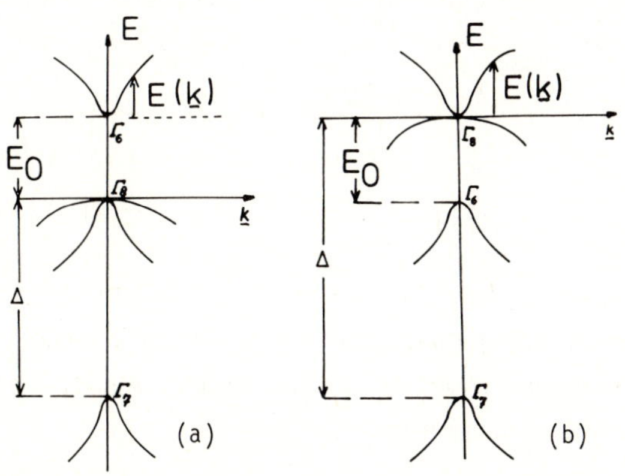

Fig. 22. Band structure of InSb type (a) and α-Sn type (b) materials at the Γ point (schematically) /85/

The model leads to an 8x8 matrix eigenvalue problem, which results in the following equations for the eigenvalues /85/ (neglecting the free electron term):

$$F(E'_{N,k_z,\pm}) = E'(E' + E_0)(E' + \Delta) - P^2\left[s(2N + 1) + k_z^2\right](E' + \frac{2}{3}\Delta) \pm \frac{1}{3}P^2\Delta\, s = 0 \quad (12)$$

for the semimetal and

$$F(E'_{N,k_z,\pm}) = E'(E' + E_0)(E' + E_0 + \Delta) - P^2\left[s(2N + 1) + k_z^2\right](E' + 2/3\Delta + E_0) \pm \frac{1}{3}P^2\Delta\, s = 0 \quad (12')$$

for the semiconducting compounds with $E' = E - \hbar^2k^2/2m$, $s = eB/\hbar$ and the momentum matrix element $P = -(i\hbar/m)\langle s|p_z|z\rangle$.

The three roots of this equation give the energy levels in the conduction, the light-hole and the split-off energy bands. The fourth eigenvalue $E'_{N,k_z,\pm} = 0$ corresponds to the band of heavy holes.

According to LAX et al. /86/ the energy expression for the conduction band can be written as

$$E'_{N,k_z,\pm} = \frac{\hbar eB}{m^*(E')}(N + \frac{1}{2}) \pm \frac{1}{2}\mu_B\, g^*(E')B \quad \text{with } E' = E'_{N,k_z,\pm}$$

$$m^*(E')^{-1} = \frac{1}{m_0^*}\cdot\frac{E_0(E_0 + \Delta)}{3E_0 + 2\Delta}\cdot\left[\frac{2}{E' + E_0} + \frac{1}{E' + E_0 + \Delta}\right] \quad (13a)$$

$$g^*(E') = g^*\cdot\frac{E_0(E_0 + \Delta)}{\Delta}\cdot\left[\frac{1}{E' + E_0} - \frac{1}{E' + E_0 + \Delta}\right].$$

For the region of energies $E' \ll 2\Delta/3$ (resp. $E' \ll E_0 + 2\Delta/3$) from (12) the resulting quadratic equation can be solved directly to give for the conduction band:

$$E'_{N,k_z,\pm} = -\frac{E_0}{2} + \left[\left(\frac{E_0}{2}\right)^2 + E_0 D(N,k_z,\pm)\right]^{1/2}$$

$$\text{with } D(N,k_z,\pm) = \hbar\omega_c(N + \frac{1}{2}) + \frac{\hbar^2k_z^2}{2m_0^*} \mp \frac{1}{2}\mu_B|g^*|B \quad (13b)$$

where $\omega_c = eB/m_0^*$ is the effective cyclotron frequency and $g^* = -m/m_0^*$ the g value at the bottom of the band. A schematic diagram of the Landau and spin levels is shown in Fig. 23 /85/. The effective mass at the band edge in (13) is given by

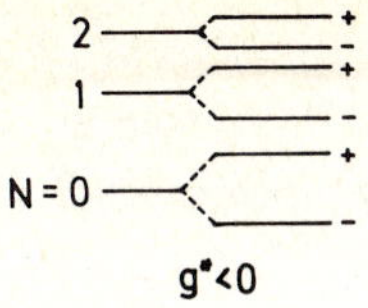

Fig. 23. Magnetic Landau- and spin levels of the conduction electrons (for $k_z = 0$) in a small-gap semiconductor (schematically) /85/

$$m_o^{*\,-1} = 4P^2/3\hbar^2 E_o \qquad \text{for the semimetal and}$$

$$m_o^{*\,-1} = 4P^2/3\hbar^2 E_o \, \frac{\Delta + 3E_o/2}{\Delta + E_o} \qquad \text{for the semiconducting compounds.} \tag{14}$$

For the latter the g factor at the bottom of the band is:

$$g^* = - \frac{m}{m_o^*} \cdot \frac{\Delta}{\Delta + 3E_o/2} . \tag{15}$$

The appropriate electron wave functions are given by ZAWADZKI. For vanishing magnetic field (12, 12') become

$$E'(E' + E_o)\,(E' + \Delta) - P^2k^2(E' + 2/3\,\Delta) = 0 \tag{16}$$

$$E'(E' + E_o)\,(E' + E_o + \Delta) - P^2k^2(E' + E_o + 2/3\,\Delta) = 0. \tag{16'}$$

The effective mass is thus obtained from (16') in the form

$$m^{*\prime}(k)^{-1} = \frac{1}{m_o^*} \, \frac{\Delta + 3E'/2}{(1 + 2E'/E_o)\,(\Delta + E') + E'/E_o(\Delta + E') - 3/4\, \frac{\hbar^2k^2}{2m_o^*}} \tag{17}$$

and the conduction electron g factor as

$$g^*(k) = 2 \left\{ 1 - \left(\frac{m}{m^*(k)} - 1\right) \frac{\Delta}{3E'(k) + 2\Delta} \right\} \tag{18}$$

$$g^*(k) = 2 \left\{ 1 - \left(\frac{m}{m^*(k)} - 1\right) \frac{\Delta}{3E'(k) + 3E_o + 2\Delta} \right\} \tag{18'}$$

with $E'(k)$ determined from (16, 16'). For $E' \ll 2\Delta/3$, the dispersion relation becomes

$$E'(k) = - \frac{E_o}{2} + \left[\left(\frac{E_o}{2}\right)^2 + E_o \frac{\hbar^2k^2}{2m_o^*} \right]^{1/2} \tag{19}$$

and the effective mass

$$m^{*\prime}(k) = \left(\frac{1}{\hbar^2 k} \cdot \frac{dE'(k)}{dk} \right)^{-1} = m_o^*\left(1 + 2\frac{E'}{E_o}\right). \tag{20}$$

It should be mentioned, that in the literature sometimes the values $E_p = (2m/\hbar^2)\, P^2$ or $Q = 2E_p/3$ are given instead of P.

In the case of very low values E_o in $Hg_{1-x}Cd_xTe$ near the semimetal-semiconductor transition from (14) and (19) follows:

$$E(k) = (2/3)^{1/2} P \cdot k \tag{21}$$

(the electron energy is proportional to its momentum) and

$$m_o^* = (3/2)\,(\hbar^2/P)\,k. \tag{22}$$

3.3.2 Statistics

It seems useful before discussing experimental results to give some more detailed formulae based on Fermi-Dirac statistics. The total electron concentration can be written as

$$n = \frac{1}{3\pi^2}\left(\frac{2m_o^*\,k_BT}{\hbar^2}\right)^{3/2}\, {}^{0}L_{0}^{3/2}\,(\eta,\beta) \tag{23}$$

with generalized Fermi integrals ${}^{n}L_{k}^{m}$ defined by the expression:

$${}^{n}L_{k}^{m}\,(\eta,\beta) = \int_0^\infty (-\,\partial f_o/\partial z)\; z^n\,(z+\beta z^2)^m\,(1+2\beta z)^k dz \tag{24}$$

where $\beta = k_BT/E_o$, $\eta = E_F/k_BT$ (see /82/). In the case of strong degeneracy (23) can be evaluated as

$$n = \frac{1}{3\pi^2}\left(\frac{2m_o^*\,k_BT}{\hbar^2}\right)^{3/2}(\eta+\beta\eta^2)^{3/2}. \tag{25}$$

It follows

$$\eta = \frac{E_F}{k_BT} = (1/2\,\beta)\,(\sqrt{D}-1) \text{ with } D = 1 + 2(3\pi)^{2/3}\frac{\hbar^2}{m_o^*\,E_o}\,n^{2/3} \tag{26}$$

(in this case the electron concentration is simply $n = (1/3\,\pi^2)\,k_\zeta^3$)

$$E_F = \frac{-E_o}{2} + \left\{\frac{E_o^2}{4} + \frac{2}{3}P^2\,(3\pi n)^{2/3}\right\}^{1/2}. \tag{27}$$

With $m^*(\eta) = m_o^*\,D^{1/2}$ it follows

$$\left(\frac{m^*/m}{1-m^*/m}\right)^2 = \left(\frac{m_o^*}{m}\right)^2 + \frac{1}{2}\left(\frac{3}{\pi}\right)^{2/3}\frac{\hbar^2 m_o^*}{E_o\,m^2}\,n^{2/3} \tag{28}$$

$$m^*/m = \left\{ 1 + \frac{4P^2}{3\hbar^2} m \, (E_o^2 + 8/3\, P^2k^2)^{-1/2} \right\}^{-1}. \qquad (29)$$

In the region of energies $E \ll E_o + 2\Delta/3$ (for $E_o > 0$) and $E \ll 2\Delta/3$ (for $E_o < 0$) the density of states in a magnetic field is given by /87/:

$$z(E'_{N,k_z,\pm}) = \frac{1}{8\pi^2} \left(\frac{2m_o^*}{\hbar^2} \right)^{3/2} \left(\frac{2E'}{E_o} + 1\right) (\hbar\omega_c)^{1/2} \sum_{n,\pm} |C_\pm - N - \tfrac{1}{2}|^{-1/2} \qquad (30)$$

$$C_\pm = \frac{E'^2 + (E' \pm \frac{1}{2}\mu_B|g^*|B)E_o}{E_o\hbar\omega_c}$$

which leads to the following expression for the free electron concentration

$$n = \int_0^\infty z(E)f_o(E)dE = \frac{1}{4\pi^2} (k_BT)^{1/2} \left(\frac{2m_o^*}{\hbar^2} \right)^{3/2} \hbar\omega_c$$

$$\sum_{N,\pm} \left(1 + \frac{4[\hbar\omega_c(N+\frac{1}{2}) \mp \frac{1}{2}\mu_B|g^*|B]}{E_o} \right)^{1/4} {}^oL_o^{1/2}(\eta_{N,\pm}, \beta_{N,\pm}) \qquad (31)$$

$\eta_{N,\pm} = \frac{1}{k_BT}(E_F + E'_{N,k_z,\pm})$, $\beta_{N,\pm} = k_BT/2(E'_{N,k_z,\pm} - E_o/2)$ with E_F Fermi level (relative to the bottom of the conduction band in the absence of a magnetic field).
In the case of a degenerate semiconductor (31) becomes

$$n = \frac{1}{4\pi^2} \left(\frac{2m_o^*}{\hbar^2} \right)^{3/2} (\hbar\omega_c)^{3/2} \sum_{N,\pm} (C_\pm - N - \tfrac{1}{2})^{1/2} \qquad (32)$$

Because of the energy dependence of the effective mass in a non-parabolic energy band, the mass values determined in transport experiments represent values averaged over the participating electrons /82/. In strongly degenerate conditions these electrons have energies very close to the Fermi energy and the averages, in principle different for different experiments, are all essentially equal to the mass value at the Fermi level. If, however, the degeneracy is not complete the mean values determined in different measurements can be quite different. According to ZAWADZKI /82/ one determines the following ratios:

free carrier optical reflectivity

$$\bar{m}_R/m_o^* = {}^oL_o^{3/2} \,/\, {}^oL_{-1}^{3/2}$$

Splitting of reflection minima in a magnetic field

$$\bar{m}_{R_H}/m_o^* = {}^oL_{-1}^{3/2} \,/\, {}^oL_{-2}^{3/2}.$$

Faraday rotation

$$\bar{m}_F/m_o^* = ({}^oL_o^{3/2} \,/\, {}^oL_{-2}^{3/2})^{1/2}$$

free carrier Voigt effect

$$\bar{m}_V/m_o^* = ({}^oL_o^{3/2} \; / \; {}^oL_{-3}^{3/2})^{1/3}$$

thermoelectric power

$$\bar{m}_{\alpha(\infty)}/m_o^* = \frac{2}{\pi^2} ({}^oL_1^{3/2})^{2/3} ({}^1L_o^{3/2} \; / \; {}^oL_o^{3/2} - \eta).$$

3.3.3 Comparison Between Theoretical and Experimental Results

3.3.3.1 The Dependence of the Energy Gap E_o on Composition x, Temperature T and Pressure p

As already shown in Fig. 19 band structure calculations in the virtual crystal approximation predict a linear dependence of the fundamental gap E_o on composition x of the alloy system. Positions of the semimetal-semiconductor transition differ only slightly /64,70/ dependent probably on the fit parameters used. Experimentally the gap energy E_o was determined in a large number of experiments such as Shubnikov-de Haas and photoconductivity measurements /88,89,90,91/, measurements of the thermoelectric power and magneto-plasmareflection /92-95/, interband magnetoreflection /96/, cyclotron resonance and non-resonante cyclotron absorption /97-99/, optical absorption edge /100-101/, photoconductivity /102-103/, temperature dependence of the carrier density /49/, pressure and temperature dependence of galvanomagnetic data /79/, plasma reflection /104/, photoluminescence /40/ and others /105/.

The most extensive studies on the dependence of E_o on composition x and temperature T have been reported by WILEY and DEXTER /98/ ($0.135 \leq x \leq 0.203$), SCOTT /100/ ($0.23 \leq x \leq 0.61$, $10\ K \leq T \leq 300\ K$) and SCHMIT and STELZER /102/ ($0.13 \leq x \leq 0.6$, $20\ K \leq T \leq 300\ K$). Their results have been fitted by the following expressions, respectively:

$$E_o(x,T) = -0.30 + 5 \cdot 10^{-4}\ T + (1.91 - 10^{-3}\ T)x \qquad /98/$$

$$E_o(x,T) = -0.303 + 1.73x + 5.6 \cdot 10^{-4}\ (1 - 2x)\ T + 0.25\ x^4 \qquad /100/ \qquad (33)$$

$$E_o(x,T) = -0.25 + 1.59x + 5.233 \cdot 10^{-4}\ (1 - 2.08x)\ T + 0.327\ x^3 \qquad /102/$$

The first three terms in (33) represent linear x and T dependencies of E_o, the last term adjusts the second and third curve to fit the gap of CdTe. A slight observed non-linearity of the T dependence in /100/ is not included in the second formula. Following the authors /100,102/ the best fit was obtained in the regions $T \geq 77$ K and $0.2 \leq x \leq 0.33$.

In Figs. 24, 25 a-d, E_o is plotted versus x and T corresponding to the expressions given above together with data points published by various other authors.

It is interesting (Fig. 25) that the E_o versus x curve seems slightly bowed downward from a straight line. This effect, observed in other alloy systems too,

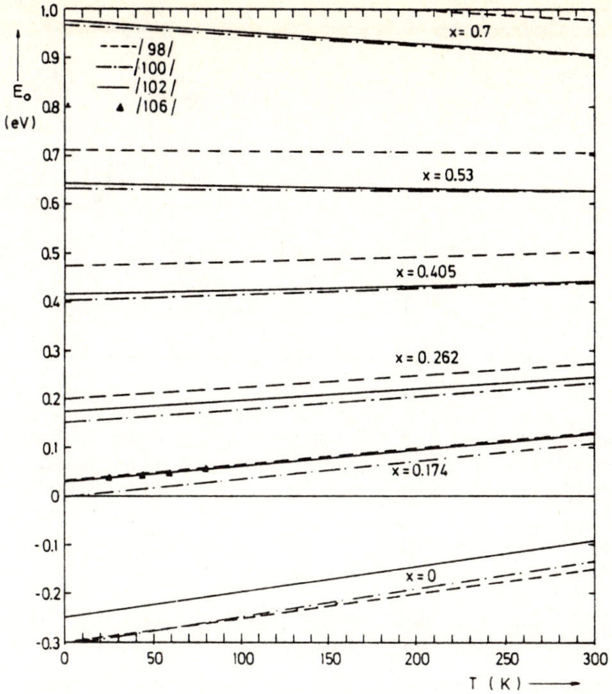

Fig. 24. Temperature dependence of the energy gap E_0 at different compositions x

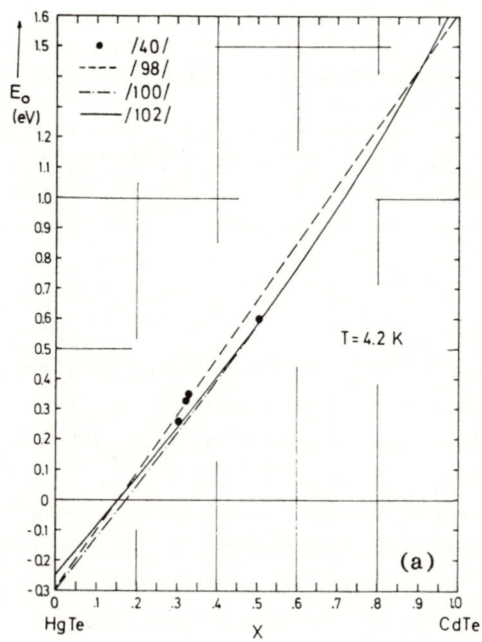

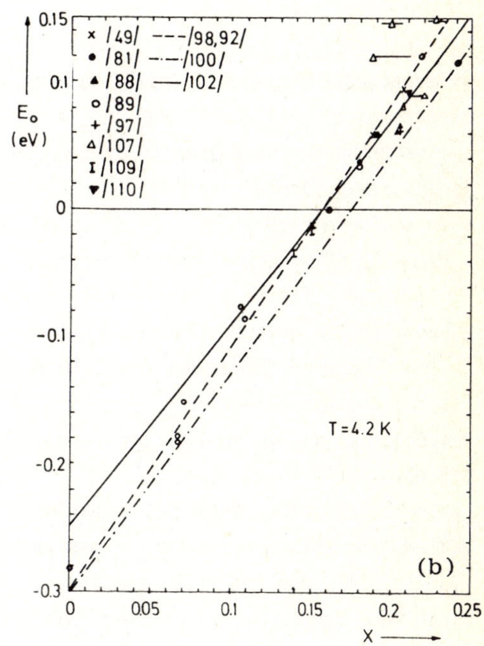

Fig. 25a-d. Composition dependence of the energy gap E_0 at (a) T = 4.2 K, $0 \leq x \leq 1$, (b) T = 4.2 K, $0 \leq x \leq 0.25$, (c) T = 77 K, (d) T = 300°K

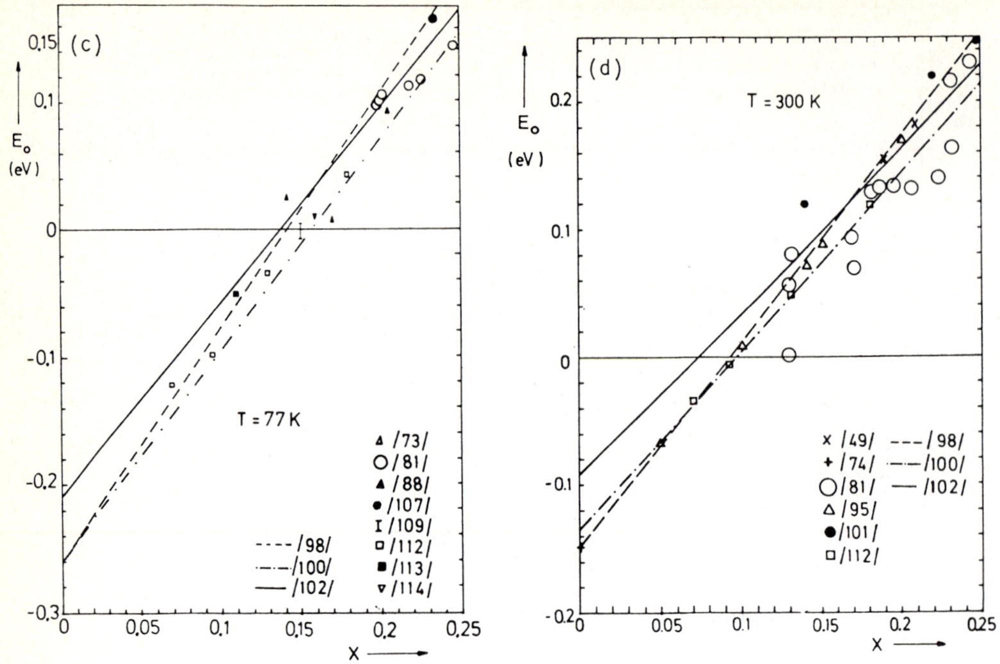

Fig. 25c,d. Figure caption see p. 159

is expected theoretically /115/ but was not taken into account in published band structure calculations on $Hg_{1-x}Cd_xTe$ (Sect. 3.6).

The temperature dependence of the energy gap dE_0/dT is plotted in Fig. 26. It is anomalous for two reasons:

- in the HgTe rich alloys dE_0/dT is positive in contrast to the negative dependence normally observed in semiconductors,
- $E_0(T)$ has a finite slope down to very low temperatures (Fig. 24) rather than becoming temperature independent (Sect. 3.4).

The change of the energy gap E_0 with pressure has been determined by STANKIEWICZ /116/ and others (Sect. 3.5). In Fig. 27 the changes in the $E_0(x)$ curve for two temperatures and an applied pressure of 8.11 kbar are shown. Values of dE_0/dP are given in Table 13.

In their band structure calculation KATSUKI et al. /63/ have computed the detailed band structures in the vicinity of Γ along three symmetry lines (in the (100), (110), and the (111) direction). Because of the usual k-linear term in zincblende compounds originating from the absence of inversion symmetry the Kramer's doublet is split. As a result of the combination of the linear and quadratic k term a valence band maximum $E_m = E_{\Gamma_8} + E_t$ away from k = 0 at $k_m = 3 \cdot 10^{-3}$ $2\pi/a$ in the (111) direction is found, where $E_t = 9 \cdot 10^{-5}$ eV for x = 0

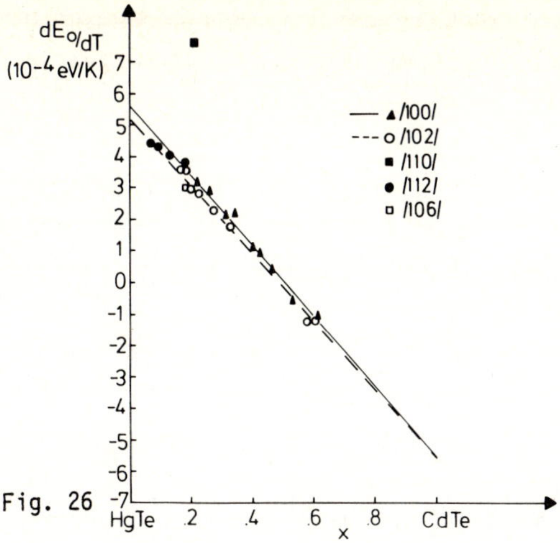

Fig. 26

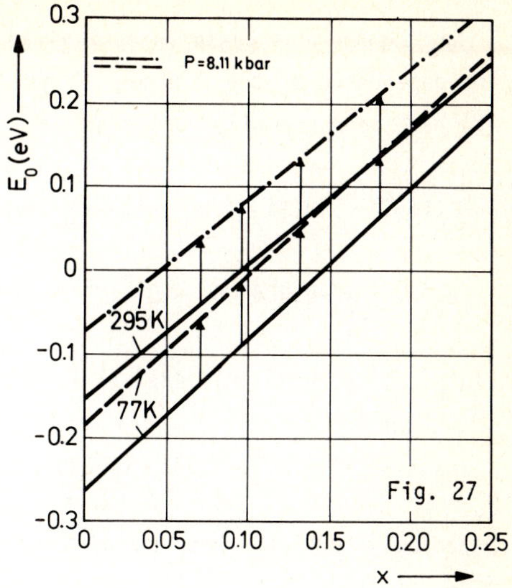

Fig. 26. Composition dependence of dE_0/dT

Fig. 27. Change in energy gap E_0 in $Hg_{1-x}Cd_xTe$ with pressure (solid lines P = 1 bar, dashed and dashed-dotted lines P = 8.11 kbar) at 77 and 295 K

Table 13. Variation of the energy gap E_0 with pressure

x	dE_0/dP [meV/kbar]	Ref.
0	12 ± 3	/53,79,113/
0	8	/39/
0	13.8	/76/
0	7	/109/
$0 \leq x \leq .2$	11.8	/78/
0.07	9.4 ± .5	/112/
0.095	10.4 ± .5	/112/
0.13	9.5 ± .5	/112/
0.18	8.4 ± .5	/112/

and 0.146 and $5 \cdot 10^{-5}$ eV for x = 1. The corresponding values given by OVERHOF /59/ are $k_m = 7.7 \cdot 10^{-3}$ $2\pi/a$ and $E_m = E_{\Gamma_8} + 2.6 \cdot 10^{-4}$ eV. OVERHOF has published plots of constant energy lines in various planes too. The values $E_m = E_{\Gamma_8} + E_t$ have to be compared with published data $E_t = 3 \cdot 10^{-3}$ eV /76/, $2 \cdot 10^{-2}$ eV /3/ for x = 0.15, $2 \cdot 10^{-3}$ eV /117/ for x = 0.2 and 0 eV /78,94/ for x = 0.15. In both calculations /59,63/ the shape of the conduction band is not in complete agreement with (19), since a small but significant anisotropy of band energies is observed.

New expressions based on optical and transport measurements /367,432,433a,b/ for the dependence of the fundamental gap E_0 (eV) on composition x and temperature T (K) have been recently published:

$$E_0(x,T) = -0.31 + 1.88x + (1-2x) \cdot 5 \cdot 10^{-4}\, T \qquad /432/ \quad (68a)$$

$$E_0(x,T) = -0.304 + 6.3 \cdot 10^{-4}\, T^2\, (1-2x)/(11+T) + 1.858x + 0.054x^2 \qquad /367/ \quad (68b)$$

$$E_0(x,T) = -0.337 + 1.948x + (1-1.89x) \cdot 6.006 \cdot 10^{-4}\, T \qquad /433a/ \quad (69a)$$

$$E_0(x,T) = -0.302 + 1.93x + (1-2x) \cdot 5.35 \cdot 10^{-4}\, T - 0.810x^2 + 0.832x^3 \qquad /433b/ \quad (69b)$$

Especially (68b) has been shown to fit a series of data from various publications at temperatures below 100 K and for compositions $x \leq 0.3$ /367/. The temperature dependence for E_0 in HgTe has also been reinvestigated in /434/. Values are given in Table 14.

Luttinger and Kane parameters as well as the corresponding effective heavy-hole masses deduced from magnetooptical measurements are described in Sect. 3.7.2.

Considerable discrepancies remain between the values for the hole effective masses determined in different investigations. This may be caused by the calculated strong temperature dependence /435,436/ or by the influence of non-locality of the potential in zero-gap materials /437-442/ (see Sect. 3.7.2.2).

Table 14. Experimental values of the gap E_0 for HgTe at various temperatures /434/

T [K]	E_0 [meV]	
10 ± 5	-(301 ± 1)	exp.
28 ± 3	-(293 ± 2)	
46 ± 3	-(282 ± 2)	
79 ± 1	-(263 ± 2)	
280	-120.9	from $E_0(T) = E_0(0)+\alpha T^2/(T+\beta)$ with
286	-116.4	$E_0(0) = -0.303$ eV
292	-112.4	$\alpha = (7.2 \pm 0.2) \cdot 10^{-4}$ eV/K
295	-110.2	$\beta = (30 \pm 10)$ K
300	-106.6	

Empirical expressions for the intrinsic carrier concentration based on corresponding relations $E_0(x,T)$ given above and several other reasonable assumptions have been published in /433a,c/:

$$n_i = 1.265 \cdot 10^{22}\, T^{3/2}\, (6+x)^{-3/2}\, E_0^{3/2} \{1+[1+19.40\, (6+x)^{-3/2} (m_h^*/m_0)^{-3/2} \cdot E_0^{3/2} \exp(E_0/k_B T)]^{1/2}\}^{-1} \; (m^{-3}) \quad . \qquad (70a)$$

which may be replaced by the parabolic expression

$$n_i = 26.54 \cdot (6+x)^{-3/4}\, [10^{20}\, E_0^{3/4}\, T^{3/2} \exp(-E_0/2k_B T)] \; (m^{-3}) \qquad (70b)$$

for temperatures below 100 K /433a/. The authors of /433a/, however, found a best fit value for $m_h^*/m_0 = 0.9$ independent of composition and temperature which deviates considerably from the average value in other investigations.

In /433c/ instead, the intrinsic carrier concentration is described by

$$n_i = (5.585-3.820x+1.753 \cdot 10^{-3}\, T-1.364 \cdot 10^{-3}\, xT) \cdot [10^{20}\, E_0^{3/4}\, T^{3/2} \exp(-E_0/2k_B T)] \; (m^{-3}) \qquad (70c)$$

based on energy gap values given by (69b) and a heavy hole mass $m_h^*/m_0 = 0.443$ /367/. This equation is quoted to provide an excellent fit to experimental values and to calculations based on the full Kane band structure (within 1%) in the range $x \leqq 0.4$ and for $T \leqq 300$ K.

A comparison of the dependence of experimentally determined effective electron masses on the band gap and the carrier concentration with the predictions from Kane theory has been undertaken in /443/. A simplified form of Kane's dispersion relation has been given in /444/.

3.3.3.2 Effective Masses

The dependence of effective masses on composition as determined from calculated energy bands and by the k . p perturbation method in /63/ is shown in Fig. 28. The computed values of effective masses at the band edge are about twice the experimental ones which are collected from various references in Fig. 29. But even the experimental values exhibit large deviations reaching a factor 2 for some compositions.

From Fig. 29 and (14) it is seen that the effective mass of electrons near the conduction band edge is extremely small for x values near the $\Gamma_8 - \Gamma_6$ crossover.

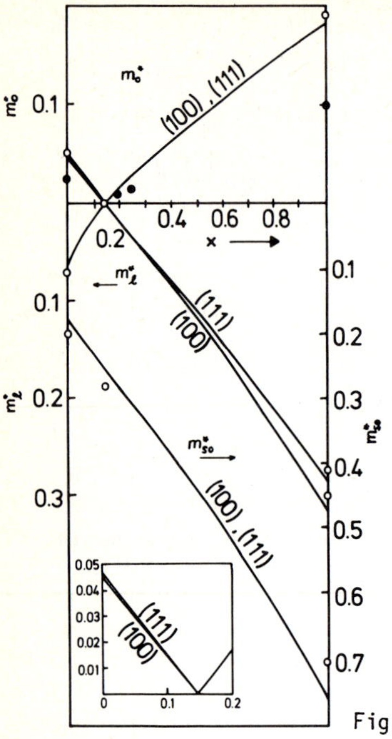

Fig. 28

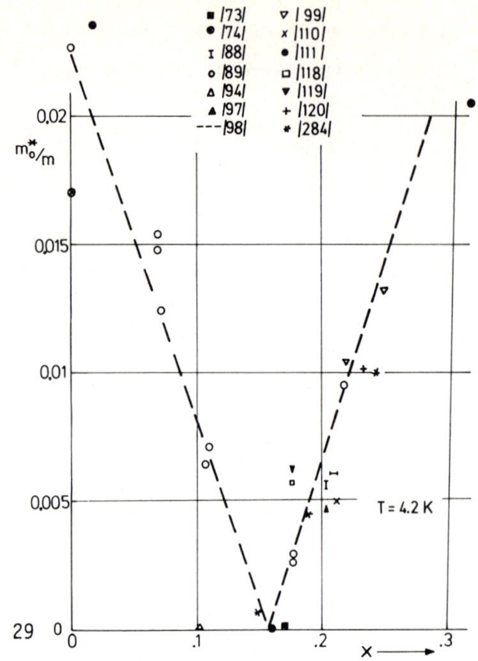

Fig. 29

Fig. 28. The effective masses of $Hg_{1-x}Cd_xTe$ calculated by the k . p perturbation method. Open circles show the masses calculated directly from the computed energy bands of /63/, closed circles some experimental values. (m_o^* conduction electron, m_ℓ^* light hole, and m_{so}^* splitt-off hole effective masses)

Fig. 29. Composition dependence of the conduction electron band edge masses

To observe these small masses, however, the carrier densities must be rather low. Because of the high density of states in the valence band, the Fermi level is pushed well up into the conduction band even at moderately low carrier densities. In **Fig. 30** calculated /98/ and experimental values of m^* as a function of x and carrier concentration n at T = 4.2 K are given (using the $E_0(x,T)$ relation given in /102/ instead of the linear dependence in /98/ in calculations).

As shown in (28) a plot of $\frac{m^*/m}{1-m^*/m}$ versus $n^{2/3}$ can be used to determine E_0 and P (see for example Fig. 31 /89/). Values $E_p = (\frac{2m}{\hbar^2})\, P^2$ experimentally determined by various authors are plotted in **Fig. 32**.

Calculated momentum matrix elements given in /59/ and /63/ are

$P = 6.4 \cdot 10^{-10}$ eVm (E_p = 10.8 eV) and
$P = 5.9 \cdot 10^{-10}$ eVm (E_p = 9.0 eV) respectively.

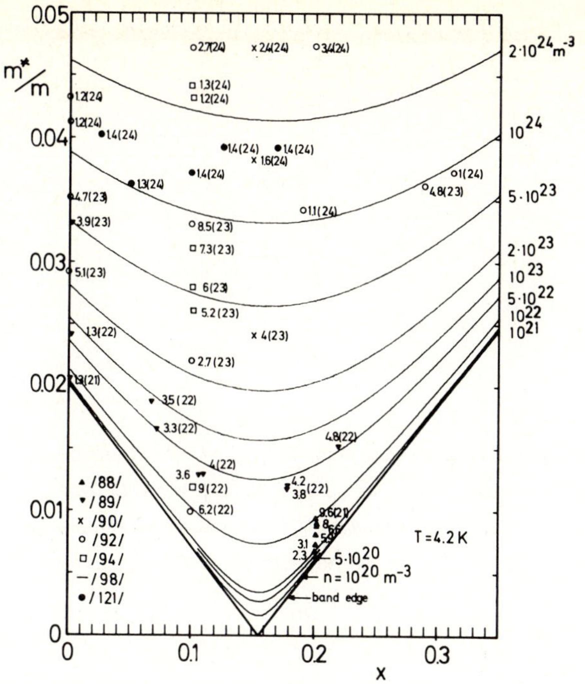

Fig. 30. Conduction band effective masses as a function of x at liquid helium temperatures for various carrier concentration n

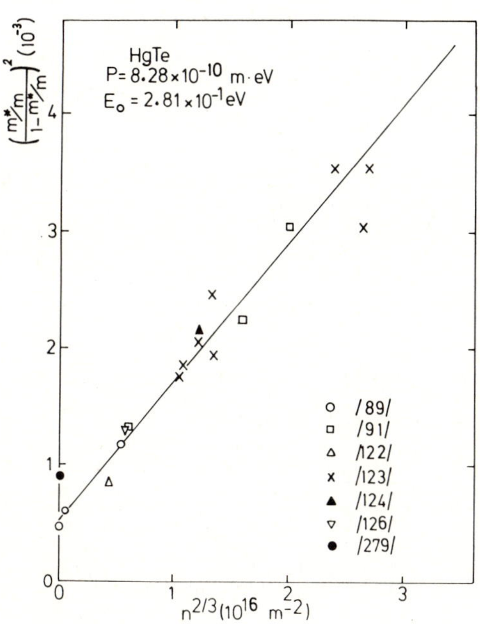

Fig. 31. Plot of $(m^*/(m_0 - m^*))^2$ versus $n^{2/3}$ for HgTe samples with various carrier concentrations /89/

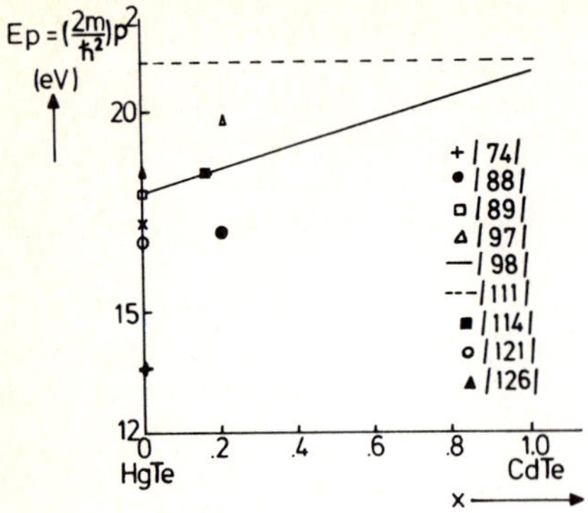

Fig. 32. $E_p = (2m/\hbar^2)P^2$ as a function of x

The masses of heavy holes have been determined from measurements of the thermoelectric power /127,128/, interband magnetoreflectance /114,126/, plasmareflection /129/, and temperature dependence of carrier concentration /98/ and the light hole mass estimated from plasma reflectance and mobility measurements /131/. Results are collected in Table 15.

Table 15. Heavy hole effective masses

x	p- or n-type $[m^{-3}]$	m^*_{hh}/m	Ref.
0	p, band edge	0.53 ± 0.04	/130/
0	p, $4.6 \cdot 10^{24}$	0.6	"
0	p, $2.1 \cdot 10^{25}$	0.7	"
0.1	p, $9.6 \cdot 10^{24}$	0.41	/127/
0.1	p, $9.0 \cdot 10^{24}$	0.5	"
0.1	p, $8.1 \cdot 10^{24}$	0.59	"
0.1	p, $7.8 \cdot 10^{24}$	0.37	"
0.1	p, $7.2 \cdot 10^{24}$	0.46	"
0.1	p, $6.5 \cdot 10^{24}$	0.53	"
0.149	n, $6.8 \cdot 10^{21}$	0.71 ± 0.05	/ 98/
0.161	n, $9.4 \cdot 10^{20}$	0.28 ± 0.1 (100)-dir.	/114/
0.161	density of states	0.35	/114/

In /126/ the anisotropic valence band cyclotron masses are given for various directions:

$$x = 0 \quad m^*_{hh}/m = 0.32\ (100) \quad 0.406\ (110) \quad 0.445\ (111)$$
$$0.185\ (100) \quad 0.241\ (110) \quad 0.266\ (111)$$

The two sets of data result from two different possible sets of fit parameters used to fit the observed $\Gamma_6 - \Gamma_8$ transitions.

3.3.3.3 g Factor of Conduction Electrons

The spin spectroscopic g factor of conduction electrons can be determined directly for instance from electric-dipole excited electron spin resonance (EDE-ESR) measurements as

$$g^*(N,B) = h \cdot \nu_s(N)/\mu_B B \qquad (34)$$

In the three-band model g^* is in principle given by (12,13), but in this form the g value is a function of the Landau quantum number N and the magnetic field strength (**Fig.** 33 /132/). In the limit of vanishing magnetic fields the g factor is given by (18).

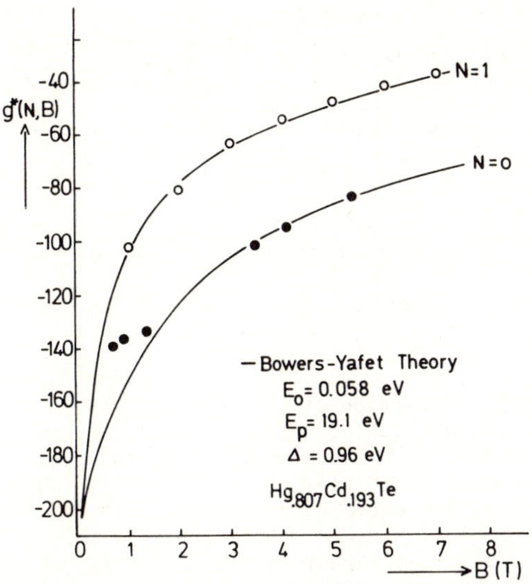

Fig. 33. Electron g values at the N = 0 and N = 1 Landau levels versus magnetic field for $Hg_{1-x}Cd_xTe$. The solid lines are calculated from the Bowers and Yafet model /132/

Zero field g values determined from Shubnikov-de Haas (SdH) measurements /89, 110,118/ and combined resonance transition measurements (EDE-ESR) /132,133/ are shown in **Fig. 34**. For compositions around the SM-SC transition the very small values of E_0 result in very large negative g values. The insert in **Fig. 34** shows the g factor of conduction electrons at the Fermi energy as a function of electron concentration in the band.

3.3.3.4 E_1 and $E_1 + \Delta$ Gaps

Electroreflectance and normal incidence reflection measurements (see Sec. 5.1) have been used by various authors to determine the transition energies $E_0 + \Delta_0$ $(\Gamma_7 - \Gamma_8)$, E_1 $(\Lambda_{4.5} - \Lambda_6)$, $E_1 + \Delta_1$ $(\Lambda_6 - \Lambda_6)$, E_0' $(\Gamma_8 - \Gamma_7)$, E_2 $(X_2 - X_6)$ and $E_2 + \delta$ $(X_6 - X_7)$ (see **Fig. 18**) and their variation with alloy composition and temperature /135-141/. The results are presented in Figs. 35 to 39. In Fig. 35

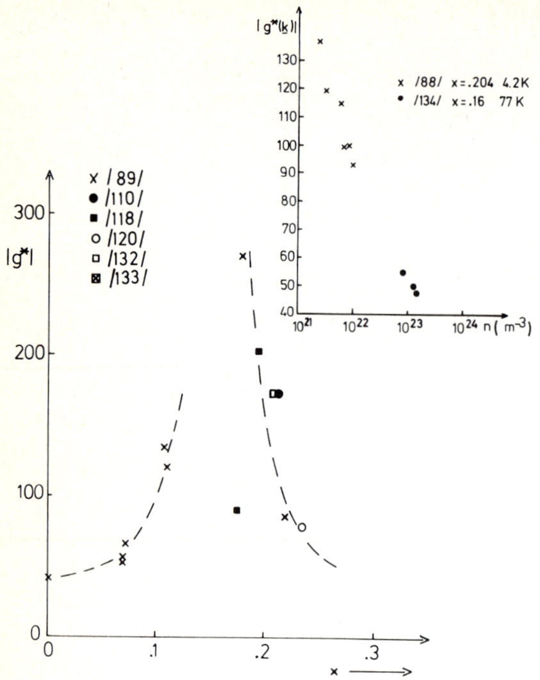

Fig. 34. Composition dependence of the band edge effective g^x values. The insert shows g values versus carrier concentration n

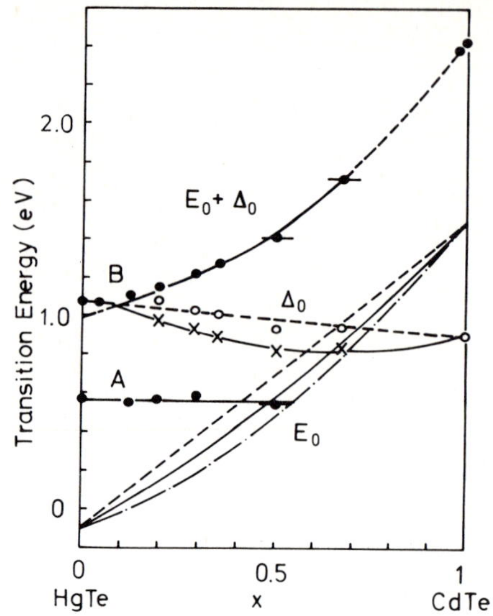

Fig. 35. Variation of E_0, $E_0 + \Delta_0$, and Δ_0 with composition. Closed circles represent the composition dependence of the energies corresponding to peaks in electroreflectance spectra. The upper circles are attributed to energies $\Delta_0 (0 \leq x \leq 0.1)$ and $E_0 + \Delta_0 (0.1 \leq x \leq 1)$. ---: linear variation for E_0, —— and x: calculated curve for E_0 with $r_V = 1.04\ r_{IV}$ (see Sec. 3.6) and deduced values of Δ_0. -.- and o: calculated curve for E_0 with $r_V = 1.08\ r_{IV}$ and deduced values of Δ_0 /135/

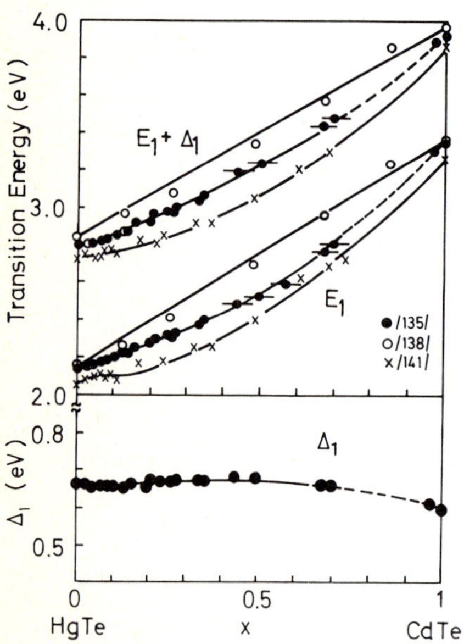

Fig. 36 Variation of E_1, $E_1 + \Delta_1$, and Δ_1 with composition x /135/

measurements of $E_0 + \Delta_0$ on the semiconducting and Δ_0 on the semimetal side are shown together with various calculations which are discussed in Sect. 3.6. A comparison of spin-orbit splitting energies Δ_0 as determined or used as fit parameters is given in Table 16.

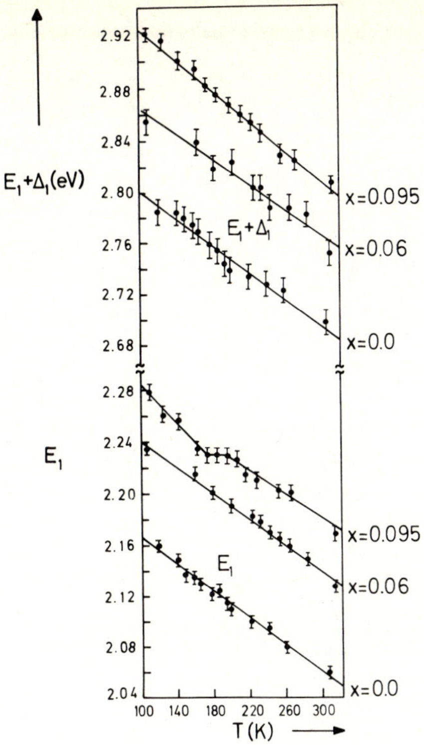

◁ Fig. 37. Temperature dependence of E_1 and $E_1 + \Delta_1$ for three values of the mole fraction x of Cd on the sample surface /140/

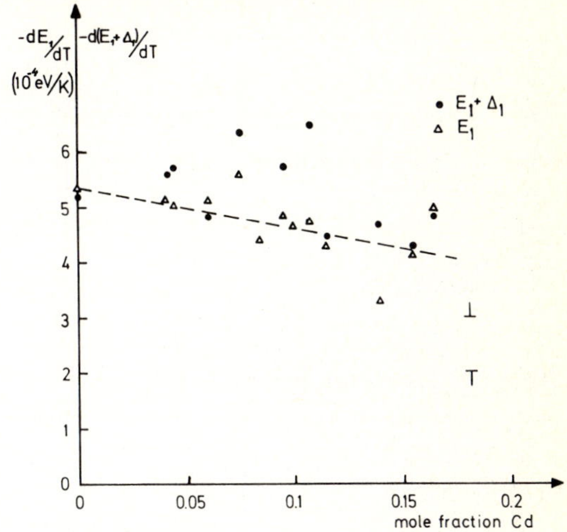

Fig. 38. Composition dependence of $-dE_1/dT$ and $-d(E_1 + \Delta_1)/dT$

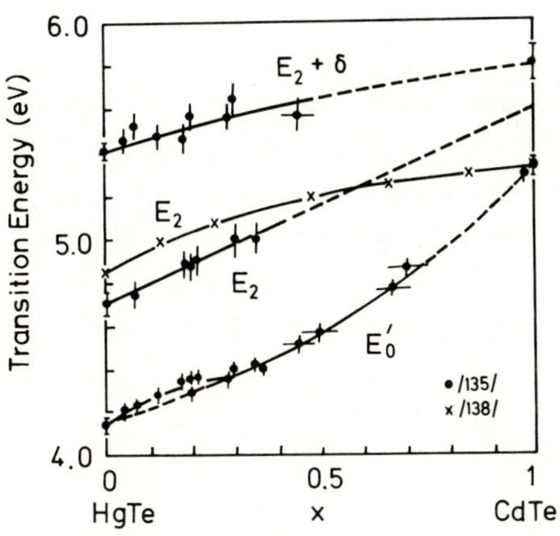

Fig. 39. Variation of $E_2 + \delta$, E_2 and E_0' with composition x /135/

Transition energies E_1 and $E_1 + \Delta_1$ and resulting Δ_1 values at room temperature as determined by /135,138,141/ are given in Fig. 36. A linear variation of transition energies with x has been observed only on epitaxial films by LUDEKE et al. The values obtained by /135,141/ as well as by /139/ could be fitted by parabolas $cx^2 + bx + a$ /139/. Parameters are given in Table 17.

Table 16. Spin-orbit splitting energy Δ_o

x	Δ_o [eV]	Ref.
0.0	1.20 (Ions) 0.98 (Atoms)	/143/
0.0	1.21	/142/
0.0	1.10	/72,74/
0.0	0.89 (77 K)	/63/
0.0	1.08 ± 0.02	/135/
0.0	1.00	/126/
0.204	0.75	/88/
	0.96	/97,110,133/

Table 17. Parameter values for $E_i = a + bx + cx^2$ (in eV)

Transition	c	b	a	Ref.
$E_1 + \Delta_1$, $0 \leq x \leq 1$	0.53	0.57	2.80	/135/
"	0.48	0.64	2.73	/139/
"	0.97	0.17	2.74	/141/
E_1, $0 \leq x \leq 1$	0.86	0.35	2.15	/135/
$0 \leq x \leq 0.16$	-3.5	1.42	2.068	/133/

The comparison may be resumed as follows /139/:

- The consistence of energies $E_1 + \Delta_1$ is sufficiently good over the whole composition range $0 \leq x \leq 1$. The slight shift in energies may be due to the difference between experimental methods.
- For energies E_1 the results are consistent in the range $0.2 \leq x \leq 1.0$ only. In the composition range $0.0 \leq x \leq 0.2$, however, drastic deviations occur. The abnormal behaviour of the $E_1(x)$ dependence for low composition as observed in /139,141/ is discussed in Sect. 3.6 as well as the bowing in the $E_o(x)$ curves.

The dependence of transition energies $E_1 + \Delta_1$ and E_1 /140/ on temperature is presented in Fig. 37, the composition dependence of dE_1/dT and $d(E_1 + \Delta_1)/dT$ in Fig. 38.

Measured values of energies E_o', E_2 and $E_2 + \delta$ /135,138/ are shown in Fig. 39. In /135/ the dependence of $E_2 + \delta$ and E_2 on x in the range $0 \leq x \leq 0.4$ seems to be almost linear. However, there are remarkable discrepancies between /135/ and /138/. In the variation of E_o' with x, a slight deviation from a quadratic curve is obtained between x = 0 and x = 0.3.

For comparison experimental values for the main gaps in the pure compounds CdTe and HgTe are given in Table 18.

Table 18. Experimental values for main gaps (in eV)

CdTe					HgTe					
E_1	$E_1 + \Delta_1$	E'_o	E_2	$E_2 + \delta$	E_1	$E_1 + \Delta_1$	E'_o	E_2	$E_2 + \delta$	Ref.
3.32	3.88	5.16	5.40	-	2.09	2.71	4.03	4.84	-	/144/
3.28	3.87	5.30	-	-	2.12	2.78	4.14	5.79	5.24	/145/
3.39	3.99	-	5.32	-	2.15	2.85	-	4.84	-	/138/
3.28	3.86	-	-	-	2.06	2.73	-	-	-	/141/
3.31	3.91	5.34	5.56	5.77	2.12	2.78	4.14	4.70	5.40	/135/

The relation between composition x and E_1-gap has been determined to $E_1 = 2.15 + 0.723x + 0.102x^2$ (eV) at 300 K. The analysis of thermoreflection spectra of $Hg_{.75}Cd_{.25}Te$ near the E_1 and $E_1 + \Delta_1$ gaps obtained at 90 and 300 K revealed the influence of exciton effects on the spectral shape. This shape could be explained allowing for the influence of electron-hole interaction which mixes the contributions of the critical points M_1 and M_2 /445/.

3.4 Temperature Dependence of the Band Gap

In Fig. 26 it is shown that the temperature coefficient dE_0/dT is of similar absolute value and of opposite sign in the two compounds CdTe and HgTe and changes almost linearly with composition. Thus $dE_0/dT = 0$ is obtained near $x = 0.5$. Though the inverted band structure at the Γ point occurs at room temperature in $Hg_{1-x}Cd_xTe$ approximately in the range $x \leq 0.1$, the positive temperature coefficient extends as far as $x = 0.5$.

A positive temperature coefficient is very unusual in semiconductors with zincblende structure and therefore a number of workers tried to explain this anomalous behaviour especially in HgTe. VERIE /72/ speculated about an unknown positive term, probably connected with relativistic effects.

$$\frac{dE_0}{dT}\bigg|_{\text{total}} = \frac{\partial E_0}{\partial T}\bigg|_{\text{dilatation}} + \frac{\partial E_0^{\text{phon.-electr.}}}{\partial T}\bigg|_{\text{classical}} + \frac{\partial E_0^{\text{phon.-electron}}}{\partial T}\bigg|_{\text{relativistic ?}}$$

$$\text{HgTe:} \quad > 0 \qquad\qquad < 0 \qquad\qquad < 0 \qquad\qquad > 0 \tag{35}$$

GUENZER and BIENENSTOCK /15,146/ carefully reexamined the results on HgTe in the light of the theories of FAN /147/ (explicit temperature dependence of the gap due to electron-phonon interaction at constant volume) and BROOKS and YU /148/. They measured X-ray diffraction intensities versus temperature for several reflections from a single crystal of $Hg_{.968}Cd_{.032}Te$ and derived Debye temperatures for Hg and Te ions in HgTe. They calculated temperature-dependent mean square displacements u_α for each ion α according to the formula

$$u_\alpha^2(T) = \frac{3\hbar^2}{M_\alpha k_B}\frac{T}{\Theta_\alpha^2}\left[\Phi\left(\frac{\Theta_\alpha}{T}\right) + \frac{1}{4}\frac{\Theta_\alpha}{T}\right] \tag{36}$$

where Θ_α and M_α denote the separate temperature-independent Debye temperature for the α ion and the ionic mass, respectively, and $\Phi(Z)$ is a Debye integral of order one. Figure 40 presents the mean square displacements as calculated with Debye temperatures

$\Theta_{Hg} = (75 \pm 1.5)$K and $\Theta_{Te} = (130 \pm 5)$K,

which are in reasonable agreement with other results. Debye-Waller factors are obtained by the time average

$$M_\alpha(\underline{G}) = \frac{1}{2} <|\underline{U}_\alpha(T) \cdot \underline{G}|^2> \quad . \tag{37}$$

The Brooks-Yu temperature dependence is calculated by reducing each Fourier component of a complete set of Fourier components of the ionic potentials or pseudopotentials $\{w_\alpha(\underline{G})\}$, which are used to determine the zero temperature energies, by the corresponding Debye-Waller factors of the ions

$$E(\underline{k},T) = E(\underline{k},\{w_\alpha(\underline{G}) \cdot e^{-M_\alpha(\underline{G})}\}). \tag{38}$$

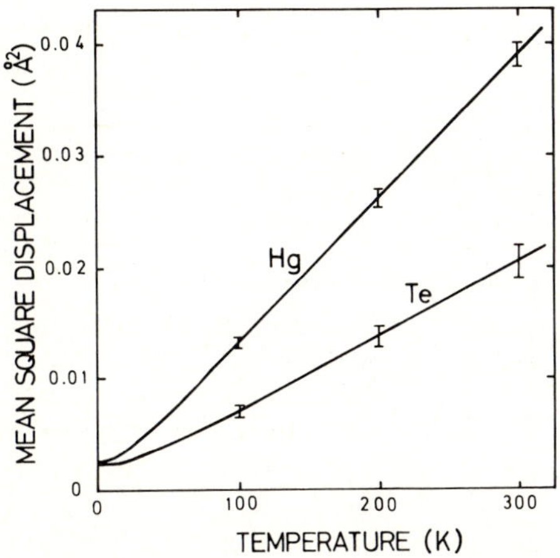

Fig. 40. Temperature-dependent mean square displacements for the two atomic species in HgTe calculated according to Debye formula with derived Debye temperatures /15/

Guenzer's and Bienenstock's calculations resulted in a temperature coefficient for HgTe of

$$\frac{\partial |E_0|}{\partial T}\Big|_V = +2.7 \cdot 10^{-4} \text{ eV/K}$$

in the temperature range 0 - 300 K. This result is about half the right magnitude but the wrong sign.

The effect of thermal expansion can be accounted for by explicity using the relation

$$\frac{dE_0}{dT} = \frac{\partial E_0}{\partial T}\Big|_V + \frac{\partial E_0}{\partial V}\Big|_T \cdot \frac{dV}{dT} . \tag{39}$$

The second term can be written as

$$\frac{\partial E_0}{\partial V}\Big|_T \cdot \frac{dV}{dT} = -3\alpha B \frac{\partial E_0}{\partial p}\Big|_T \tag{40}$$

with the lattice temperature coefficient α and the bulk modulus B. Inserting measured values /9,143,10/, a maximum value of $+7\cdot10^{-5}$ eV/K is obtained. The deviation from measured values is obvious.

Though there is some concern about the validity of the pseudopotential used in the calculation of the Brooks-Yu contribution to the temperature coefficient, GUENZER and BIENENSTOCK turn to another contribution and make an attempt to calculate the self energy terms of an electron in phonon field. They show, generalizing Fan's approach, that interband contributions must be included. Taking a simple four band model their calculation yields a sign agreeing with experiment. Because of a lack of knowledge of acoustic phonon modes in HgTe, they are not able to calculate the exact magnitude of the self energy contributions. A sketch of their results is given in **Fig. 41**.

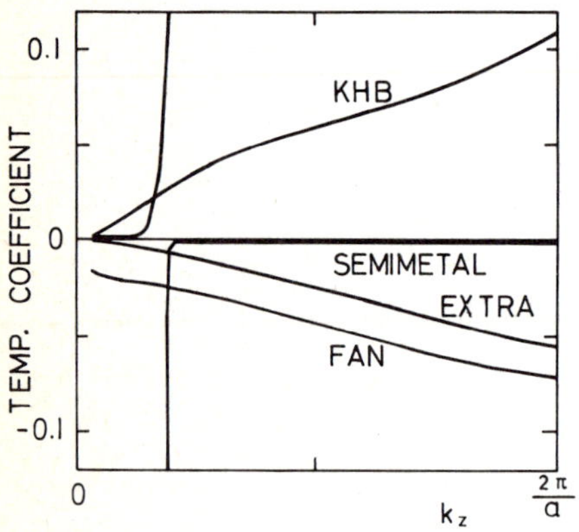

Fig. 41. Temperature coefficient (in $(eV)^{-1}$) for magnitude of the HgTe band gap calculated along the Γ - X symmetry axis /15/

The problem of the dependence of the fundamental gap E_0 in HgCdTe on temperature has been addressed in several papers /446-448/.

The temperature coefficient $dE_0(x)/dT$ is a sum of lattice dilatation and phonon-electron interaction terms. The values of the dilatation term may be computed from

$$(\partial E_0/\partial T)_{Dil} = -3\alpha c_e(\partial E_0/\partial P)_T \tag{71}$$

using measured parameter values to give approximately /447/

$$(\partial E_0/\partial T)_{Dil} = \begin{cases} -(0.029-0.018x)\cdot10^{-5} \text{ eV K}^{-1} \text{ at } 4.2 \text{ K} \\ -(2.10-1.73x)\cdot10^{-5} \text{ eV K}^{-1} \text{ at } 77 \text{ K} \\ -(6.426-1.331x)\cdot10^{-5} \text{ eV K}^{-1} \text{ at } 300 \text{ K} \end{cases} . \tag{72}$$

The main contribution to the dE_0/dT values, however, is expected to result from the electron-phonon interaction. In their calculations the authors of /446/ took into account only the intraband self-energy term, neglecting the interband contribution. The interband electron-phonon interaction in narrow-gap semiconductors leads to band mixing in addition to Kane mixing and, in the same order in the coupling constant, to an additional renormalization of the energies of the electron bands; these effects have been studied within a two-band model in various cases in /448/.

3.5 Pressure Effects on the Band Structure

It is known that in all diamond and zincblende type compounds examined so far the Γ_6 level shifts upwards with hydrostatic pressure faster than Γ_8. Thus $\frac{d}{dP}(E_{\Gamma_6} - E_{\Gamma_8})$ is positive and of the order of 10^{-5} eV/kbar. Early investigations of the dependence of the thermoelectric power and the Hall coefficient of HgTe on pressure /76,150/ have experimentally confirmed the existence of the inverted, gray-tin-like band structure of HgTe, indicated by the fact that results could only be interpreted assuming a value $d|E_0|/dp < 0$, hence $E_0 = E_{\Gamma_6} - E_{\Gamma_8} < 0$. Similarly measurements of the intrinsic carrier concentration and the mobility of $Hg_{1-x}Cd_xTe$ single crystals in the range $0 \leq x \leq 0.28$ and P up to 4.5 kbar supported the given interpretation of a negative value E_0 for low mole fractions x /78,151,152/.

According to numerous data /109,112,116,153/ hydrostatic compression leads to a linear change of the energy E_0 at a rate dE_0/dP given in Table 13 for HgTe and $Hg_{1-x}Cd_xTe$ alloys. At a certain critical pressure P_c which depends on x and T the alloys exhibit the semimetal semiconductor transition (see Fig. 27). No clearcut dependence of dE_0/dP on x or T has been observed in the region $0 \leq x \leq .3$.

An early direct evidence of bandcrossing at the SM-SC transition has been given by VERIE and MARTINEZ /113/, who submitted a p-n junction of semimetallic $Hg_{.89}Cd_{.11}Te$ (E_0 = -50 meV, 77 K) to a pressure up to 14 kbars. The I-V characteristic remained ohmic for P < 6 kbars, but became rectifying at higher pressures. The contact potential, which is equal to E_0 in p-n junctions with degenerate electron and hole gases, increases linearly with pressure up to 70 meV for 14 kbars. The opening of the band gap by applying a hydrostatic pressure is clearly visualized by this experiment /152/.

The pressure dependence of the carrier concentration of n-type polycrystalline /112,116,153/ and p-type single crystals /109/,53,154/ as well as the mobility dependence /53,109,116,153/ will be briefly mentioned in Chap. 4. Some general features of the pressure dependence of μ_H and R_H in n-type samples are demonstrated e.g. in Fig. 42 /79/.

Anomalies in the transport properties of HgTe and $Hg_{1-x}Cd_xTe$ alloys versus pressure and temperature in the low temperature region have been reported by various authors /79,38,53,109,153,39,54/, see, e.g. Fig. 43. In most cases these anomalies have been attributed to resonant acceptor states in the conduction band and are discussed in more detail in Sect. 2.3.

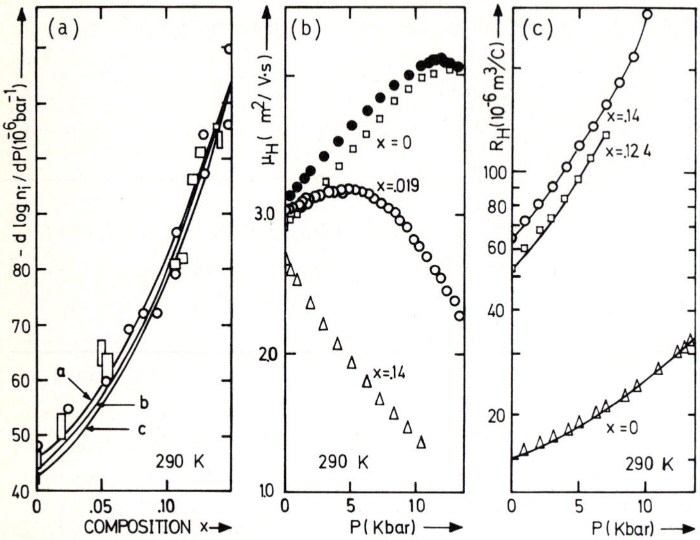

Fig. 42. (a) Fit of the pressure variation rate, extrapolated at P = 0, of the intrinsic carrier concentration as a function of the alloy composition. (b) Experimental values of Hall mobilities versus P. (c) Fits of Hall coefficient as a function of P with the following band parameters:

x	E_0 [meV]	E_p [eV]	m^*_{hh}/m	dE_0/dP [meV/kbar]	
0	-125	13.98	0.6	10	
0.124	+ 63	15.15	0.6	10	
0.14	+ 82	15.15	0.6	9.5	/79/

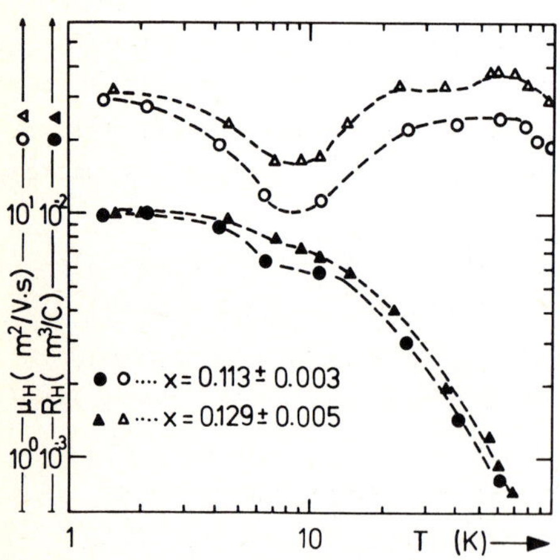

Fig. 43. Hall mobilities and Hall coefficient (B = 0.03 T) in the low temperature range /79/

The effects of uniaxial stress are quite different from those of hydrostatic pressure. As the degeneracy of conduction and valence band at Γ_8 in the semimetallic region as well as the double degenerate Γ_8 state itself are results of the cubic symmetry of the crystal, these degeneracies can be removed by destroying the cubic symmetry with uniaxial stress /155/.

The piezoresistance of $Hg_{1-x}Cd_xTe$ was investigated by KIREEV et al. /156/ in the range $x \le 0.45$ and $P \le 300$ bar.

Their results can be summarized as follows:

- At room temperature the pressure does not affect the resistivity of most of the samples, at 77 K the coefficient $\pi_\ell = \frac{1}{\rho_o''} \left(\frac{\rho'' - \rho_o''}{-p}\right)$ is positive for p-type and negative for n-type samples (see Table 19), the stress dependence is mostly linear.
- At 77 K in the range $x \le 0.2$ the application of uniaxial compression alters the Hall coefficient more strongly, in the range $0.2 \le x \le 0.45$ the resistivity more strongly.
- π_ℓ depends on orientation of the samples.
- The increase in the resistance of the n-type samples with uniaxial compression is related to the change in the effective mass and the deformation potential constant /155/, the changes in ρ and R_H of the p-type samples can best be explained by the splitting of the Γ_8 band.

Piezoresistive properties of n- and p-type HgTe in the temperature range 4.2 - 200 K have been investigated in /157/ on heavily doped samples. Magnetoresistance measurements on HgTe under uniaxial pressure conditions at low temperatures /158/ have been reported with pressures up to 1 kbar and magnetic fields up to 4.6 T. The observation of a large negative magnetoresistance has been inter-

Table 19. Dependence of the resistivity and the Hall coefficient on uniaxial compression /156/

Type of cond.	Orient.	x	$\pi_\ell \cdot 10^6$ [bar^{-1}]		$(1/R \cdot \partial R/\partial(-P))_T 10^6$ [bar^{-1}]	
			77 K	293 K	77 K	293 K
p	Unorient.	0.14	80	0	100	0
p	"	0.16	79	0	230	0
p	"	0.17	84	0	(500)	-
p	"	0.22	78	-	145	0
p	[100]	0.18	120	0	0	-
p	[111]	0.28	77	-	-	-
p	[111]	0.33	150	27	0	-
p	[111]	0.42	156	32	0	0
n	[111]	0.28	-19	-13.5	-20	0

preted with the lifting of the Γ_6 - Γ_8 degeneracy and the occurrence of a small band gap of 2 - 3 meV /155/. The dependence of the transverse and longitudinal magnetoresistance seems to be governed by the complicated character of the Landau levels in HgTe /159/.

Various pressure coefficients and compressibilities of the two parent compounds are given in Table 20 /160/.

Table 20. Energy-band pressure coefficients, pressure dependence of ln (n) and of the ionization potentials I, and compressibilities K as calculated with the dielectric theory of the chemical bond (in 10^{-6} eV bar^{-1}, except d ln(n)/dP and K in 10^{-6} bar^{-1})

	dE_0/dP	dE_1/dP	dE_2/dP	$dE_{\Gamma L}/dP$	$dE_{\Gamma X}/dP$	d ln(n)/dP	dI/dP	K
CdTe	8.1	3.9	3.1	1.9	-1.9	0.1 ±0.1	2.9	2.36
HgTe	12.0	5.4	3.4	-	-		-	2.40

Measurements of the thermal conductivity of $Hg_{.72}Cd_{.28}Te$ under hydrostatic pressure have been reported in /449/.

The influence of uniaxial compressional stress and the opposite effect of tensile strain on symmetry-induced zero-gap materials with zincblende structure has recently been reviewed in /450/. It can be described using the formalism of BIR and PIKUS /155/.

The main effect of compressional stress is to pull apart conduction and valence band and to open an energy gap. The maximum of the valence band is shifted from the original band-edge point along a direction perpendicular to the stress axis. Thus, the new maxima may be situated along a ring around the stress axis and the constant-energy surfaces for holes become toroidal in shape. The region near the center of the toroidal surface is an anomalous region in which the hole-mass component along a direction perpendicular to the stress axis becomes negative.

Effects of uniaxial stress on HgCdTe have recently been reported in /451-454/.

TAKITA et al. /454/ have additionally applied a strong magnetic field and calculated the Landau level structure for uniaxially compressed HgTe. They found a crossing between the lowest Landau levels of the conduction and valence bands, giving rise to a pronounced minimum of the longitudinal magnetoresistance at a pressure-dependent magnetic field value.

Tensile strain has been shown to induce a band overlap in HgTe /455/. This overlap leads to segments in the conduction and valence bands with reversed curvature or to carriers with negative masses which, in turn, led the authors predict a negative differential resistance effect in HgTe.

3.6 The Disorder Problem in Pseudo Binary Alloys

In recent years there has been considerable interest in the electronic properties of pseudo binary alloys. In addition to purely theoretical interest as part of the problem of disordered structures, a better understanding would also be of major practical value, especially in connection with the task of creating materials to prior specification such as $Hg_{1-x}Cd_xTe$. It is therefore appropriate to conclude this chapter with some general remarks on the methods developed over the last few years for predicting the behaviour of "random alloys".

Many III-V or II-VI semiconducting compounds with either a common cation or anion may be alloyed in arbitrary concentrations. This implies the introduction of an isovalent substitutional impurity (an impurity from the same group and subgroup of the periodic table as the substituted atom) into the crystal, which usually leads to continously varying properties intermediate between the parent compounds. Crystal structure (e.g. zincblende structure) is preserved; although size effects produce strains, there is no topological disorder.

The influence of a substitutional "impurity" on the energy spectrum of an electron in a crystal can show itself in the following ways:

1) the electronic potential of the impurity differs from the potential of the substituted atom; the difference gives rise to a local disturbance of the potential;
2) the lattice of the initial crystal is deformed by the introduction of the impurity, which leads to a change in the average interatomic distance and in the volume of the crystal as a whole;
3) the impurity atom perturbs the potential of the nearby atoms in the lattice.

A very large number of investigations has been devoted to the problem both from a theoretical and from a more empirical point of view. We refer to some review articles on various aspects of this problem /161-164/.

The fact that in contrast to topologically disordered materials, such as glasses, the optical spectra and electronic behaviour of substitutional alloys are so similar to those of the ordered, pure parent compounds together with the assumption that the atoms of the alloy lie on periodic lattice sites has led to the so-called virtual crystal approximation (VCA) /165-167/. In this approximation, which is widely used in band structure calculations (Sect. 3.1), the randomly fluctuating cation potential in an $A_x^N B_{1-x}^N C^{8-N}$ alloy is replaced by a suitable weighted average cation potential $\bar{V}$ defined by

$$\bar{V}_x = x\, V_A + (1-x)\, V_B \quad . \tag{41}$$

However, to avoid misinterpretation, it must be pointed out that the average should include only the residual potential seen by the valence electrons outside the atomic core, which themselves remain intact. Even so some caution is needed in using the above averaging formula /168,169/. As usually the volume Ω_x of the alloy unit cell is different from that of the parent compounds Ω_{AC} and Ω_{BC}, pseudopotential form factors

$$V^P_{A,B}(\underline{G}) = \frac{2}{\Omega_{AC,BC}} \int V^P_{A,B}(\underline{r}) \exp(-i\underline{G}\underline{r}) d^3\underline{r}$$

of the compounds must be renormalized to get the alloy form factors:

$$\bar{V}^P_x = (\Omega_{AC}/\Omega_x)\, x\, V^P_A + (\Omega_{BC}/\Omega_x)\,(1-x)\, V^P_B \qquad (42)$$

With form factors obtained in this way the energy bands of the $A^N_x B^N_{1-x} C^{8-N}$ alloy may be calculated as if the latter were a pseudocompound with cation pseudopotential $\bar{V}^P_x$, and anion pseudopotential V^P_C.

It is not expected even in this simple approximation that the energy gap E_0 varies linearly with x because it is not a linear function of the potential for any of the commonly used methods of band structure calculation. Conventionally the variation of gaps E_i with x is fitted to a quadratic:

$$E_i\,(x) = a_i + b_i x + c_i x^2 \qquad (43)$$

where $a_i = E^{AC}_i$, $b_i = (E^{BC}_i - E^{AC}_i) - c_i$, $c_i = 4 \cdot \left[(0.5\, E^{AC}_i + 0.5\, E^{BC}_i) - E_i(0.5)\right]$ and E^{AC}_i, E^{BC}_i are the band gaps of the parent compounds. The bowing parameter c_0 is always positive, but c_i may be negative for higher gaps /170/.

As was pointed out by VAN VECHTEN and BERGSTRESSER /171/, even small errors in defining $\bar{V}$ in the virtual crystal model can reverse the sign of c_0 in calculated band structures.

In the VCA model as described it has been implicitly assumed that the atoms of each compound sit exactly on the alloy lattice sites. This, however, may not be the exact situation. PHILLIPS /172/, RICHARDSON /168/, and VAN VECHTEN /173/ have argued from thermodynamic considerations that displacements from the lattice sites are expected in order to minimize the perturbation in the crystal. These self-consistent rearrangements may include adjustments in bond lengths and angles, diffusion of the ions and screening of the electrons. As a result in substitutional alloys grown under nearly equilibrium conditions and allowed to equilibrate in the cooling process or during very long annealing times no fluctuations in the pseudopotential may occur within several k_BT of the band gap. An influence of the conditions of formation of alloys on the bowing parameter is indicated experimentally in /174/ on $InSb_xAs_{1-x}$. A more definite theoretical estimate has been given in /175/.

Another approach in calculating the band structure of semiconductor alloys has been presented by VAN VECHTEN and BERGSTRESSER /171,176/. It is based on the dielectric two-band method /177/ of calculating the band structure of tetrahedrally coordinated semiconductors. In this theory the band gap energies $E_i (i = 0.1)$ are given by

$$E_i = \left[E_{ih} - (D_{av} - 1)\,\Delta E_i\right]\left[1 + (C_{AB}/E_{ih})^2\right]^{1/2} \qquad (44)$$

with E_{ih} being the average homopolar energy gap (= gap for the corresponding group IV material)

C_{AB} the average ionic energy gap $C_{AB} = e^2 bZ|1/r_B + 1/r_A|e^{-ksR}$ (= Phillips electronegativity difference)

$R = 1/2\left[r_C + (1 - x)\, r_B + x r_A\right]$ the average covalent radius

(for alloy $A^N_x B^N_{1-x} C^{8-N}$)

$b \approx 1.5$

k_s linearized Thomas-Fermi screening wave number

r_α covalent radius of element α ($\approx$ half the nearest-neighbour distance of the corresponding group IV element in the diamond type crystal)

$Z = Z_A = Z_B$ valence number

ΔE_i Parameter, only dependent on the nearest-neighbour distance

D_{av} average valence band to d-band f-sum.

In the spirit of the virtual crystal approximation the values E_{ih}, D_{av}, C_{AB} and the lattice constant are assumed to vary linearly with composition x between the values of the pure compounds. The total bowing parameter c is proposed to be the sum of the "intrinsic" bowing parameter c_i, determined from the above virtual crystal approximation, and an "extrinsic" bowing c_e due to the effects of a short-range aperiodic part of the crystal potential neglected in the VCA. The value c_e is estimated as

$$c_e = C_{AB}^2/A \qquad (45)$$

with A fitted (for all compounds) to 0.98 eV.

The bowing parameters c_e, c_i and c_{calc} have been calculated for $Hg_{1-x}Cd_xTe$ according to the model presented above by MORITANI et al. /135/ and have been compared with their experimentally determined values and with the data of SCHMIT and STELZER /102/. Using a value for the covalent radius $r_{Hg} = 1.04\, r_{Sn}$ they found good agreement between Schmit and Stelzer's results and the calculated curve $E_0(x)$, but not with their own measurements on E_1 and E_0' (Fig. 44 and Table 21). A good fit to their measurements was obtained assuming a value of $r_{Hg} = 1.08\, r_{Sn}$, deviating from Phillips' and Van Vechten's proposal.

However, as is shown in Table 21, a large fraction of the observed non-linear variation of the gaps is attributed to c_e, and thus to the influence of fluctuations rather than to the virtual crystal variation. Considering the sharp features of the spectra an explanation seems difficult and must include deviations of atomic positions from the sites of a perfect lattice as mentioned above. To determine the

Table 21. Bowing parameters (in eV) and parameter values used in calculation /135/

	c_i		c_e		$c_{calc.}$		$c_{exp.}$
r_V/r_{IV}	1.04	1.08	1.04	1.08	1.04	1.08	
E_0	0.22	0.22	0.22	0.67	0.44	0.89	
E_1	0.16	0.16	0.22	0.67	0.38	0.83	0.84 ± 0.20
E_0	0.02	0.02	0.22	0.67	0.24	0.69	0.70 ± 0.20

	Phillips ionicity	b	C [eV]	E_h [eV]	Row No.	$a_{eff.}/a_B$	D_{av}	D
CdTe	0.675	1.55	4.90	3.08	IV-IV	12.25	1.43	1.303
HgTe	0.65	-	4.0	2.92	V-IV	12.16	1.60	-

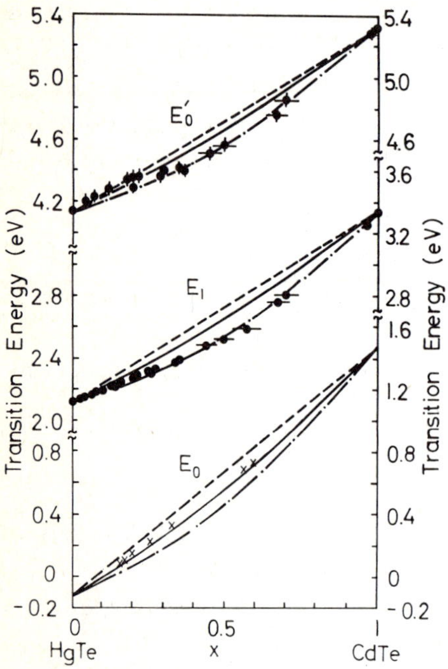

Fig. 44. Variation of E_0, E_1, and E_0' with composition. (Closed circles: experimental data from /135/, crosses: experimental data from /102/, dashed lines: linear variation, solid lines: calculated curves with $r_V = 1.04\ r_{IV}$, dashed-dotted lines: calculated curves with $r_V = 1.08\ r_{IV}$)

extent of disorder-induced mixing, VAN VECHTEN, BEROLO and WOOLLEY examined the variation of spin orbit splittings Δ_0 and Δ_1 in III-V alloys /178/. In the VCA, $\Delta_0(x)$ and $\Delta_1(x)$ deviate from linearity only to the small extent that the overlap of the relevant wave functions with the nuclei vary. As in $Hg_{1-x}Cd_xTe$ (see Figs. 35 and 36) there is in III-V alloys an upwards bowing (negative c) in the $\Delta_1(x)$ curve and a downward bowing (positive c) in the $\Delta_0(x)$ curve. The authors suggested /179/ that, because of the breaking of the crystal symmetry by the aperiodic terms, mixing of the valence and conduction band states occurs. They suggested further that observed deviations in the effective mass values m_0^* for intermediate compositions from predictions of a simple Kane model might be attributed to the same kind of mixing of valence and conduction bands.

SIGGIA /175/ pointed out too, in generalising the Kane model to alloys, that it might be necessary to go back to the experimental data and to examine how a proper treatment of alloy scattering will affect the interpretation of these data. This seems to be worthwhile in the case of $Hg_{1-x}Cd_xTe$ too.

In recent years several attempts have been made to get a more realistic description than that provided by the rigid band or virtual crystal models discussed above. As a result of different approaches /164,180-183/ an approximation superior, both from a physical and mathematical point of view, has been obtained, the coherent potential approximation (CPA). The essential idea is analogous to the molecular field theory. On each lattice point except one, an atom having a "yet unknown" coherent potential is located. An atom on the chosen point can be either A or B with respective probabilities x and 1 - x. Thus the problem reduces to that of a single impurity embedded in the effective medium and solved exactly once the effective medium defined by the coherent potential is specified. At the same time the solution obtained characterizes the coherent potential, which thus is determined self-consistently (see Fig. 45).

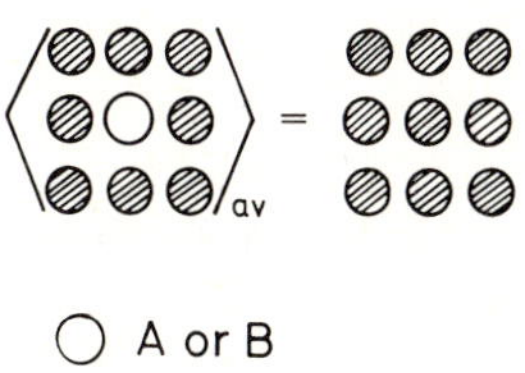

Fig. 45. Schematical representation of CPA

The formalism is described in great detail elsewhere /161,164/ and will not be considered here. However, some remarkable points should be mentioned:

i) The CPA has been shown to be the best single site approximation and to be at least as good as the VCA at all concentrations x.

ii) The CPA is exact in several limiting conditions, e.g. in the dilute limit $x \ll 1$, the virtual crystal limit (weak scattering) $V/w \ll 1$ (V being the difference of the atomic levels and w being the band width), the split band (atomic) limit $V/w \gg 1$ etc..

iii) Because the CPA is essentially a single site theory and all the higher-order effects are averaged to provide a leveled out environment of a one-site scatterer, the theory neglects effects due to the clustering of atoms in a disordered alloy. Such clustering, which will occur statistically even in alloys with no short-range order, can be significant whenever the difference of atomic levels of constituent atoms is large compared with the band width (Sect. 5.2). Another restriction is the assumption that the Hamiltonian expressed in the tight-binding representation contains only the site-diagonal randomness, the off-diagonal terms being regular and possessing the periodicity of the crystal. For a disordered binary alloy this is equivalent to the assumption that the density of states of the pure metal of each constituent is the same except for the shift of the energy level.

A number of investigations on disordered alloys based on the CPA and its extensions have been published concerning the band structure /184,185/, the density of states and the electric conductivity /185,187/, galvanomagnetic effects in quantizing magnetic fields /188/, the vibrational spectrum /189/, k . p theory in semiconductor alloys /175/, etc. For a more complete list see /161,164/. Various other approaches to the alloy problem are also available /190-192/, including short-range order properties /193/ as well as accurate numerical solutions /194/ for large three-dimensional models.

We have considered the problem of disorder in alloys in some detail as we feel that the observed nonlinearities in the concentration dependence of almost all properties demand a more rigorous theoretical treatment than the conventional more or less empirical and qualitative reasoning in the spirit of the virtual crystal approximation.

A more rigorous theory could possibly explain some reported anomalies in the temperature and concentration dependence of the E_1 gap, which seem to be connected with the transition from the simple to the inverted structure in $Hg_{1-x}Cd_xTe$ and which were tentatively ascribed to interactions between the Γ and Λ points (see Figs. 37, 102b /138-141/).

As has been pointed out before, most theoretical calculations of the band structure of HgCdTe have been performed within the framework of the virtual crystal approximation (VCA) and, indeed, these theories have been very successful in predicting specific band quantities (band gap, effective masses) and describing optical and transport properties. The high mobility in HgCdTe with small band gap, for example, indicates highly non-localized electron wavefunctions characteristic for a periodic rather than a random lattice for states near the band edge.

The relative ease of forming a Hg vacancy, on the other hand, suggests that the bonding differs for Hg and Cd and that the electronic structure may reflect the difference in potential of the Hg and Cd sites in the cation sublattice /430c/.

Ultraviolet photoemission spectroscopy (UPS) has recently been used /430c,d/ to probe the complete valence band of the pure compounds HgTe and CdTe and several HgCdTe alloy compositions. While the upper, mainly p-like valence states within 3.5 eV from the valence band maximum were found to be bandlike and qualitatively similar for all compositions studied, pronounced differences have been observed for states near 5 eV below the valence band maximum. The latter states are mainly s electron derived and much more localized than the upper states. They distinctly reflect the difference in Hg- or Cd-cation-to-Te bonds.

Recently calculations treating the disorder problem in HgCdTe within the coherent potential approximation scheme /430e,f,h/ indeed showed that the principal disorder in the HgCdTe alloy system arises from the decrease of the Hg s-state energy compared to that of Cd caused by relativistic terms. The density of states obtained in these calculations is quite similar to the VCA case for the upper p-like valence bands, but exhibits important differences in the mainly s derived region, when compared to the VCA, in accordance with the UPS results. Higher conduction bands more than 2 eV above the band-edge are predicted to show similar shifts and broadening associated with strong disorder. Bowing parameters for several gaps as well as the alloy scattering contribution to the total mobility have been calculated within CPA as well /430h/.

3.7 Band Structure and Dielectric Function

HgCdTe is a narrow-gap or, for a range of compositions, a zero-gap semiconductor. Contrary to, e.g. the lead salts, in HgCdTe the degeneracy of the conduction and the valence bands is not accidential, but symmetry induced. Hence the zero-gap can be destroyed only by perturbations which reduce the symmetry. It has recently been demonstrated (for a review see /450/) that many physical properties are crucially affected by this particular structure of the bands. Among these properties are the dielectric function, the symmetry character of the electron wavefunctions, etc., which in turn strongly influence the optical and transport properties.

In this section a number of recent theoretical and experimental investigations concerning the dielectric function and optical properties (Sect. 3.7.1), magneto-optics and band structure in a magnetic field (Sect. 3.7.2) and finally impurity states (Sect. 3.7.3) will be briefly reviewed.

3.7.1 Dielectric Function

In a zero-gap semiconductor one might expect that the Coulomb interaction between electrons and holes which are not separated by an energy gap may destabilize the ground state of the system and thus the latter might collapse into a new ground state of bound electron-hole pairs /456/. This, however, is not observed due to the anomalous dielectric screening associated with Γ_{8v}-Γ_{8c} excitations. The dielectric function describes this dielectric screening and thus strongly influences optical and transport properties.

Published theoretical work concerning the dielectric function is based on the random-phase approximation. Although there have been some doubts about the violation of the adiabatic approximation when excitation energies of plasmons, phonons and band-to-band transitions are of the same order of magnitude /457/, it seems that experimental results /458/ do not support this view. Then the complex dielectric function can be separated into different parts,

$$\varepsilon(\omega) = \varepsilon_\infty + \Delta\varepsilon_{inter} + \Delta\varepsilon_{intra} + \Delta\varepsilon_{ph} \qquad \text{with} \tag{73}$$

ε_∞ a high-frequency dielectric constant due to all interband transitions except Γ_8-$\Gamma_{8(6)}$;

$\Delta\varepsilon_{inter}$ a contribution due to Γ_8-Γ_8 interband transitions in the zero gap ($E_0 \leqq 0$) and Γ_8-Γ_6 transitions in the case $E_0 > 0$;

$\Delta\varepsilon_{intra}$ a contribution due to intraband carriers in the $\Gamma_{8(6)}$ band; and

$\Delta\varepsilon_{ph}$ a contribution due to phonons.

In the mixed crystal $Hg_{1-x}Cd_xTe$ all contributions are dependent on the composition x.

The general expressions for $\Delta\varepsilon_{inter}$ and $\Delta\varepsilon_{intra}$ are

$$\Delta\varepsilon_{inter}(\omega,q,x) = -\frac{e^2}{\pi^2}\frac{1}{q^2}\int d^3\underline{k}\,\{|\langle\underline{k},v|e^{-i\underline{q}\underline{r}}|\underline{k}+\underline{q},c\rangle|^2$$

$$\cdot[f(E_{\underline{k}+\underline{q},c})-f(E_{\underline{k},v})] + (v \rightleftarrows c)\}/(E_{\underline{k}+\underline{q},c}-E_{\underline{k},c}-\hbar\omega-i\hbar/\tau_s) \tag{74}$$

[$f(E_{\underline{k}})$ is the Fermi-Dirac distribution, τ_s denotes scattering lifetimes],

$$\Delta\varepsilon_{intra}(\omega,q,x) = -\frac{e^2}{\pi^2}\frac{1}{q^2}\int d^3\underline{k}\,|\langle\underline{k},c|e^{-i\underline{q}\underline{r}}|\underline{k}+\underline{q},c\rangle|^2$$

$$\cdot[f(E_{\underline{k}+\underline{q}})-f(E_{\underline{k}})]/E_{\underline{k}+\underline{q}}-E_{\underline{k}}-\hbar\omega-i\hbar/\tau_n) \quad . \tag{75}$$

3.7.1.1 Static Dielectric Function $\varepsilon(0,q)$

For a pure material at T = 0 K $\Delta\varepsilon_{intra}$ is zero. For $\Delta\varepsilon_{inter}$, expression (74) gives /459/

$$\Delta\varepsilon_{inter}(0,q,x) = \varepsilon_0(x) + k_0/q + O(q^n) \quad , \tag{76}$$

with $k_0 = f[m_0^\star(x), m_v(x)]$ in a parabolic two-band approximation. Thus the static dielectric function has an anomalous q^{-1} divergence, which is due to zero-energy excitations. With free carriers in the band, however, this divergence is removed. Now for small q in the degenerate limit and for parabolic bands $\Delta\varepsilon_{inter}$ is /460/

$$\Delta\varepsilon_{inter}(0,q,x) = \varepsilon_0(x) + \frac{8e^2m_0^\star}{\pi h^2 k_F}\left(1 - \frac{1}{12}\cdot\frac{q^2}{k_F^2}\right) \tag{77}$$

with k_F being the Fermi momentum.

For $\Delta\varepsilon_{intra}$ the usual Thomas-Fermi result is modified to /461/

$$\Delta\varepsilon_{intra}(0,q,x) = (k_{TF}^2/q^2)\ f_1(q) \tag{78}$$

with k_{TF} the Thomas-Fermi momentum at the temperature T and the intraband screening function $f_1(q)$ given in the parabolic approximation by

$$f_1(q) = 1 - 0.47(q/k_F)^2 + 0.075(q/k_F)^4 \quad . \tag{79}$$

In the case of non-parabolic bands the deviation of the electronic dielectric function from the Thomas-Fermi result has been calculated in /462/ and expressed in terms of $\phi_F(\lambda,t)$ with $\lambda = 4\ k_F^2/E_0$, $t = q/k_F$, a function which is equal to the filter function $\phi(q)$ usually introduced to describe the deviation from the Thomas-Fermi approximation

$$\Delta\varepsilon(0,q) = 1 + \frac{1}{q^2L^2}\ \phi(q) \quad , \tag{80}$$

L being the Thomas-Fermi screening length.

Non-parabolicity leads to quantitative changes of ϕ and even has a qualitative influence on the filter function for high q values.

3.7.1.2 Dynamic Dielectric Function

For an infinite lifetime and perfect degeneracy the real and imaginary parts of $\Delta\varepsilon_{inter}(\omega,0)$ are /463/

$$\Delta\varepsilon'_{inter}(\omega,0) = \frac{2\sqrt{2}e^2}{\pi\hbar^{3/2}}\left(\frac{m_0^*}{1+\gamma}\right)^{1/2} \frac{1}{\omega^{1/2}}\left[\frac{\pi}{2} - \tan^{-1}\left(\frac{\omega_F}{\omega}\right)^{1/2} + \ln\left|\frac{1+(\omega_F/\omega)^{1/2}}{1-(\omega_F/\omega)^{1/2}}\right|\right]$$

$$\Delta\varepsilon''_{inter}(\omega,0) = \frac{\sqrt{2}e^2}{\hbar^{3/2}}\left(\frac{m_0^*}{1+\gamma}\right)^{1/2} \frac{\Theta(\omega_F-\omega)}{\omega^{1/2}} \tag{81}$$

with

$$\gamma = m_0^*/m_v^*,\ \hbar\omega_F = \begin{cases} (1+\gamma)E_F & \text{n-type} \\ (1+\gamma^{-1})E_F & \text{p-type} \end{cases}, \qquad \Theta(\omega_F-\omega) = \begin{cases} 1 & \omega > \omega_F \\ 0 & \omega < \omega_F \end{cases} .$$

Thus, $\varepsilon'(\omega,0)$ has a singularity and $\varepsilon''(\omega,0)$ a finite discontinuity for $\omega = \omega_F$. For finite lifetime or finite temperature these structures are smoothed out.

Using wavefunctions given in /464/ the matrix element in (74) can be calculated to give /458/

$$|\langle \underline{k}+\underline{q},c|e^{-i\underline{qr}}|\underline{k},v\rangle|^2 = \frac{3}{8}\,\frac{\eta-E_0}{\eta}\cdot\frac{q^2\sin^2\theta}{|\underline{k}+\underline{q}|^2} \tag{82}$$

where θ is the angle between $\underline{k}$ and $\underline{q}$ and

$$\eta^2 = E_0^2 + \frac{8}{3}\,\frac{\hbar^2k^2}{2m_0}\,E_p \quad . \tag{83}$$

Integration then gives

$$\Delta\varepsilon''_{inter}(\omega,0,x) = 2\left(R\,\frac{3}{2E_p}\right)^{1/2}\left(\frac{\hbar\omega-E_0}{\hbar\omega}\right)^{1/2}(f_v-f_c) \tag{84}$$

with $R = m_0e^4/2h^2 = 13.6$ eV. Equation (84) generalizes the parabolic band model for both semimetallic and semiconducting compositions. The real part $\Delta\varepsilon'_{inter}(\omega,0,x)$ can be obtained by a Kramers-Kronig inversion /457/ (Fig. 46).

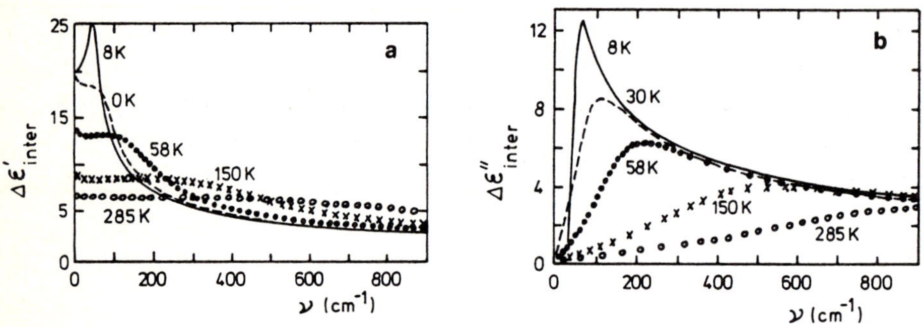

Fig. 46 a,b. Calculated real (a) and imaginary (b) part $\Delta\varepsilon_{inter}$ of the Γ_8-Γ_8 interband contribution to dielectric function

The intraband part of ε is due to normal free carriers and may be described according to the classical Drude-Zener theory as

$$\Delta\varepsilon_{intra} = -\,\omega_p^2/(\omega^2+i\Gamma\omega) \quad , \tag{85}$$

where Γ is the damping parameter for free carriers.

3.7.1.3 Phonon Contribution $\Delta\varepsilon_{ph}$

In the framework of the classical theory of independent harmonic oscillators, the phonon contribution to the dielectric function is given by

$$\Delta\varepsilon_{ph} = \sum_j \frac{S_j\omega_{TO,j}}{\omega_{TO,j}^2-\omega^2-i\Gamma_j\omega} \tag{86}$$

where $\omega_{TO,j}$ is the wavenumber of the transverse mode for each oscillator. S_j are the oscillator strengths, and Γ_j the damping parameters.

In /457/ the authors fitted their experimental results on HgTe using two independent lattice oscillators, one (ω_{TO} = 117 cm^{-1}) being the HgTe lattice oscillator. The origin of the second oscillator, which is fairly strong ($S_2 > 1$) at high temperatures, has been discussed in /457/ but remains highly speculative (see also Sect. 5.2).

In the mixed crystal regime the phonon contribution $\Delta\varepsilon_{ph}$ is much more difficult to describe since HgCdTe exhibits a two-mode behaviour (Sect. 5.2). This occurence of two long-wavelength optical modes has usually been analysed within the framework of the so-called REI (random elements isodisplacement) or MREI (modified REI) models. These do, in fact, describe the dependence of the optical mode frequencies of HgCdTe on the composition x satisfactorily in the range $1 \geqq x \geqq 0.2$ /465/. On the mercury-rich side ($x \leqq 0.2$) one has to take the interband transitions and the plasmon contribution into account in a consistent way to calculate the phonon spectrum /458/.

Even in the range where isodisplacement models seem to describe the phonon spectrum correctly, there remains an inherent difficulty: REI models usually use three force constants, two between nearest neighbours and one between next nearest neighbours. A parameter fit to experimental measurements then results in values for these force constants of approximately /465/

$$F_{CdTe} \cong 92 \text{ N/m}$$
$$F_{HgTe} \cong 101 \text{ N/m}$$
$$F_{HgCd} \cong 92 \text{ N/m} \quad ,$$

that is, the second-neighbour forces between ions of the same sign should be of the same order of magnitude as forces between nearest neighbours with opposite signs, which seems quite unjustified.

3.7.2 Magnetooptics and Band Structure in a Magnetic Field

In recent years the results of a number of magnetooptical investigations in HgTe and HgCdTe have been published /432,466-475/. These investigations differ in many cases not only in their experimental results but sometimes also in the theoretical models used to interpret these results and in the approximations made using the same basic model. In addition, some of the published papers contain (printing?) errors or notational inconsistencies. Consequently, several material parameters derived from these investigations span a rather wide range if compiled from the literature. In the following sections we would like to summarize the main results on the band structure of HgCdTe in a magnetic field and the results of magnetooptical investigations. We shall point out some discrepancies and open questions.

3.7.2.1 Band Structure in a Magnetic Field

The problem of calculating the band structure of HgCdTe for compositions corresponding to a small or zero-energy gap in a magnetic field is an intricate one because

i) due to the small energy distance Γ_8-Γ_6, the interaction between these bands induces a strong non-parabolicity,

ii) the inversion asymmetry of the zincblende lattice induces k-linear terms,

iii) as the Γ_8-Γ_6-Γ_7 energy separations are not small compared to the energy distance to higher bands, their influence must be taken into account in the description of heavy-hole mass and the warping of both bands,

IV) in the zero-gap case the conduction and valence bands are degenerate at the Γ_8 point giving rise to Luttinger effects, and

V) in a gapless semiconductor a non-local potential due to the exchange interaction can lead to a strong modification of the hole energy spectrum /437-442/.

The first approach to this problem using a three-level (Γ_8-Γ_6-Γ_7) model has been discussed in detail in Sect. 3.3.1.

The theory of Landau levels at the Γ point in small-gap materials with zincblende lattice was established by PIDGEON and BROWN /476/ within the effective mass approximation /159,477/. By applying a procedure of quasi-degenerate states, the closely spaced Γ_8, Γ_6, Γ_7 bands are treated exactly, and the effects of the higher bands upon the closely spaced states are accounted for to order k^2.

The Luttinger effects in Γ_8 are included as well as the non-parabolicity effects induced by the k·p interactions between s- and p-states.

According to GERHARDTS /478/ the Luttinger Hamiltonian in the Pidgeon and Brown basis for the valence bands reads

$$\frac{1}{\hbar^2} m_0 H_L = \begin{pmatrix} D_a & N_{ab} \\ N_{ab}^+ & D_b \end{pmatrix} \qquad \begin{aligned} \Delta &= \frac{\hbar^2}{m_0}\delta \\ k_x^2 + k_y^2 &= k_\perp^2 \\ k_\pm &= \frac{1}{\sqrt{2}}(k_x \pm i k_y) \\ k^2 &= k_z^2 + k_\perp^2 \\ s &= \frac{eB}{\hbar} \end{aligned} \tag{87a}$$

$$N_{ab} = -3\gamma_3 k_z \begin{Vmatrix} -i\sqrt{\frac{2}{3}}\,k_- & 0 & +\frac{1}{\sqrt{3}}\,k_- \\ 0 & +i\sqrt{\frac{2}{3}}\,k_- & +k_+ \\ +k_+ & +\frac{1}{\sqrt{3}}\,k_- & 0 \end{Vmatrix} \tag{87b}$$

$$D_a = \begin{Vmatrix} -\frac{1}{2}(\gamma_1+4\gamma_2)k^2+3\gamma_2(k_z^2+\frac{1}{2}k_\perp^2)-\frac{1}{2}s3\kappa & \frac{\sqrt{3}}{2}\{-\gamma_2(k_+^2+k_-^2)+\gamma_3(k_+^2-k_-^2)\} & -i\sqrt{\frac{3}{2}}\{-\gamma_2(k_+^2+k_-^2)+\gamma_3(k_+^2-k_-^2)\} \\ \frac{\sqrt{3}}{2}\{-\gamma_2(k_+^2+k_-^2)-\gamma_3(k_+^2-k_-^2)\} & -\frac{1}{2}(\gamma_1+4\gamma_2)k^2+\gamma_2\{k_z^2+\frac{5}{2}k_\perp^2\}+\frac{1}{6}s3\kappa & -i\sqrt{2}\gamma_2(k_z^2-\frac{1}{2}k_\perp^2)-i\frac{\sqrt{2}}{6}s(3\kappa+3) \\ -i\sqrt{\frac{3}{2}}\{-\gamma_2(k_+^2+k_-^2)-\gamma_3(k_+^2-k_-^2)\} & i\sqrt{2}\gamma_2(k_z^2-\frac{1}{2}k_\perp^2)+i\frac{\sqrt{2}}{6}s(3\kappa+3) & -\delta-\frac{1}{2}(\gamma_1+4\gamma_2)k^2+2\gamma_2k^2+\frac{1}{3}s(3\kappa+\frac{3}{2}) \end{Vmatrix} \tag{87c}$$

$$D_b = \begin{Vmatrix} -\frac{1}{2}(\gamma_1+4\gamma_2)k^2+\gamma_2(k_z^2+\frac{5}{2}k_\perp^2)-\frac{s}{6}3\kappa & \frac{\sqrt{3}}{2}\{-\gamma_2(k_+^2+k_-^2)+\gamma_3(k_+^2-k_-^2)\} & -i\sqrt{2}\gamma_2(k_z^2-\frac{1}{2}k_\perp^2)+i\frac{\sqrt{2}}{6}s(3\kappa+3) \\ \frac{\sqrt{3}}{2}\{-\gamma_2(k_+^2+k_-^2)-\gamma_3(k_+^2-k_-^2)\} & -\frac{1}{2}(\gamma_1+4\gamma_2)k^2+3\gamma_2(k_z^2+\frac{1}{2}k_\perp^2)+\frac{s}{2}3\kappa & -i\sqrt{\frac{3}{2}}\{-\gamma_2(k_+^2+k_-^2)-\gamma_3(k_+^2-k_-^2)\} \\ i\sqrt{2}\gamma_2(k_z^2-\frac{1}{2}k_\perp^2)-i\frac{\sqrt{2}}{6}s(3\kappa+3) & i\sqrt{\frac{3}{2}}\{-\gamma_2(k_+^2+k_-^2)+\gamma_3(k_+^2-k_-^2)\} & -\delta-\frac{1}{2}(\gamma_1+4\gamma_2)k^2+2\gamma_2k^2-\frac{s}{3}(3\kappa+\frac{3}{2}) \end{Vmatrix} \tag{87d}$$

which is identical to the Hamiltonian given in /84/ with the following identifications:

$$-\frac{1}{2}\gamma_1 = \frac{1}{2}+\frac{1}{3}(2C+B)\frac{m}{\hbar^2}$$

$$\gamma_2 = \frac{1}{3}(C-B)\frac{m}{\hbar^2} \tag{88}$$

$$\gamma_3 = -\frac{1}{3}(C-\frac{1}{2}B)\frac{m}{\hbar^2}$$

$$\kappa = -\frac{1}{3}+\frac{1}{3}(C+\frac{1}{2}B)\frac{m}{\hbar^2} \quad .$$

If one neglects the anisotropy in Γ_8 (which means setting $\gamma_2 = \gamma_3 = :\bar{\gamma}$) one obtains

$$E_g = \frac{\hbar^2}{m_0}\frac{1}{2}q_g^2 \qquad q_g^2 = \frac{m_0}{m_c^*}k_g^2 \cong 10^2 k_g^2$$

$$\frac{1}{\hbar^2}m_0 H = \begin{pmatrix} D_a & N_{ab} \\ N_{ab}^+ & D_b \end{pmatrix} \qquad Pk_\mu = \frac{\hbar^2}{m_0}pk_\mu \qquad \Delta = \frac{\hbar^2}{m_0}\delta \tag{89a}$$

$$p^2 = \left(\frac{m_0}{\hbar^2}\right)^2 P^2 = \left(\frac{m_0}{\hbar^2}\right)^2 \frac{\hbar^2}{2}\frac{E_p}{m_0} = \frac{m_0}{2}\frac{E_p}{\hbar^2}$$

$$N_{ab} = \begin{Vmatrix} 0 & \sqrt{\frac{2}{3}}pk_z & 0 & \frac{i}{\sqrt{3}}pk_z \\ 0 & i\sqrt{6}\,\bar{\gamma}k_z k_- & 0 & -\sqrt{3}\,\bar{\gamma}k_z k_- \\ \sqrt{\frac{2}{3}}pk_z & 0 & -i\sqrt{6}\,\bar{\gamma}k_z k_- & -\sqrt{3}\,\bar{\gamma}k_z k_+ \\ -\frac{i}{\sqrt{3}}pk_z & -3\,\bar{\gamma}k_z k_+ & -\sqrt{3}\,\bar{\gamma}k_z k_- & 0 \end{Vmatrix} \tag{89b}$$

$$D_a = \left\| \begin{matrix} \frac{1}{2}(q_g^2+k_z^2+k_\perp^2+s) & ipk_+ & \frac{i}{\sqrt{3}}pk_- & \sqrt{\frac{2}{3}}pk_- \\ -ipk_- & -\frac{1}{2}(\gamma_1-2\bar{\gamma})k_z^2-\frac{1}{2}(\gamma_1+\bar{\gamma})k_\perp^2-\frac{3}{2}s\kappa & -\sqrt{3}\,\bar{\gamma}k_-^2 & i\sqrt{6}\,\bar{\gamma}k_-^2 \\ -\frac{i}{\sqrt{3}}pk_+ & -\sqrt{3}\,\bar{\gamma}k_+^2 & -\frac{1}{2}(\gamma_1+2\bar{\gamma})k_z^2-\frac{1}{2}(\gamma_1-\bar{\gamma})k_\perp^2+\frac{1}{2}s\kappa & -i\sqrt{2}\,\bar{\gamma}(k_z^2-\frac{1}{2}k_\perp^2)-\frac{i}{\sqrt{2}}s(\kappa+1) \\ \sqrt{\frac{2}{3}}pk_+ & -i\sqrt{6}\,\bar{\gamma}k_+^2 & i\sqrt{2}\,\bar{\gamma}(k_z^2-\frac{1}{2}k_\perp^2)+\frac{i}{\sqrt{2}}s(\kappa+1) & -\delta-\frac{\gamma_1}{2}(k_z^2+k_\perp^2)+s(\kappa+\frac{1}{2}) \end{matrix} \right\| \qquad (89c)$$

$$D_b = \left\| \begin{matrix} \frac{1}{2}(q_g^2+k_z^2+k_\perp^2-s) & \frac{i}{\sqrt{3}}pk_+ & ipk_- & \sqrt{\frac{2}{3}}pk_+ \\ -\frac{i}{\sqrt{3}}pk_- & -\frac{1}{2}(\gamma_1+2\bar{\gamma})k_z^2-\frac{1}{2}(\gamma_1-\bar{\gamma})k_\perp^2-\frac{1}{2}s\kappa & -\sqrt{3}\,\bar{\gamma}k_-^2 & -i\sqrt{2}\,\bar{\gamma}(k_z^2-\frac{1}{2}k_\perp^2)+\frac{i}{\sqrt{2}}s(\kappa+1) \\ -ipk_+ & -\sqrt{3}\,\bar{\gamma}k_+^2 & -\frac{1}{2}(\gamma_1-2\bar{\gamma})k_z^2-\frac{1}{2}(\gamma_1+\bar{\gamma})k_\perp^2+\frac{3}{2}s\kappa & i\sqrt{6}\,\bar{\gamma}k_+^2 \\ \sqrt{\frac{2}{3}}pk_- & i\sqrt{2}\,\bar{\gamma}(k_z^2-\frac{1}{2}k_\perp^2)-\frac{1}{\sqrt{2}}s(\kappa+1) & -i\sqrt{6}\,\bar{\gamma}k_-^2 & -\delta-\frac{1}{2}\gamma_1(k_z^2+k_\perp^2)-s(\kappa+\frac{1}{2}) \end{matrix} \right\| \qquad (89d)$$

Eigenvectors of H have the form

$$\begin{matrix} c_1\phi_n \\ c_2\phi_{n-1} \\ c_3\phi_{n+1} \\ c_4\phi_{n+1} \\ c_5\phi_{n+1} \\ c_6\phi_n \\ c_7\phi_{n+2} \\ c_8\phi_n \end{matrix}$$

With $k_z = 0$ this reduces further to

$$k_+\phi_n = \sqrt{s(n+1)}\phi_{n+1}, \quad k_-\phi_n = \sqrt{sn}\phi_{n-1}, \quad k^2\phi_n = s(2n+1)\phi_n \quad .$$

With

$$f_a = \begin{pmatrix} a_1\phi_{n-1} \\ a_3\phi_{n-2} \\ a_5\phi_n \\ a_7\phi_n \end{pmatrix} \text{follows } D_a f_a = \begin{pmatrix} \tilde{a}_1\phi_{n-1} \\ \tilde{a}_3\phi_{n-2} \\ \tilde{a}_5\phi_n \\ \tilde{a}_7\phi_n \end{pmatrix} \text{ and } \begin{pmatrix} \tilde{a}_1 \\ \tilde{a}_3 \\ \tilde{a}_5 \\ \tilde{a}_7 \end{pmatrix} = \tilde{D}_a \begin{pmatrix} a_1 \\ a_3 \\ a_5 \\ a_7 \end{pmatrix} , \tag{90}$$

$$\tilde{D}_a = \begin{Vmatrix} \frac{1}{2}q_g^2+sn & ip\sqrt{s(n-1)} & \frac{i}{\sqrt{3}}p\sqrt{sn} & \sqrt{\frac{2}{3}}p\sqrt{sn} \\ -ip\sqrt{s(n-1)} & -(\gamma_1+\bar{\gamma})s(n-\frac{3}{2})-\frac{3}{2}s\kappa & -\sqrt{3}\bar{\gamma}s\sqrt{n(n-1)} & i\sqrt{6}\bar{\gamma}s\sqrt{n(n-1)} \\ -\frac{i}{\sqrt{3}}p\sqrt{sn} & -\sqrt{3}\bar{\gamma}s\sqrt{n(n-1)} & -(\gamma_1-\bar{\gamma})s(n+\frac{1}{2})+\frac{1}{2}s\kappa & i\sqrt{2}\bar{\gamma}s(n+\frac{1}{2})-\frac{i}{\sqrt{2}}s(\kappa+1) \\ \sqrt{\frac{2}{3}}p\sqrt{sn} & -i\sqrt{6}\bar{\gamma}s\sqrt{n(n-1)} & -i\sqrt{2}\bar{\gamma}s(n+\frac{1}{2})+\frac{i}{\sqrt{2}}s(\kappa+1) & -\gamma_1(n+\frac{1}{2})s+s(\kappa+\frac{1}{2})-\delta \end{Vmatrix} \tag{91}$$

A similar expression holds for D_b and is easily derived from (89d). In the parabolic limit the corresponding states a(n) reduce to $\sigma_z = +1/2$ spin sublevels for Γ_6 bands, whereas b(n) states are the equivalent of $\sigma_z = -1/2$ spin sublevels. The determinant equations (91) depend on six parameters: the interaction gap $E_0 = E_{\Gamma_6}-E_{\Gamma_8}$, the Kane matrix element through $E_p = (2/m)|\langle s|p_z|z\rangle|^2$, the spin-orbit splitting Δ, and the modified Luttinger parameters $\gamma_1,\bar{\gamma},\kappa$ describing the interaction between Γ_8 and higher bands. They are related to the Luttinger parameters $\gamma_1^L,\bar{\gamma}^L,\kappa^L$ by

$$\gamma_1 = \gamma_1^L-E_p/3E_0, \quad \bar{\gamma} = \bar{\gamma}^L-E_p/6E_0, \quad \kappa = \kappa^L-E_p/6E_0$$

(which essentially means adding the contribution of the Γ_6 level).

Equation (91) corresponds to the results in /466/ after several notational inconsistencies are removed. It also corresponds to the results in /467/ if $n \rightarrow n+1$, $\gamma = \bar{\gamma}$ and the wrong $\kappa+3/2$ in $\tilde{D}_{a,b}$ (4.4) of /467/ is corrected to $\kappa+1/2$.

In this derivation the F and G matrix elements defined by KANE /83/ are neglected. In Fig. 47 the dependence of the lowest conduction band Landau levels on the applied magnetic field is compared, calculated with expressions (13a,b, 91).

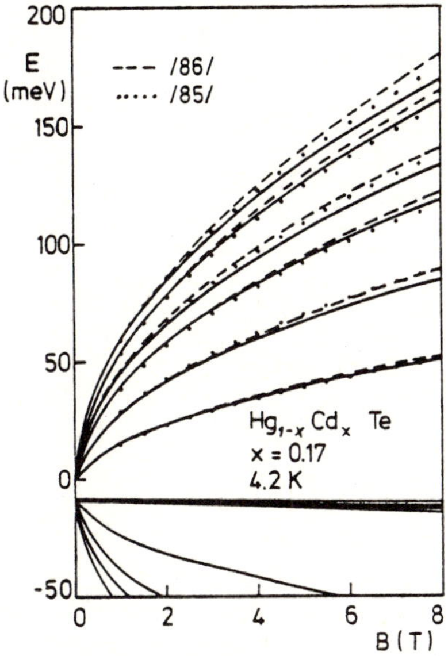

Fig. 47. Comparison of the dependence of the lowest Landau levels on the magnetic field for $Hg_{.83}Cd_{.17}Te$ at 4.2 K calculated from (91) and after /85,86/

3.7.2.2 Magnetooptics

The description given above has been used in /466,467/ to derive selection rules and transition probabilities for magnetooptical transitions and to interpret magnetooptical spectra of HgCdTe alloys on both sides of the SM-SC transition.

This description, of course, neglected the non-sphericity of the Γ_8 bands. It is clear, however, as already pointed out in /298/, that due to the warping of Γ_8 bands the structure of magnetic subbands does not remain the same as the magnetic field is rotated with respect to the crystal axes (Fig. 48). Although this anisotropy is more pronounced, e.g. in α-Sn than in HgCdTe, it can be important for an unambiguous assignment of magnetooptical transitions.

WEILER et al. /432,472/ have recently interpreted their magnetoreflectance data on HgCdTe using determinantal equations for the energies of Landau levels given in /473/. There expressions are given including the four Luttinger parameters $\gamma_1^L, \gamma_2^L, \gamma_3^L, \kappa^L$ and also the Kane parameter F. For arbitrary spherical polar angles θ, ϕ of the applied magnetic field with respect to the cubic crystal

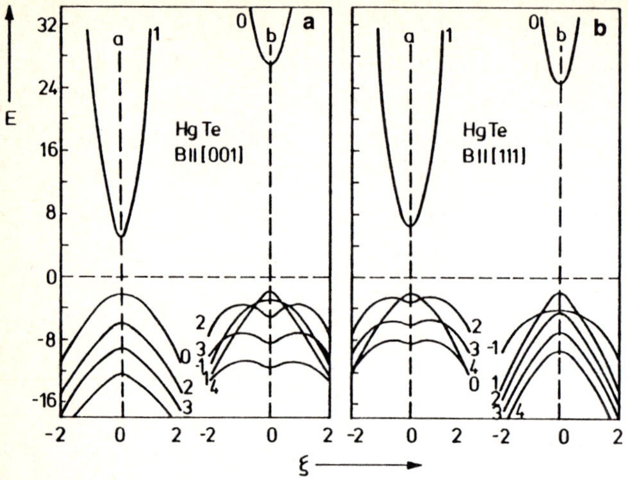

Fig. 48 a,b. Calculated k_z dependence of magnetic subbands in HgTe for magnetic fields (a) B‖ [001] and (b) B‖ [111]. $\zeta = k_z(\hbar c/eH)^{1/2}$

axes, expressions

$$\gamma' = \gamma_3-(\gamma_2-\gamma_3)f(\theta,\phi) \tag{92}$$

$$\gamma'' = \frac{1}{3}\gamma_2+\frac{2}{3}\gamma_3+(\gamma_2-\gamma_3)f(\theta,\phi) \tag{93}$$

$$f(\theta,\phi) = \frac{1}{4}(3\cos^2\theta-1)^2+\frac{3}{4}\cos^2 2\phi\ \sin^4\theta \tag{94}$$

describe the influence of orientation on the Landau level energies.

In recent papers on the band structure of InSb the same authors have even gone further in the refinement of the quasi-Ge model /474,475/, extending the number of parameters by another three or four. It seems doubtful, however, whether the introduction of more fit parameters to interpret magnetooptical data on HgCdTe is very meaningful. The authors of /432,472/ have admitted that their data were not sufficient to unambiguously determine the Kane parameter F in many cases.

An excellent recent review on magnetooptical results and the different theoretical approaches to interpret these results has been published in /367/. Also a thorough discussion of parameter fitting techniques and material parameters obtained for HgCdTe can be found there.

PASTOR et al. /471/ have included anisotropy effects on the Landau level structure of HgTe by using second-order perturbation theory for the calculations.

In /437-442/ it has been shown that the non-locality of the potential due to the electron-electron exchange interaction may lead to a renormalization of the electron and hole energy spectrum in a "gapless semiconductor" [e.g. if $m_h >> m_e$ and $k \tilde{<} a_b^{-1}$, where $a_b = \hbar^2 \varepsilon_0 / m_e e^2$]. SZYMANSKI /442/ has included the screening effects due to the intraband (free-carrier) contributions to the dielectric function and performed a numerical analysis for HgTe on the dependence of electron and hole effective masses on energy and temperature (Fig. 49). The result is a strong dependence of the hole mass on different factors (type of sample, hole energy, temperature), which may be responsible for the serious spread of hole masses reported in different investigations.

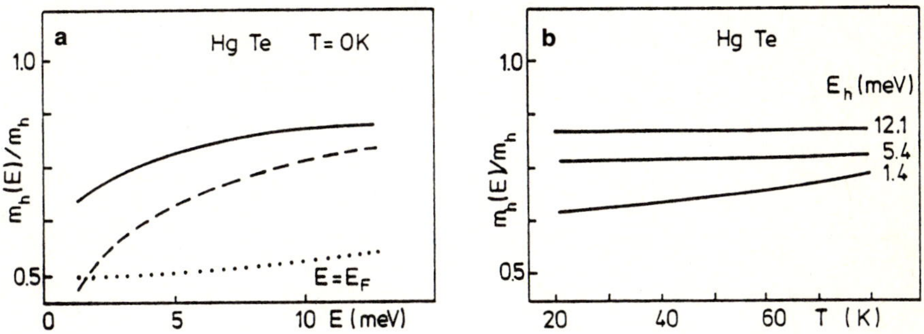

Fig. 49. (a) Hole momentum mass vs energy with screened (———) and unscreened (-----) exchange corrections for intrinsic HgTe at T = 0 K. (.....) hole mass at the Fermi level vs Fermi energy for p-type HgTe. (b) Hole momentum mass vs temperature for intrinsic HgTe for different hole energies E_h in meV

SHIMIZU et al. /470/ have analysed magneto-absorption data on HgTe using both a non-local /438/ and the local potential approximation, which resulted in totally different assignments of many transitions and a differing set of fit parameters. They judged that the non-local approach gave a more reasonable interpretation of their experimental data.

Parameters of different investigations are collected in Tables 22, 23.

Changes in selection rules for magneto-absorption at all higher harmonics of cyclotron frequency in mixed semiconductors due to microscopic composition fluctuations have been theoretically predicted by MYCIELSKI et al. /479/.

3.7.3 Impurity States

Since data and experimental results on imperfections such as native defects and foreign dopants observed in HgCdTe have already been discussed and collected in Sect. 2.3, this section concentrates on resonant acceptor levels in zero-gap compositions and their observation mainly in magnetooptical investigations. Theory and experimental results on resonant states have been recently reviewed in several excellent articles /450,484,485/, and we only repeat the main results here.

<u>Table 22.</u> Luttinger parameters and corresponding effective masses from magneto-optical measurements on HgTe and HgCdTe

(a) HgTe:	γ_1^L	γ_2^L	γ_3^L	κ^L	
	-12.8	-8.4	-8.4	-10.5	/289/
	-16.8	-10.4	-9.2	-11.2	/480/
	-14.8	-8.6	-8.6	-10.2	/481/
	-14.8	-10.9	-8.7	-11.7	/482/
	-14.8	-9.0	-8.2	-10.2	/468/
	-17.0	-10.6	-8.8	-9.5	/471/
	-14.0	-7.8	-7.4	-10.0	non-local potential } /470/
	-15.6	-9.6	-8.6	-11.2	local potential } /470/
	-18.3	-11.2	-10.0	-12.0	/298/

(b) HgCdTe:	γ_1	γ_2	γ_3	κ	F	E_p	
	1.5±1.0	-1.3±.7	.5±.7	-.2±.8	0	16.8	/471/
	2.5± .4	-.3±.2	.5±.1	-1.2±.1	-.7±.3	17.9	/432/
	3.3± .2	.1±.1	.9±.1	-.8±.1	-.8±.3	19.0	/367/
	4.5± .5	1.0		-1.0±.5	0	19.0	/467/
	5.2	1.4		-.2	0	18.0 (HgTe)	/466/
		$\bar{\gamma}$					

(c)	m^*_{hh}/m	[001]	[011]	[111]	
		0.40±0.07	0.45±0.09	0.53±0.11	/432/
		0.3		0.7	/471/
		0.37	0.45	0.48	/468/
		0.40	0.40	0.40	/466/
	m^*_c/m	0.027		0.028	/471/
	(HgTe)	0.031	0.032	0.032	/468/
	g^*_c	-25	-22	-21	/468/
	g^*_v	+6	+4	+3	/468/

Table 23. Fit parameters for $E(k_z)$ dependence of magnetic subbands of HgTe: $E = Ak_z^2 + B|k_z| + C$ /483/

	Subband	A [eV m^2]	B [eV m]	C [eV]
valence band	+1	$1.98 \cdot 10^{-20}$	$8.25 \cdot 10^{-12}$	$2.41 \cdot 10^{-4}$
	0	$4.4 \cdot 10^{-20}$	$5.01 \cdot 10^{-12}$	$3.4 \cdot 10^{-4}$
	2	$1.5 \cdot 10^{-19}$	$2.18 \cdot 10^{-11}$	$2.8 \cdot 10^{-4}$
	-1	$1.5 \cdot 10^{-19}$	0	$4.25 \cdot 10^{-4}$
conduction band	e_1	$1.4 \cdot 10^{-18}$	0	$6.9 \cdot 10^{-4}$
	e_2	$1.4 \cdot 10^{-18}$	0	$2.53 \cdot 10^{-3}$

The theory of impurity states in materials like HgCdTe, where the energy gap is identically zero for certain compositions, is a rather intricate problem. In these symmetry-induced "zero-gap semiconductors" impurity levels can only be superimposed on the continuous spectrum corresponding to either the conduction or the valence band. Thus, bound states cannot exist in a strict sense, but a change in the density of states may occur for energies of the order of the Bohr energy of the impurity state (Fig. 50). The energy spectrum remains continuous. What the impurity does is to phase-shift the running free-electron waves. A resonant level will be produced any time the phase-shift $\delta(\varepsilon)$ passes through $\pi/2$. To be observable the density-of-states peak should be reasonably narrow, which

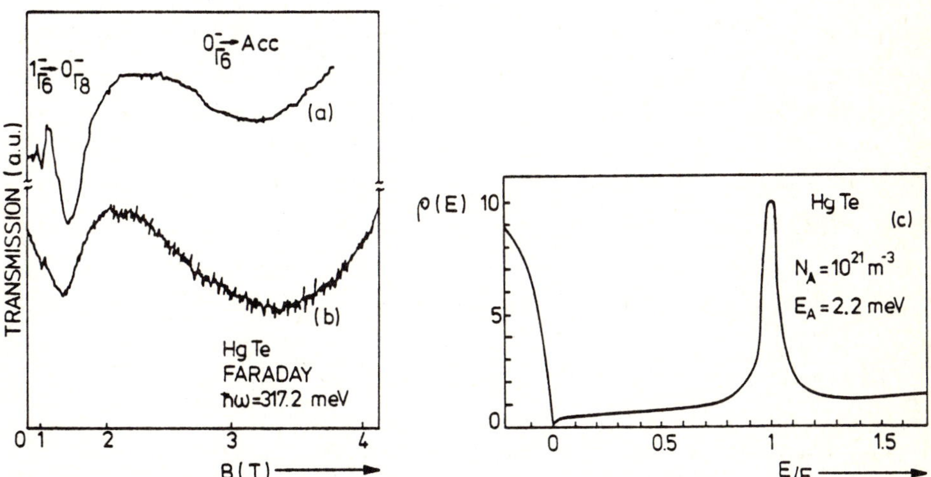

Fig. 50. Comparison between impurity magnetoabsorption observed in annealed (a) and unannealed bulk HgTe samples (b). Density of states of HgTe including resonant acceptor level (schematically) (c)

means the phase-shift has to vary rapidly around the resonance. It has been shown that the phase-shift or the width of the resonant state are related to the ratio of conduction to valence band masses and hence the pronounced asymmetry between these masses (due to the proximity of the Γ_6 s band $m_c/m_v << 1$) is a key feature for the occurrence of only acceptor resonant states in zero-gap HgCdTe.

The first attempt to understand the structure of impurity-induced resonant states in zero-gap materials was made by LIU and BRUST /486/, but their work neglects entirely the detailed Γ_8 band-edge structure. Similarly MAUGER and FRIEDEL /39/, using a perturbing potential entirely localized (which might be appropriate for a vacancy type of defect), did not take into account the proper symmetry of the bands.

A more quantitative analysis of resonant impurity states has subsequently been made using an effective-mass Hamiltonian, in which the electrons and holes are described by the k•p perturbation expansion about the Γ_8 point in a spherical approximation /41,52,487/.

GELMONT and DYAKONOV /52/ have considered this Hamiltonian for an unscreened Coulomb potential and in the limit of zero electron mass and predicted a binding energy for the acceptor level of 4/9 hole Rydberg (6.4 meV for HgTe). JOOS et al. /487/, however, using the full RPA screened Coulomb potential, demonstrated that the singular q dependence of the dielectric function decreases this value down to 1-2 meV depending on the exact value of the heavy-hole mass. The relative width of the resonance proved to be only weakly dependent on screening effects.

Owing to the large screening in zero-gap materials several authors used a short-range Slater-Koster type potential to treat the problem of a Hg vacancy in HgCdTe /405c,420,484,488/.

Similar to the results cited above, the position of the acceptor resonance ε_A was shown to depend mainly on the valence band (for a given effective impurity potential), whereas the acceptor width Γ_A is controlled by the conduction band mass /485/,

$$\frac{1}{V_{eff}} = \frac{\sqrt{2}}{4\pi\hbar^3} m_v^{3/2} \sqrt{\varepsilon_A} \tag{95}$$

$$\frac{\Gamma_A}{2\varepsilon_A} = (m_c/m_v)^{3/2} \quad . \tag{96}$$

Conversely, donor resonances ($\varepsilon_D < 0$), which would be superimposed on the valence-band background, are always overdamped since $\Gamma_D/2|\varepsilon_D| = (m_v/m_c)^{3/2} >> 1$.

Obviously, as most of the latter models contain the potential matrix elements as adjustable parameters, one can not determine quantitatively the location of the resonance states, but one is able to make some qualitative predictions, e.g. on the variation of the resonance energy with composition in HgCdTe. On the zero-gap side an approximate $\varepsilon_A \propto E_0^2$ law results /418/.

Theory described so far is carried out in the approximation of the infinite dilute system, i.e. taking into account only one scattering centre. Because of the very low effective mass near the Γ_8-Γ_6 crossover in HgCdTe the electron wavelength might become comparable or greater than the mean distance between impurities $N_A^{-1/3}$ even for moderate doping. BASTARD /420/ took into account multiple scattering effects and DUGAEV et al. /419/ extended these calculations to the case of an acceptor impurity band (Fig. 51).

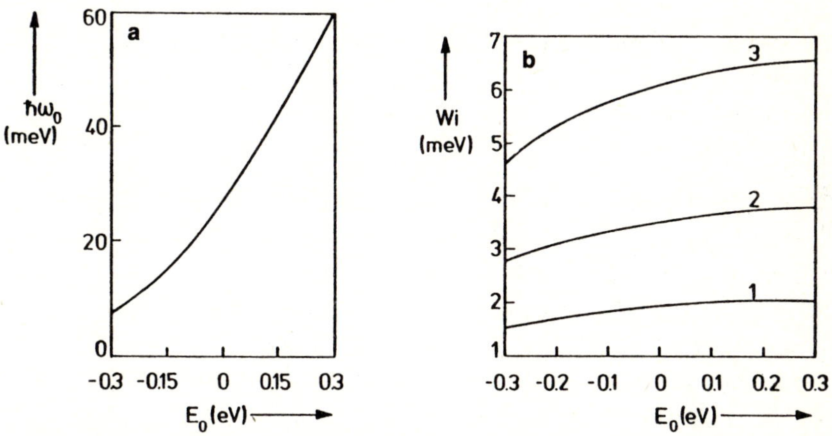

Fig. 51 a,b. Position of the impurity band centre as a function of energy gap (a). Impurity band half-width as a function of energy gap (*1*) $N_a = 3\cdot10^{22}\ m^{-3}$, (*2*) $10^{23}\ m^{-3}$, (*3*) $3\cdot10^{23}\ m^{-3}$ (b). /419/

Evidence for resonant states in zero-gap semiconductors was first found in transport measurements in the extrinsic regime ($k_BT \lesssim \varepsilon_A$) (see the references in Sects. 2.3, 4.1,2 of the original article and /483,489-493/). One may distinguish schematically two kinds of effects:

- a population effect due to the extra density of states in the conduction band,
- a scattering effect; the scattering of electrons is enhanced and consequently the conduction relaxation time lowered near the resonance.

Magnetooptical /466-471/ and optical /493/ measurements provide a more direct way to observe resonant states through the occurrence of impurity transitions. For HgTe the existence of at least two resonant acceptor states at $\varepsilon_{A_0} \cong 0.7 - 0.8$ meV and $\varepsilon_{A_1} \cong 2.2$ meV (extrapolated to zero-magnetic field) seems to be firmly established now from transport and magnetooptical measurements. Their origin, however, is still a matter of debate (for a discussion see, e.g. /456,484,492/ and Sect. 2.3). Also, while there seems to be some agreement on their composition dependence for zero-gap HgCdTe (Fig. 52), the magnetic field dependence is still controversial. Assuming a linear variation of the binding energy with field B, $\varepsilon_A(B) = \varepsilon_A(0)+b$ with values for b between +0.25 meV/T /467/ and -0.2 meV/T /468/ for the A_1 resonance have been reported (see also /494/).

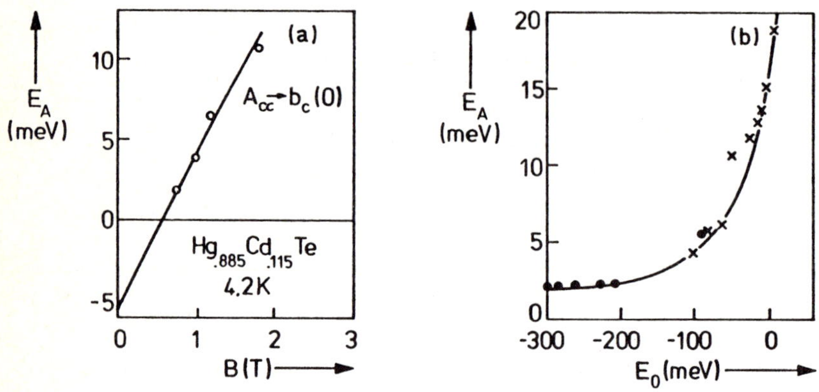

Fig. 52. (a) Impurity transition observed in the submillimetre spectrum. (b) Evolution of the acceptor E_A binding energy with the interaction gap obtained from magnetooptical data (o) and transport (x)

Also the existence and the nature of other transport anomalies and possible acceptor states is still much disputed. These include a proposed state A_2 at ε_{A_2} = 9.5 - 10.5 - 13.2 meV in HgTe /38,495/, levels at 3.1 meV /493/ or 5.4 meV /496/ in $Hg_{.85}Cd_{.15}Te$ and 0.58 and -0.05 meV in HgTe /489/.

Data for semiconducting compositions close to the SC-SM transition are scarce; a resonant acceptor level at about 9 meV has been proposed for $Hg_{.8}Cd_{.2}Te$ /45, 493/.

4. Transport Properties

In this chapter we report and discuss various transport properties of the $Hg_{1-x}Cd_xTe$ alloy system. In some cases properties of pure HgTe and CdTe are also mentioned for comparison. Data on band edge characteristics such as E_0, m_0^*, m_{hh}^*, P, etc. obtained from measurements of transport properties have already been included in Chap. 3 and will not be repeated.

As the topics of the first two sections have already been reviewed elsewhere, the discussion will in places be very short and confine itself to the more recent investigations. However, we have attempted to give almost all the references available.

Various investigations have been published in recent years reporting structures and anomalies in the temperature dependence of various transport properties /396, 497-501/ mostly interpreted as due to resonant acceptor states or other impurity states or bands. The effect of ionized intrinsic point defects on the current carrier concentration in HgCdTe has been determined in /502/. Fabrication and characterization of galvanomagnetic devices such as Hall generators have been described in /503-505/.

4.1 Hall Coefficient, Intrinsic Carrier Density

The main topic of this section has been the subject of a large number of experimental investigations /1,20,30,38,40,44,45,53,73,77-79,94,112,127,197-215/ and has already been reviewed in earlier publications (see, for example /2,3,4/). So we may restrict ourselves to a brief report of the main results and a short discussion of some peculiarities.

SCHMIT /199/ has calculated the intrinsic carrier concentration n_i by determining the value of the reduced Fermi energy for which the electron concentration n from the Kane-model is equal to the hole concentration p, thus yielding $n_i = n = p$. The electron concentration n has been obtained, taking into account the effects of non parabolic bands and degeneracy, from the relation

$$n = \frac{3}{4\pi^2} \left(\frac{3}{2}\right)^{1/2} \left(\frac{k_B}{P}\right)^3 T^3 \int_0^\infty \frac{X^{1/2}(X + X_g)^{1/2} (2X + X_g)\, dX}{1 + \exp(X - \eta)} \tag{46}$$

with

$$X = \left\{ - E_o + \left[E_o^2 + \frac{1}{3} (8P^2\underline{k}^2)\right]^{1/2} \right\} /2k_BT, \quad X_g = 1/\beta = E_o/k_BT, \quad \eta = E_F/k_BT$$

and P being the momentum matrix element. In the calculations it has been assumed that the Kane model, strictly valid only at 0 K, still applies at higher temperatures. The variation of E_o with x and T has been taken from /102/, values $P = 9\cdot10^{-6}$ eVm and $m^*_{hh} = 0.55$ m have been used.

The density of free holes in the valence band has been taken from

$$p = 4\pi \left(\frac{2k_B}{h^2}\right)^{3/2} m^{*\,3/2}_{hh}\, T^{3/2}\, F_{1/2}\left[-(\eta + X_g)\right] \tag{47}$$

assuming a parabolic shape of the heavy hole Γ_8 band ($F_{1/2}$ is the Fermi-Dirac integral). The calculated temperature dependence of the intrinsic carrier concentration of $Hg_{1-x}Cd_xTe$ for various x values is shown in Fig. 53. The temperature dependence of a parabolic equivalent effective mass ratio (an effective mass which is obtained assuming a parabolic band shape and fitting the resulting values of n_i to the ones given by the above described procedure) is given in Fig. 54, the reduced Fermi energy versus T of $Hg_{1-x}Cd_xTe$ with 10^{21} m^{-3} free electrons added to the conduction band is shown in Fig. 55. Some uncertainties in the curve presented above arise from uncertainties in the values used for the parameter P and m^*_{hh}: taking a value of m^*_{hh} = 0.3 m, for example, would reduce all n_i curves by a factor of about 0.65, leaving the shape nearly unchanged.

It was pointed out by LEONARD et al. /200/ that the integral form for the electron density used by SCHMIT is valid only under the condition $\Delta > E_0$, which is not true for higher values of x and T. Numerical results showing the x dependence of n at 80 K under the assumptions $\Delta > E_0$, $\Delta = E_0$ and $\Delta < E_0$ are presented in Fig. 56. It is clearly demonstrated that the values obtained by SCHMIT may have a large error for large band gap compositions x.

In /2,44,78/ curves of Hall coefficient R_H versus temperature have been shown for very pure high mobility samples, having the expected shape for a narrow gap

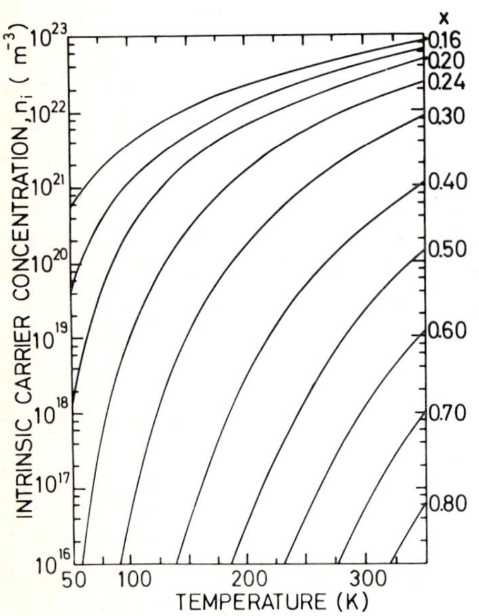

Fig. 53. Calculated temperature dependence of the intrinsic carrier concentration of $Hg_{1-x}Cd_xTe$ for $0.16 \leq x \leq 0.80$ /199/

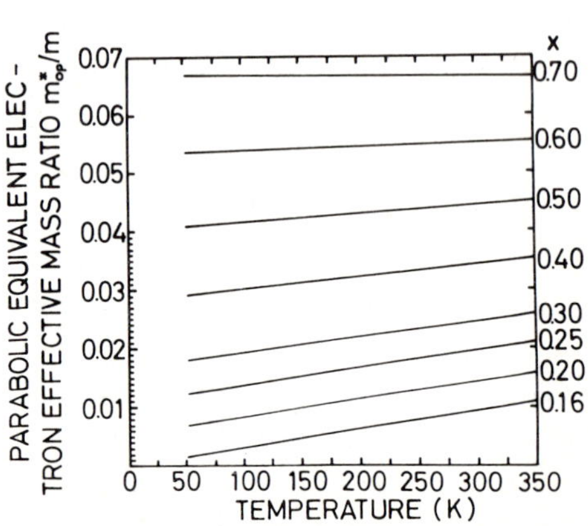

Fig. 54. Temperature dependence of the electron effective mass of $Hg_{1-x}Cd_xTe$ calculated from n_i /199/

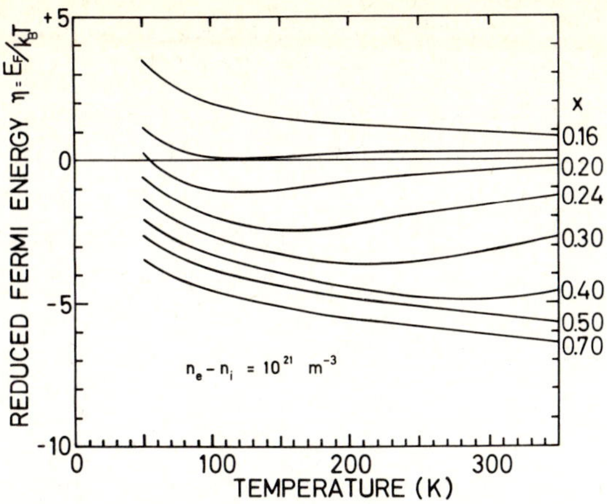

Fig. 55. Temperature dependence of the reduced Fermi energy of $Hg_{1-x}Cd_xTe$ with 10^{21} free electrons/m^3 added to the conduction band /199/

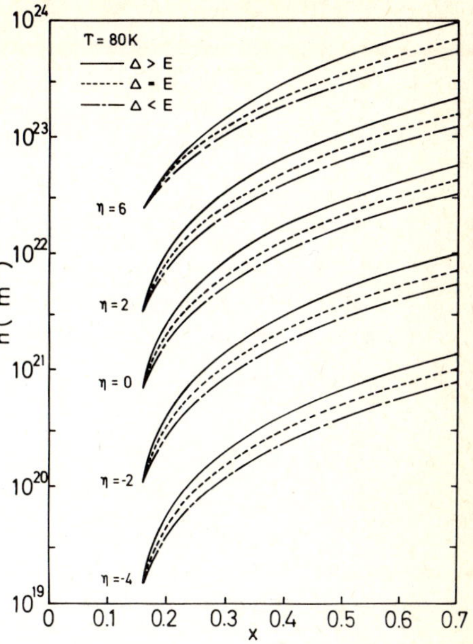

Fig. 56. Carrier density versus composition as a function of the spin-orbit splitting energy for different reduced Fermi levels /200/

InSb-like semiconductor. These samples have been n type in the extrinsic range at low temperatures and remain n type in the intrinsic range because the electron mobility is greater than the hole mobility by a factor of about 200 (at 77 K). No carrier freeze-out into discrete impurity or donor levels has been observed even at the lowest temperatures (40 mK in /44/), which may be explained by the overlap of the electronic wave functions according to (1), expected even at small impurity densities of $\sim 10^{20}$ m^{-3}. However, in high magnetic fields the wave functions become compressed and the ionization energy is increased, so, at least qualitatively, a magnetic freeze out is expected /216/. In high-mobility samples no magnetic freeze-out has been observed in fields up to 5 T (1.4 K) either /44/. DORNHAUS et al. /44/ proposed to take the extrinsic electrons to be due to Te vacancies. Similarly to PbTe /46,217,218/ (Sect. 4.2), the Te vacancy is assumed not to have a bound state.

A number of investigations have been published showing quite a complex behaviour in the temperature dependence of the Hall coefficient. The effects of doping on the electrical properties of HgTe /213/ and $Hg_{.9}Cd_{.1}Te$ /20/ have been measured (see Sect. 2.3.2). The influence of sample preparation and annealing procedures (Figs. 57 and 58) has been investigated /205,211,212,214/ and measurements have been interpreted assuming mixed conduction /205/ or even electronic conduction in the valence band /211,212/. Different models of donor and acceptor levels have been proposed /204,53,109/ to explain the observed phenomena (Fig. 59, see also Sect. 2.3.1).

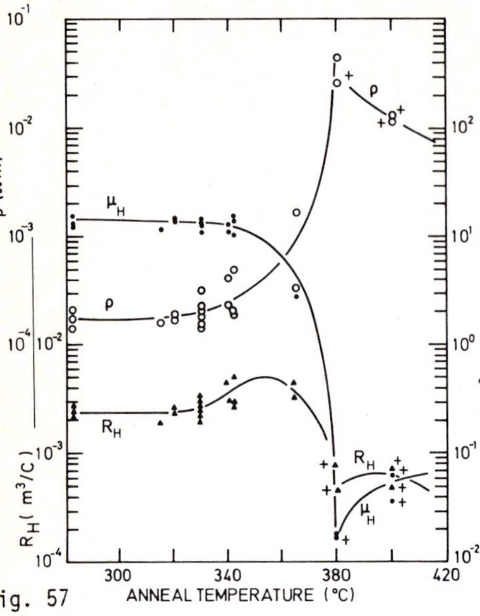

Fig. 57

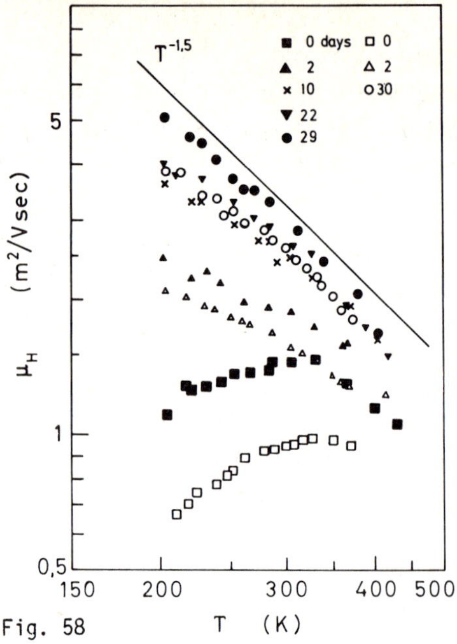

Fig. 58

Fig. 57. Hall coefficient, resistivity and Hall mobility versus annealing temperature for samples from one $Hg_{.730}Cd_{210}Te$ ingot /205/

Fig. 58. Hall mobility versus temperature for HgTe samples with different annealing times
(Crystal 1: ■ not annealed, ▲ 2 days annealed, x 10 days, ▼ 22 days, ● 39 days; Crystal 2: □ not annealed, Δ 2 days, o 30 days) /214/

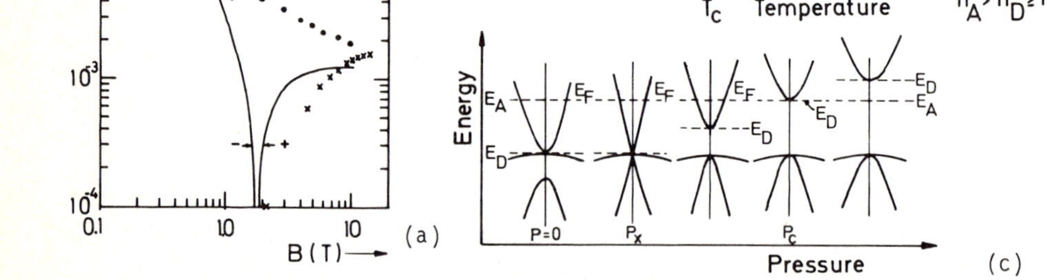

Fig. 59. (a) Hall coefficient as a function of magnetic field at 77 K and 4.2 K. Solid line - fit with a two-band expression for the Hall coefficient at 77 K (HOWARTH, JONES, PUTLEY, Proc. Phys. Soc. 70, 124 (1957)) /204/. (b) Schematic diagram of proposed donor and acceptor levels /204/ versus temperature. (c) Schematic diagram of proposed donor and acceptor levels versus pressure /40/

The temperature dependent behaviour of R_H in p-type samples has been studied in /204,205,210,40,127,198/ as well as its pressure dependence /53,109,154/. Whereas in some investigations, showing several changes of sign in the $R_H(T)$ curves, surface effects have obviously influenced the measurements /210/, other curves show the shape expected for a p-type InSb-like sample with R_H positive in the low temperature, extrinsic range and, because of the higher mobility of the electrons, reversing sign to become negative in the intrinsic range /127,198/. However discrepancies between the calculated and measured intrinsic carrier concentrations have been reported (**Fig. 60**), /198/. Whereas in earlier experimental investigations no evidence for carrier freeze-out even in p-type samples with a relative high valence band mass was reported, ELLIOTT et al. /40/ found a carrier freeze-out in p-type samples with x in the range 0.21 - 0.50 (**Fig. 61**) and deduced acceptor energies of the order of 20 meV in agreement with comparable energy levels observed in photoluminescence measurements (Sect. 2.3.1).

The pressure dependence of the carrier concentration in p-type samples of different compositions has been reported in /53,109,154/ (Sect. 2.3.1), the effect of electron irradiation on the electrical properties in /49/.

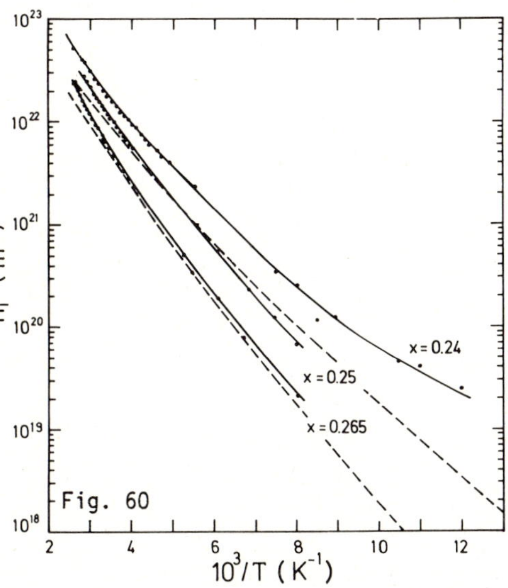

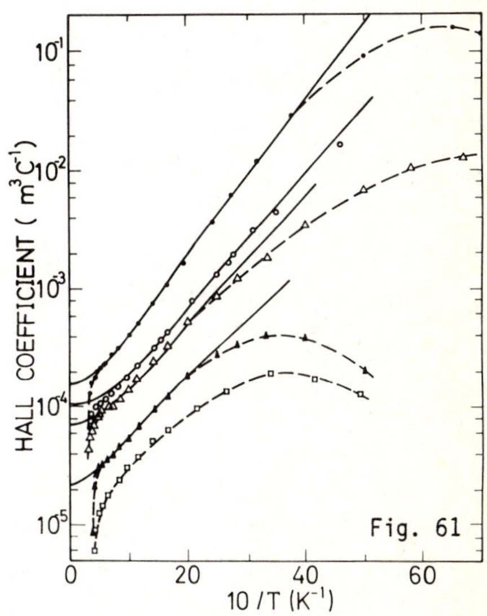

Fig. 60. Intrinsic concentration as a function of reciprocal temperature for p type $Hg_{1-x}Cd_xTe$ samples with x = 0.24, 0.25, and 0.265. Dashed lines - calculated curves for the sample with x = 0.24 assuming $E_0(x,T)$ following the relation of WILEY and DEXTER /54/ (lower curve) and SCOTT (upper curve) and m^*_{hh} = 0.3 m /198/

Fig. 61. Hall coefficient as a function of reciprocal temperature for p type $Hg_{1-x}Cd_xTe$ samples of different compositions. Solid lines show computed fits to the freeze-out region /40/

The magnetic field and temperature dependence of the Hall coefficient and the conductivity of HgTe in the temperature range 4.2 to 300 K and in magnetic fields up to 3 T /208/ (Fig. 62) or 6 T /209/ have been measured. It may be described using a model involving three types of carriers: high- ($\mu = 25 m^2/Vs$, $n \approx 2 \cdot 10^{21} m^{-3}$ at 4.2 K) and low- ($\mu = 0.1 m^2/Vs$, $n \approx 10^{23}$-$10^{24} m^{-3}$ at 4.2 K) mobility electrons and low-mobility holes. Low-mobility electrons have been observed at low temperatures (< 30 K) and strong magnetic fields. The conductivity resulting from these carriers exhibits a very weak temperature and magnetic field dependence and may be ascribed to resonant acceptor states /208/ (For further information on the problem - the influence of resonant states, impurity band or electrons ascribed to a valence band on conductivity - see references cited in /208/, see also STANKIEWICZ and GIRIAT in Phys. Rev. B 13, 665 (1976)).

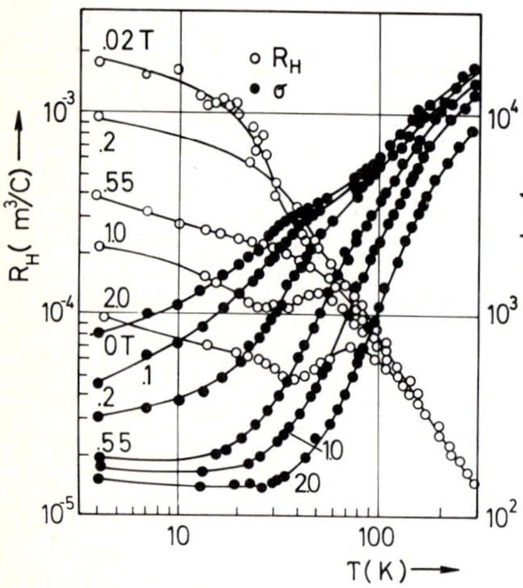

Fig. 62. Hall coefficient and conductivity versus temperature and magnetic field for HgTe /208/

The anomalous Hall effect and other spin-dependent transport properties have attracted considerable theoretical and experimental interest in recent years (see for example /219,220/). A material ideally suited for investigations on these effects should be $Hg_{1-x}Cd_xTe$ because of its very large spin-orbit interaction. Some preliminary investigations have been reported in /221/.

4.2 Carrier Mobility, Scattering Mechanisms

A large amount of experimental and theoretical work has been devoted to the elucidation of the dominant scattering mechanisms in the alloy system $Hg_{1-x}Cd_xTe$ for various x values and different temperature regions /38,44,78,79,94,111,116,152,153, 202,204,222-250/. While it is believed that ionized impurity scattering controls

the mobility at low temperatures /111/ (in very pure, high mobility samples, however, this may not be true /44/), several different mechanisms have been proposed for the principal scattering mechanisms at ambient and high temperatures /231,234/.
As was pointed out by ELLIOTT and SPAIN /234/, it is difficult to obtain reliable values for the exponent in temperature of the mobility, since measurements at high temperatures are restricted by evaporation of mercury and at low temperatures by the onset of extrinsic phenomena in p-type samples and ionized impurity scattering in n-type samples. The analysis of the temperature dependence on the other hand has to take into account several peculiarities:
(i) Especially at the small gap end of the concentration range x ($x \lesssim 0.3$) the non-parabolic structure of the conduction band strongly affects the transport properties not only via the energy-dependent effective mass and hence the density of states but also through the electronic wave-function admixture and hence the scattering probability /222-225,231/.
(ii) In the semimetallic (gapless) region of compositions the electron-electron interaction seems to play an important role in transport properties /225,226,235, 246,248-250/.
(iii) The strong positive temperature coefficient for $x < 0.5$, in contrast to other zincblende materials, leads to a pronounced temperature dependence of the band-edge effective mass.
(iv) $Hg_{1-x}Cd_xTe$ shows a two-mode behaviour (see Chap. 5), therefore both longitudinal optical (LO) phonon modes have to be included in the scattering probabilities /229,231,238/. Whereas SCOTT /229/ in his analysis only averaged the reciprocal mobilities $1/\mu_{op} = x/\mu(CdTe) + (1-x)/\mu(HgTe)$, NIMTZ et al. /238/ modified the expressions for polar-optical phonon scattering given in /222,223/ to include the two-mode behaviour. They calculated the individual mode strengths S_j of the HgTe- and CdTe-like phonon mode from the modes of the pure compounds according to

$$S_1(x) = (1-x)\, S_{HgTe} \text{ and } S_2(x) = x\, S_{CdTe} \tag{48}$$

Using

$$E_o^j = m^* e\, S_j(x)/\omega_{LO}^j \quad ; \quad j = 1,2$$

which characterizes the polar electron phonon coupling for each phonon mode, the scattering rates of both phonon modes were calculated and added with the same weight.
NAG /230/ has presented an iteration method to solve the Boltzmann equation also in the case of two-phonon mode scattering. He has used a modified collision operator L of the form:

$$\begin{aligned} L\Phi = {} & S_o(E)\Phi(E) - S_{1a}(E)\Phi(E-k_B\Theta_1) - S_{1e}(E)\Phi(E+k_B\Theta_1) \\ & - S_{2a}(E)\Phi(E-k_B\Theta_2) - S_{2e}(E)\Phi(E+k_B\Theta_2) \end{aligned} \tag{49}$$

where Θ_1, Θ_2 are the corresponding Debye temperatures for the two phonons, S_e and S_a denote the inscattering rate due to emission and absorption, respectively, S_o is the total outscattering and Φ is the perturbation of the distribution function f_o. However, in calculating the polar-optical scattering rate /231/ two mode strengths have been used with $S_1/S_2 = (1-x)/x$ which is not justified in $Hg_{1-x}Cd_xTe$ (Sect. 5.2 and Fig. 108).

(v) At intermediate x values alloy scattering /232/ seems to contribute significantly to the total scattering rate /231/.

(vi) Experiments on oscillatory photoelectric effects in $Hg_{.76}Cd_{.24}Te$ /236/ indicate that electrons may be coupled to the transverse-optical vibration of HgTe, since these vibrations have a polar character in the mixed crystal /231/.

Experimental investigations on electrical properties of $Hg_{1-x}Cd_xTe$ have usually been restricted to a fairly small range in the composition with $x \leq 0.20$. Early measurements in this range /203,78,152/ have been interpreted by assuming that acoustic mode scattering is predominant at room temperature. Most of the other workers, however, concluded that the scattering arises predominantly from optical phonons. SCOTT /229/ has measured the temperature dependence of the Hall mobility in $Hg_{1-x}Cd_xTe$ for a variety of compositions ($0 \leq x \leq 0.6$). His results are shown in Fig. 63. He has estimated the contributions of various scattering mechanisms to the electron mobility and pointed out that the measured mobilities are at least 100 times too small to be accounted for by acoustic deformation potential or piezoelectric scattering using realistic values for the appropriate constants. Although the estimates are somewhat preliminary as they did not take into account some of the features described above, they are not expected to give values more than one order of magnitude too high /234/ (Fig. 64). SCOTT has also pointed out that at higher temperatures electron-hole scattering becomes increasingly important as the number of carriers increases, but does not seem to be the dominant scattering mechanism. His conclusions are confirmed by recent calculations of CHATTOPADHYAY and NAG /231/. They emphasized, however, that discrepancies remain between calculations of the room temperature mobility taking into account scattering by LO phonons and additionally by deformation-potential acoustic phonons, ionized impurities and heavy holes. These discrepancies could be partly removed allowing for the contributions described under (v) and (vi). But it should be mentioned that the values ε_∞ and ε_o, determining the polar optical coupling constant $\propto (1/\varepsilon_\infty - 1/\varepsilon_o)$ and hence the mobility, are not known very precisely (Sect. 5.2, Fig. 107). So only a comparison of the temperature dependence of the calculated and measured mobilities seems meaningful, not a comparison between absolute values.

LONG /111/ has calculated the ionized impurity scattering mobility of conduction electrons in n type $Hg_{1-x}Cd_xTe$ at 4.2 K for x values below 0.3 under the condition of high degeneracy. His results are shown in Fig. 65. He concludes that ionized impurity scattering is dominant at helium temperatures.

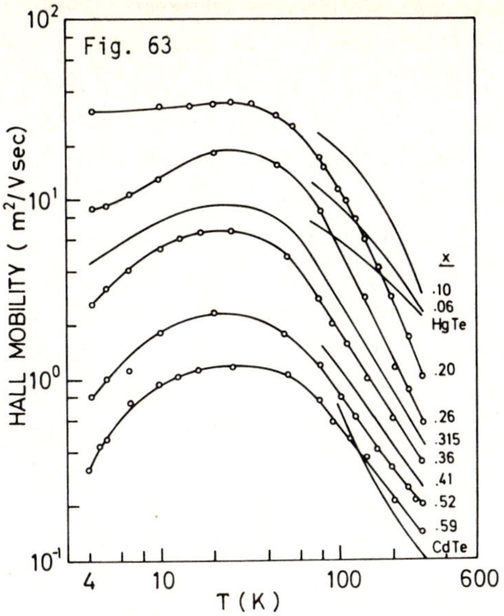

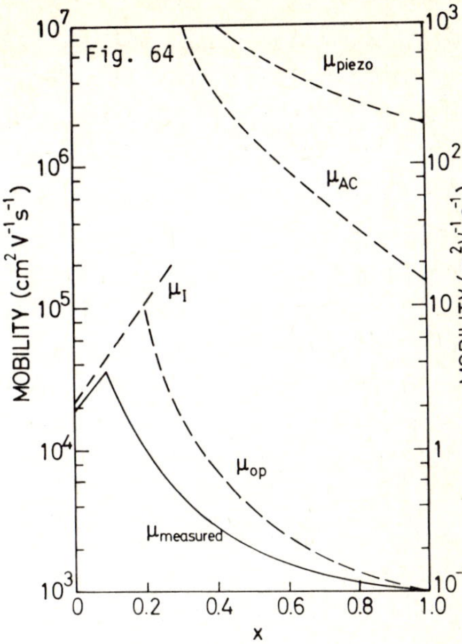

Fig. 63. Temperature dependence of Hall mobility in $Hg_{1-x}Cd_xTe$ for a variety of compositions /229/

Fig. 64. Room temperature mobility in $Hg_{1-x}Cd_xTe$, $\mu_{meas.}$ (SCOTT 1972 /229/), as a function of x compared to the calculated mobility assuming different scattering mechanisms /229/

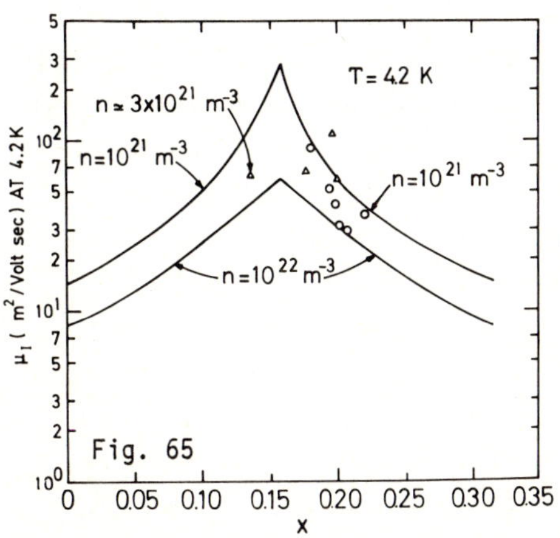

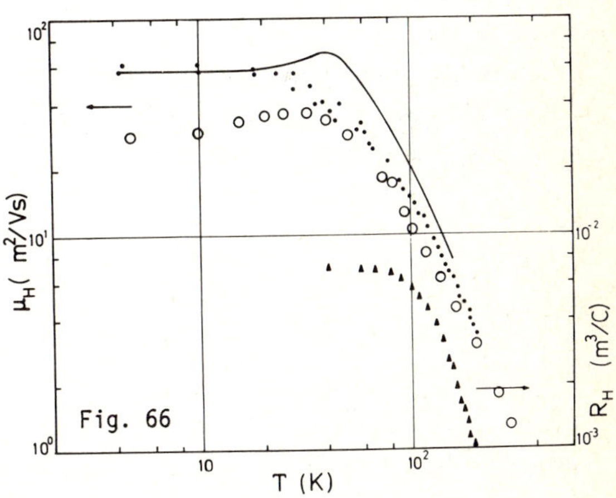

Fig. 65. Ionized impurity scattering mobility versus alloy composition x in $Hg_{1-x}Cd_xTe$ at 4.2 K. The curves are theoretical and the points represent experimental data ($0.5 \cdot 10^{21}\ m^{-1} \leq n \leq 2 \cdot 10^{21}\ m^{-3}$) /111/

Fig. 66. Temperature dependence of Hall mobility in $Hg_{.8}Cd_{.2}Te$ (● exp. points from /238/, o from /111/, — calculated /238/) and low field Hall coefficient (▲ /238/)

Calculations on the dependence of the mobility on temperature, including ionized impurity scattering and polar optical phonon scattering, and taking the number of impurities equal to the free carrier density $n = 9\cdot10^{20}m^{-3}$ /44,238/ are shown in Fig. 66. The calculated data show a maximum in the range 40 - 50 K, which results from the decreasing scattering rate of ionized impurities with increasing T and increasing rate of polar optical scattering with T. Such a maximum has been found in many experimental investigations with a somewhat lower mobility, as shown by the typical curve of LONG, but not in samples with a very high mobility /44,238/. DORNHAUS et al. /44/ have shown that in these samples neither thermal (at temperatures down to 50 mK) nor magnetic (magnetic fields up to 5 T) freeze-out of carriers takes place. From the observed discrepancies between theoretical and experimental values (Fig. 66) they concluded that in these high-purity, high-mobility samples ionized impurity scattering is not the dominant scattering process, which is supported by hot carrier experiments carried out with the same samples at 4.2 K (see Sect. 4.5). They proposed a model of Te vacancies (similar to that on PbTe /46,217, 218/) without a net charge. For an essentially repulsive vacancy potential the total scattering cross section, similar to a neutral hydrogen-like impurity, is expected to be proportional to 1/v, thus leading to a velocity-independent momentum relaxation time τ /240-243/. Such a velocity-independent scattering mechanism would explain the low temperature mobility which has been observed to be independent of temperature and electric field.

The dependence of the 4.2 K mobility on the doping level in $Hg_{1-x}Cd_xTe$ has been investigated by SCOTT /229/. He assumed that there is no compensation (all the published data seem to indicate that the degree of compensation is usually very low in $Hg_{1-x}Cd_xTe$ /234/) and that the donors are always fully ionized. Treating the screened ions as independent scattering centres (which does not seem justified for material with $x < 0.2$ /229/), mobilities as a function of donor concentration have been calculated and compared with experimental data (Fig. 67).

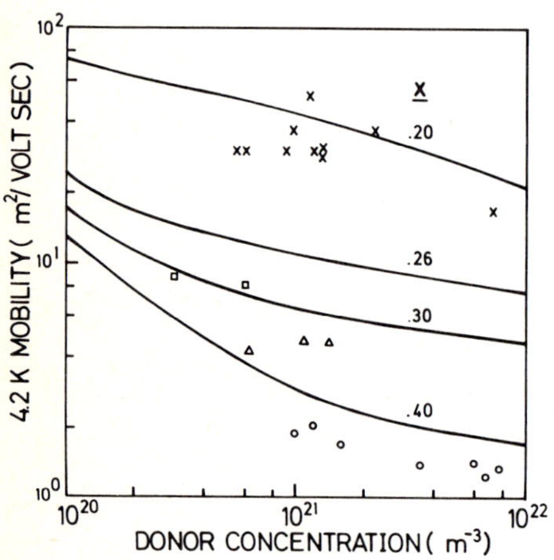

Fig. 67. Dependence of 4.2 K mobility on the doping level in $Hg_{1-x}Cd_xTe$ for a few compositions: x 0.195 - 0.206, □ 0.26 - 0.27, Δ 0.30 - 0.32, o 0.38 - 0.40 /229/

For the small and negative band-gap region LIU and BRUST /249/ have suggested significant changes in the dielectric constant with carrier concentration, which lead to variations of the mobility with band gap at low carrier concentrations /235/, (Fig. 68). This effect was not taken into account by LONG, but seems to be confirmed in a number of investigations /153,246,38/ especially in those on the pressure dependence of the mobility.

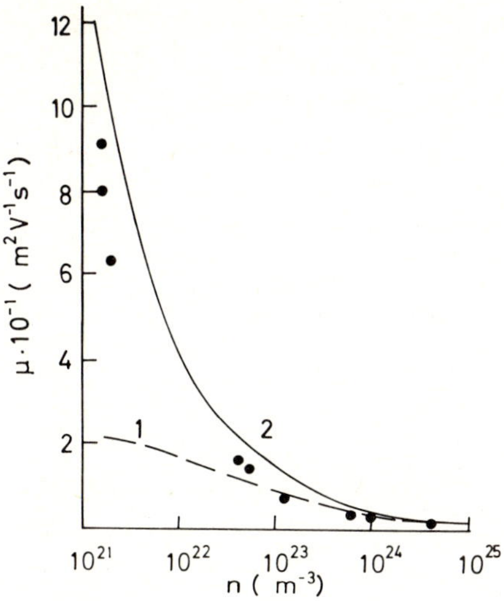

Fig. 68. Dependence of the electron mobility on the electron density in HgTe at 4.2 K. Dots - experimental dependence, dashed line - calculated dependence $\varepsilon(n)$ neglected, solid line - $\varepsilon(n)$ taken into account

In spite of the large efforts in recent years to understand the physics of narrow-gap and zero-gap semiconductors, the quantitative description of the basic galvanomagnetic and thermomagnetic properties of HgCdTe is still incomplete, and there are still conflicting conclusions about the dominant scattering mechanisms. Theoretically this is due to the fact that a complete calculation of electron scattering in these materials has to include peculiarities of the nonparabolic bands both in the density of states and the structure of the wavefunctions, the temperature dependence of the band parameters, the temperature and momentum transfer dependence of the dielectric function and several other problems (see Sect. 4.2).

Experimentally the incomplete quantitative understanding is, at least partially, due to the still relatively poor knowledge and control of native defects in the materials.

The most complete theoretical description has probably been given in a series of papers by SZYMANSKA et al. /506,507/ (see references therein), and results on ionized-impurity scattering, polar optical phonon scattering (see also /508/). acoustic-phonon scattering (see /509,510/), and electron-hole scattering have recently been shortly reviewed in /450/. Several other scattering mechanisms and their possible influence on the mobility of HgCdTe under certain conditions have also been investigated; these include electron-electron scattering /511/, disorder scattering /512,513/, spin-flip electron scattering caused by acoustic phonons through the deformation-potential interaction, which is possible because of the spin-mixing (Bloch functions for conduction electrons are mixtures of p- and s-like components) /514,515/, resonant scattering /516,517/, and other effects /518,519/.

Very detailed experimental and theoretical investigations of the electron mobility in HgCdTe with $0 < x < 0.33$, 10^{21} $m^{-3} < n < 10^{24}$ m^{-3}, 4.2 K < T < 300 K have been reported recently /520,521/ (see also Figs. 69-71 and Table 24). Data on hole mobility in HgCdTe ($0.2 \leq x \leq 0.4$) are given in /396,522/ (see also Fig. 72). A theory for low-temperature electron mobility in $Hg_{1-x}Cd_xTe$ using a variational method to solve the Boltzmann equation was given by MEYER and BARTOLI /523/. The scattering mechanisms considered in that study include ionized and neutral impurity scattering, acoustic, piezo-electric and optical-phonon scattering, and disorder scattering. It has been found that low-temperature mobility calculations require some type of correction for multi-ion effects before they can be used to accurately predict experimental mobilities.

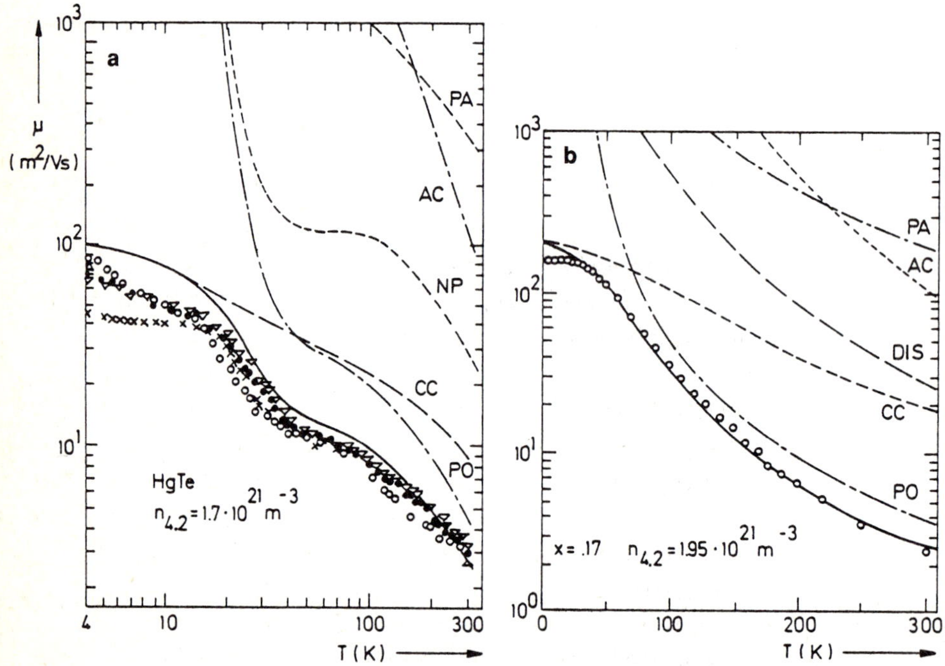

Fig. 69 a,b. Electron mobility vs temperature for undoped HgTe (a) and lightly doped $Hg_{.83}Cd_{.17}Te$ (b) with experimental points from various investigations. Dashed lines represent contributions from charged center (CC), polar (PO), nonpolar (NP), acoustic (AC), disorder (DIS) and piezo-acoustic (PA) scattering modes; the solid line is the theoretical result for mixed scattering modes /520/

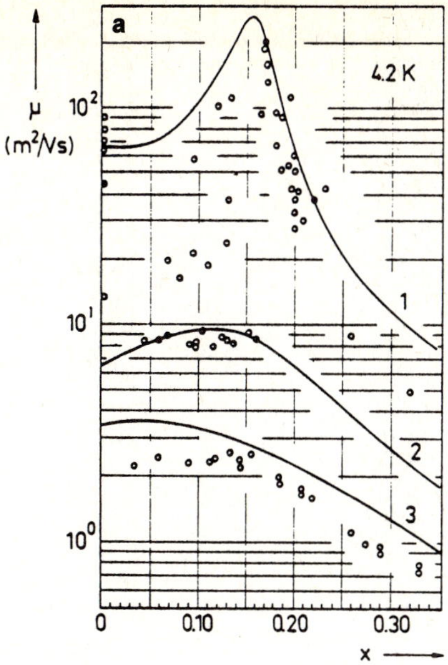

Fig. 70a

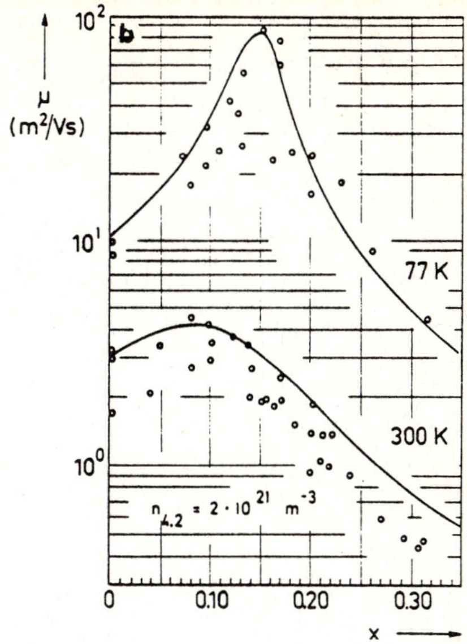

Fig. 70b

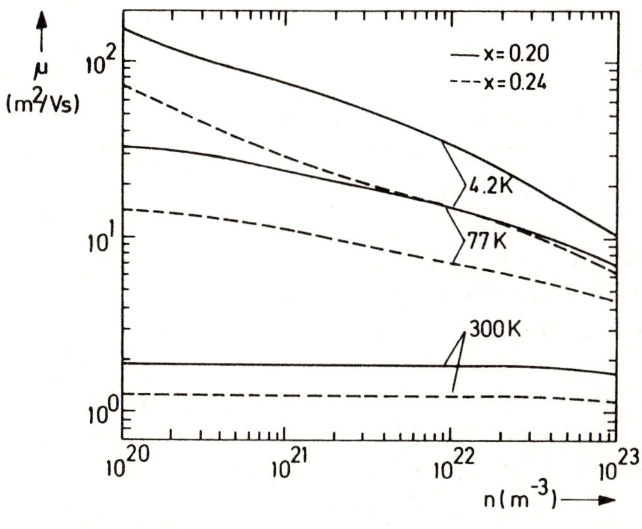

Fig. 71

Fig. 70a,b. Theoretical and experimental composition dependence of electron mobility in HgCdTe mixed crystals at 4.2 K (a), 77 and 300 K (b). Curves 1-3 in (a) are calculated with electron concentrations at 4.2 K of $2 \cdot 10^{21}$, $2 \cdot 10^{23}$ and $9 \cdot 10^{23}$ m^{-3}, respectively. Experimental points for approximately corresponding concentrations are from various investigations /520/

Fig. 71. Calculated electron mobility vs electron concentration at 4.2, 77 and 300 K for HgCdTe with x = 0.20 and x = 0.24 /520/

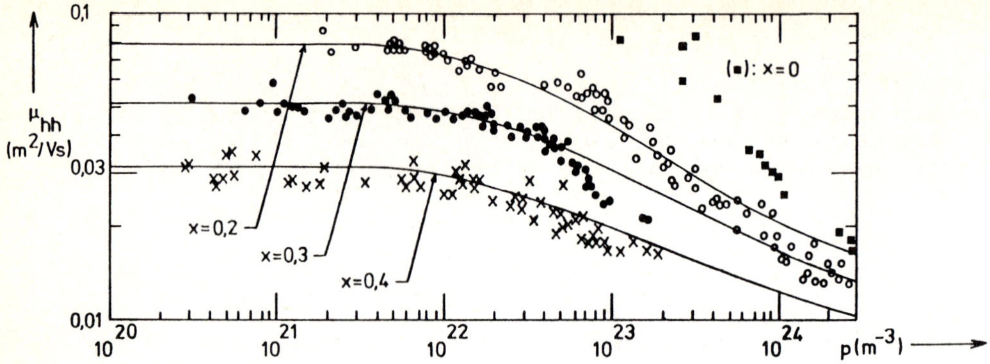

Fig. 72. Hole mobility μ_{hh} as a function of hole concentration for $Hg_{1-x}Cd_xTe$ (x = 0, 0.2, 0.3, 0.4) at 77 K. The solid lines represent calculated data considering combined lattice and ionized impurity (doubly ionized) scattering /522/

Table 24. Parameters of HgCdTe necessary for mobility calculations /520/

Material parameter	Symbol	Value, composition dependence	Units
Higher band parameters	A'	-1.0	
	M	-4.4	
	L'	-5.0	
	L'-M-N'	6.0	
Γ_8 bands overlap	$\Delta\varepsilon$	1.0	meV
Background dielectric constants	ε_∞	10.4 - 3x	
Lattice dielectric constants	ε_{L1}	4.7·(1-x)	
	ε_{L2}	3.1x	
Transverse optical phonon energy	$\hbar\omega_{TO1}$	14.5 + 1.5x	meV
	$\hbar\omega_{TO2}$	17.3 + 1.1·(1-x)	meV
Averaged sound velocities	$v_\parallel$	$2.8\cdot10^3$	m/s
	$v_\perp$	$1.6\cdot10^3$	m/s
Deformation potential constants	b	-1.3	eV
	d	-2.2	eV $d = \sqrt{3}\,b$
	C+a	-5.6	eV
	a	2.7	eV
Non-polar deformation potential constants	d_0	29.8	eV
	d_1	5.0	eV
Alloy potential matrix element	V	$9\cdot10^{-29}$	eV m^3
Piezo-electric tensor component	e_{14}	$1.9\cdot10^4$	$N^{1/2}$/m

The main conclusions which can be derived from the references given above may be summarized as follows:

- in pure samples above 50 K polar optical-phonon scattering prevails over heavy-hole and ionized donor scattering;
- non-polar optical-phonon scattering gives an approximately 10% contribution in HgTe above 50 K;
- in samples with $x > 0.1$, disorder scattering has to be considered; its magnitude, however, seems to be smaller than previously suggested /231/;
- disorder scattering gives a contribution only for strongly doped samples;
- the previously suggested polar coupling to transverse phonons /231,450/ remains a matter of debate;
- the influence of acceptor levels below $T \cong 20$ K in zero-gap HgCdTe has been firmly established; while even "anomalies" in the temperature dependence of the mobility around 30 K are a result of the intrinsic scattering mechanisms, an additional acceptor contribution seems questionable;
- besides acceptors, two other types of defects seem to occur which lower the mobility in HgCdTe: one type is characterized by a momentum-independent relaxation time and has already been described in /44,45/, the other type seems to be present in samples investigated in /229/ and is characterized by a relaxation time which decreases with electron momentum.

Transport properties of photo-excited carriers in $Hg_{.785}Cd_{.215}Te$ have been investigated in /524,525a-c/. A CO_2 laser was used to produce high-electron plasma densities. The mobility exhibited a maximum as a function of photo-excited carrier density and decreased, at high excitation levels, monotonically with increasing density. In order to explain this behaviour, the effects of electron-hole scattering, dynamic screening by photo-excited electrons and holes, multiple scattering and disorder scattering had to be considered.

4.3 Magnetoresistance

In this section we deal with experimental results on the magnetoresistance of $Hg_{1-x}Cd_xTe$ samples /251-253/ and discuss them in the light of some very general theoretical investigations on galvanomagnetic phenomena /255-272/. As measurements of the Hall coefficient and the mobility have already been discussed in the previous sections and as Sect. 4.4 is dedicated to magnetic quantum effects, we will restrict the discussion to measurements and calculations of the longitudinal (LMR) and transverse magnetoresistance (TMR) omitting the region of SdH oscillations at low temperature.

Measurements of the magnetoresistance should be interesting from a theoretical point of view since, as a consequence of the very low effective mass in $Hg_{1-x}Cd_xTe$ with $x \leq 0.2$, the extreme quantum limit can be easily reached with the normally available magnetic field strengths.

However, in comparing the experimental data with theoretical calculations some caution is needed for the following reasons:

(i) Experimentally the main problem is the homogeneity needed in experiments on the magnetoresistance in high magnetic fields. It has been pointed out by various authors that inhomogeneities may have a considerable influence on the behaviour in magnetic fields. Additionally contact effects have to be carefully avoided.
It has been pointed out in /251/ that the mentioned effects possibly account for the anomalous behaviour, such as oscillations in the temperature dependence of the Hall coefficient, more than one change of sign, etc., which is observed in some investigations.
(ii) Theoretically one has to consider the non-parabolic dispersion law and resulting effects:

- The effective mass of the electron in non-parabolic semiconductors in the presence of a quantizing magnetic field is given by

$$m^*(B) = m_o^* \, a_o \left(1 + \frac{2X_z}{X_g \, a_o^2}\right) \qquad (50)$$

with $X_z = \frac{E_z}{k_B T} = \frac{1}{k_B T}\left(-\frac{E_o}{2} + \frac{1}{2}E_o a_o + \frac{\hbar^2 k_z^2}{2m_o^* a_o}\right)$,

$X_g = 1/\beta = E_o/k_B T$ and $a_o = \left(1 + \frac{2\hbar\omega_c}{E_o}\right)^{1/2}$ /264/

thus leading to a relative change in the plasma frequency

$$\frac{\omega_p(B)}{\omega_p(o)} = \frac{1}{a_o^{1/2}} \; .$$

- The dependence of electronic mass on magnetic field alone gives rise to a non-zero LMR which increases with decreasing X_g and increasing $\hbar\omega_c/k_B T$ /270/.
- The degeneracy condition may be appreciably altered with increasing magnetic field.

The band non-parabolicity has been taken into account only in a few theoretical investigations, see e.g. /260,261,271/ and references cited therein.

Calculations of the behaviour of the magnetoresistance in the extreme quantum limit have been made by various authors, some results are given in Table 25 and compared with experimental results on $Hg_{1-x}Cd_xTe$ and InSb.

A negative longitudinal magnetoresistance (NLM) has been theoretically predicted for ionized impurity scattering under degenerate /266/ and non-degenerate /265/ conditions as well as for scattering by acoustic phonons /267,268/. Experimentally a NLM has been found in InSb /254/ and in $Hg_{1-x}Cd_xTe$ under different conditions /251,253/ (Fig. 73). However, to obtain a quantitative agreement further investigations are needed.

Table 25. Behaviour of the resistivity in the extreme quantum limit ($\rho \propto B^s$)

Ref.	Type of calculation		Scattering mech.	Statistics	s Trans.	s Long.
/257/	lowest order Born approximation parabolic bands	exact power law	point impur.	degenerate	5	2
			screened ionized imp.	"	3	1
			high temp. optical	"	4	1
/266/	"	no exact power law $2 \lesssim \omega_c/E_F \lesssim 4$	screened ionized imp.	"	-	2.2
		$\omega_c/E_F \gtrsim 8$			-	1
/271/	parabolic bands		polar optical phonon	"	-	0.6
	strong nonparabolic		(inelast. sca.)		-	0.9
/259/	self consistent theory, parabolic bands	no exact power law $2.5 \lesssim \omega_c/E_F \lesssim 4$	point impur.	"	2.9-3.6	1.0-1.5
/257/	lowest order Born approximation		high temp. optical	nondegenerate	1	0
/255/			polar optical phonon	"	1	-

Ref.	Material	Statistics	s Transv.	s Longit.
/251/	$Hg_{.8}Cd_{.2}Te$	degenerate	2.4	1.2
/252/	$Hg_{.8}Cd_{.2}Te$	"	-	1.27
Bliek et al. in /122/ p. 710	InSb	"	3.3	2.2
/251/	$Hg_{.8}Cd_{.2}Te$	nondegenerate	1	-

It was shown by LIU and TAN /272/ that in the zero-gap region the static dielectric constant at infinite wavelength and zero temperature contains a magnetic field dependent term of the form $B^{-1/2}$. Because of this anomalous term both the LMR and TMR for a p-type sample are expected to show a different field dependence from samples in the semiconducting region.

At frequencies ω satisfying the condition

$$(\omega_p^2/\omega) >> \omega_c >> \omega,\ \tau^{-1}$$

where ω_p and ω_c are the plasma and cyclotron frequencies and τ is the effective relaxation time, the excitation of helicon waves in degenerate semiconductor plasmas becomes possible. WILEY and DEXTER /98/ (see also Sect. 4.4.1) have used microwave helicons of 23 and 70 GHz to measure the carrier densities, effective masses and mobilities of electrons in $Hg_{1-x}Cd_xTe$ for $0.135 \leq x \leq 0.203$. Results have been mentioned in Chap. 3. An experimental curve illustrating the sources of the experimental data is shown in Fig. 74.

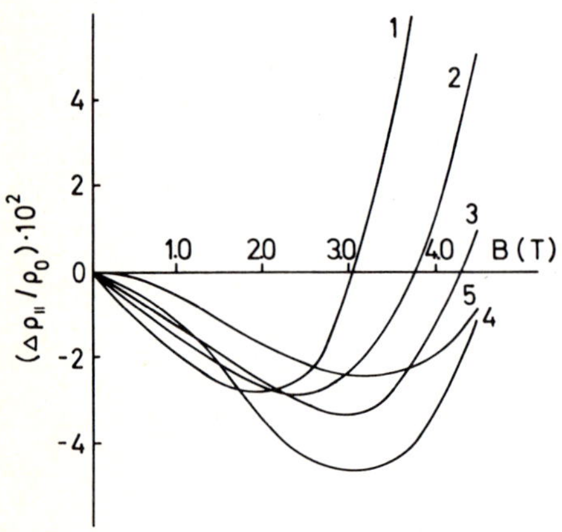

Fig. 73

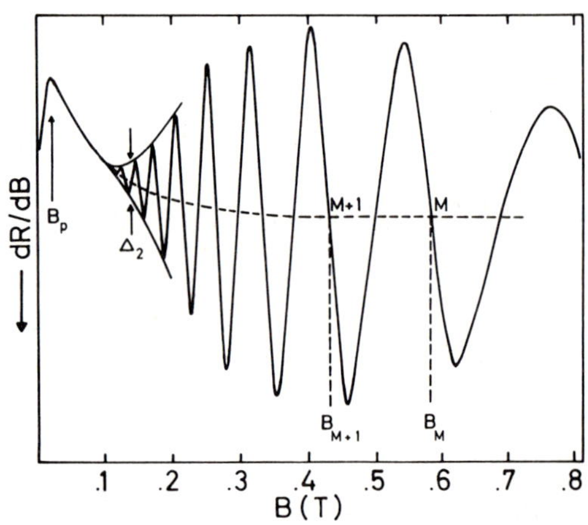

Fig. 74

Fig. 73. Dependence of the longitudinal magnetoresistance on the magnetic field applied to a sample with x = 0.18 at various temperatures T: (1) 77 K, (2) 110 K, (3) 130 K, (4) 160 K, (5) 220 K.

Fig. 74. Field derivative of the reflected microwave signal versus B (23.7 GHz, 77 K, sample thickness 0.88 mm). The low-field peak B_p, the opening part of the amplitude envelope Δ_2, and the field positions of the interference fringes B_M contain information about n, m^*, μ and E_p /98/

The influence of a non-parabolic dispersion relation on the magnetoresistance of semiconductors has been investigated both theoretically /526-529/ and experimentally /530-533/ in a number of recent papers.

In the extreme quantum limit the longitudinal and transverse magnetoresistance are found to obey asymptotic power laws /529/. The exponents in these power laws are found to be larger than in the case of parabolic energy dispersion, in reasonable agreement with experiment /532/ at 4.2 K. Screening of impurities in strong magnetic fields and its influence on the transverse conductivity have been investigated in /533/.

Samples of n-$Hg_{.8}Cd_{.2}Te$ with low carrier densities and at temperatures below ~2 K exhibit some anomalous transport properties in the extreme magnetic quantum limit which are in qualitative agreement with predictions made for strong free carrier Coulomb correlations. The Coulomb correlations are expected to produce a charge density wave transition or a Wigner condensation /534-540/. Above certain critical magnetic fields and below certain critical temperatures dependent on the composition and carrier concentration of the samples sudden pronounced changes in transverse and longitudinal magnetoresistance and in the IU characteristic have been observed. Results of Hall effect measurements are in contradiction to the "freeze-out" interpretation given by RAYMOND et al. /541-543/. The dramatic increase of resistivity which has been observed in pure undoped crystals showing a high electron mobility is caused by a drop in carrier mobility and not by a carrier freeze-out in the magnetic field /538/. The observed drop in mobility is related to a pinning of the charge density wave at lattice imperfections.

A summary of the work on the observed phase transition has recently been presented in /536/.

The effect of electron spin on the plasma frequency of non-parabolic semiconductors in quantizing magnetic fields has been investigated in /544/.

4.4 Magnetic Quantum Effects

In small-gap materials like $Hg_{1-x}Cd_xTe$ for $0 \leq x \leq 0.25$ the cyclotron frequency $\omega_c = eB/m_0^*$ is large because of the small effective mass of electrons. Hence, the range of non-quantizing magnetic fields is severely restricted and the domain of observable Landau oscillations $\hbar\omega_c < \zeta$ and $\hbar\omega_c > k_BT$ (ζ = chemical potential) and the quantum limit $\hbar\omega_c \gtrsim \zeta$, $\hbar\omega_c > k_BT$ are reached for much lower values of a magnetic field than in large-gap materials. The extreme quantum limit $\hbar\omega_c >> \zeta$ can be easily reached.

Considering for example $Hg_{.8}Cd_{.2}Te$ and some typical parameter values

$$4.2\ \text{K} : \mu = e\tau/m^* \approx 50 m^2V^{-1}s^{-1}$$

$$77\ \text{K} : \mu \approx 20 m^2V^{-1}s^{-1}$$

$$m_o^* \approx 5\cdot 10^{-3} m \qquad g^* = -164 \qquad n = 10^{21} m^{-3}$$

a rough estimation gives:

$$\hbar\omega_c > k_BT \iff mB/m_o^*T > 0.75 \iff B > 1.6\cdot 10^{-2}\,T \text{ at } 4.2\ K$$
$$B > 2.9\cdot 10^{-1}\,T \text{ at } 77\ K$$

$$\zeta \approx \hbar\omega_c \iff n^{2/3}\cdot 10^{-14}/3.18 \approx B \iff B \approx 0.31\ T$$

Additional conditions to observe quantum oscillations and a spin splitting in this oscillations are

$\omega_c\tau >> 1$ ($\mu B >> 1$) and $\mu_B|g^*|B > k_BT$ respectively. Taking the above given parameter values yields:

$$B >> \mu^{-1} \iff B >> 2\cdot 10^{-2}\,T \text{ at } 4.2\ K$$
$$B >> 5\cdot 10^{-2}\,T \text{ at } 77\ K$$
$$B > k_BT/\mu_B|g^*| \iff B > 3.8\cdot 10^{-2}\,T \text{ at } 4.2\ K.$$

In this section experimental investigations on the Shubnikov-de Haas effect /89-91, 118,273-282/, the magnetophonon effect /110,283/, cyclotron- /97,99,119,284,285/, electron-spin- /133,275,286/ and combined resonance /132,133,287/, spin-flip Raman scattering /120/, quantum oscillations in the photoconductivity /237,288/, interband magnetoreflection and -absorption /114,289/, and Auger recombination /290/ are reported.

While the dependence of the effective electron mass on the magnetic field in the extreme quantum limit had been studied earlier /270,264,526,529/, it was shown in /545/ that the effective mass in degenerate n-HgCdTe shows an oscillatory magnetic field dependence at quantizing fields with an amplitude significantly influenced by the alloy composition. The period, of course, is only dependent on the carrier concentration in this model calculation.

4.4.1 Shubnikov-de Haas Effect

The Shubnikov-de Haas effect (SdH) is a powerful tool for investigations of degenerate semiconductors and semimetals, since it leads, at least in principle, not only to effective masses of carriers but also to other band parameters of substances.

Work on SdH effect in HgTe is reported in /89,91,273,275,276,278,279,281/. These measurements have been carried out with HgTe samples in the concentration range $n = 7.4\cdot 10^{20} m^{-3} - 2.8\cdot 10^{24} m^{-3}$ in both longitudinal and transverse configurations and in steady and pulsed magnetic fields between 0.1 and 30 T. There are large discrepancies between different experimental results and disagreements in the interpretation: some authors observe a distinct spin splitting in the transverse /89,273,275, 276/ and in the longitudinal /273,275,276/ magnetoresistance, whereas others do not. In /273,275/ even a 0^- maximum in the longitudinal configuration has been observed in contrast to the usual behaviour in InSb-type semiconductors. In /89,278/ additional peaks have been found which have been attributed to SdH oscillations of a second kind of electrons /278/ or to plasma oscillations /89/ (see below). As was pointed out in /273/ the discrepancies, especially in considering the oscillations associated with the lower electron Landau levels in samples with low electron concentrations, could possibly be blamed on an incorrect symmetry of the conduction band in HgTe used in the interpretation of most experimental results.

As a consequence of degeneracy of the top of the Γ_8 valence band with the bottom of the Γ_8 conduction band, there is a mixing of the wave functions of electrons and holes near $\underline{k} = 0$. In this case the quantum energy levels in the Γ_8 bands in the presence of a magnetic field are unevenly spaced for low quantum numbers. Here the concepts of effective mass and g^* factor, adequate for describing the levels in the high quantum number limit, are meaningless /285/. At low energies, when the non-parabolicity of the conduction band of HgTe can be ignored, the Landau levels can be deduced from the Luttinger theory /159,273,285/. A comparison with results from /275,276,279/ has been given in /285/.

Investigations on the SdH effect in $Hg_{1-x}Cd_xTe$ have been reported in /90/ for $x = 0.15$,in /134/ for $x = 0.16$ ($n = 3\cdot10^{22} - 1.4\cdot10^{23}m^{-3}$, $T = 77 - 120$ K), in /118/ for $x = 0.176$, in /88,277/ for $x = 0.20$ and in /89,274/ for $x = 0, 0.07, 0.11$, 0.18 and 0.22. Experimental results on m^*, g^*, E_0, E_F, P have already been discussed in Chap. 3.

In Fig. 75 recorder traces of the transverse (TM) and longitudinal (LM) magnetoresistance voltages versus magnetic field for a sample $Hg_{.822}Cd_{.178}Te$ with $n = 3.79\cdot10^{22}m^{-3}$ are shown, in Fig. 76 longitudinal magnetoresistance curves for samples of $Hg_{1-x}Cd_xTe$ with various x-values and carrier concentrations $n \sim 10^{22}m^{-3}$ are compared /89/. Following the authors' interpretation spin splittings of the SdH oscillations are clearly observed in TM for the samples on both sides of the semimetal-semiconductor transition. In LM, however, on the semimetal side the oscillatory peaks corresponding to the spin-down level of N=0, B_0^- and the spin-up levels of N = 1, 2, 3 ..., B_N^+ are completely missing. For the semiconducting side in LM B_0^- is missing, but pair peaks for N = 1, 2, 3 ... can be observed. In contrast to the semimetal side here the B_N^+ peaks are appreciably shifted from those in TM. The experimental results are explained in terms of spin-orbit couplings. "New type oscillations" are additionally observed (dotted arrows in Figs. 75 and 76) and tentatively attributed to plasma oscillations.

There are only two experimental investigations on the SdH effect in very pure $Hg_{1-x}Cd_xTe$ ($n < 10^{21}m^{-3}$) /118,277/. In both cases anomalies occur: In /118/ the

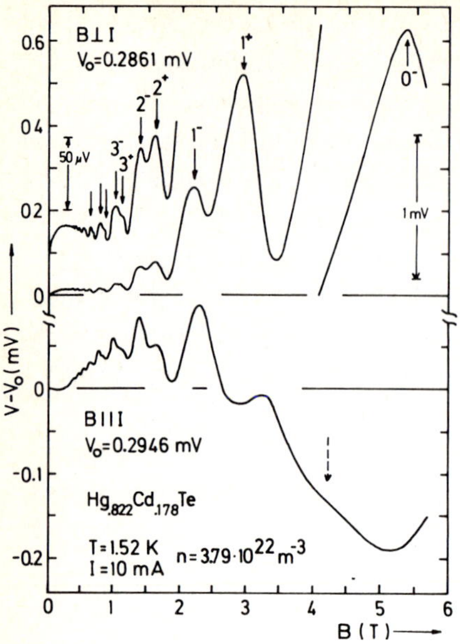

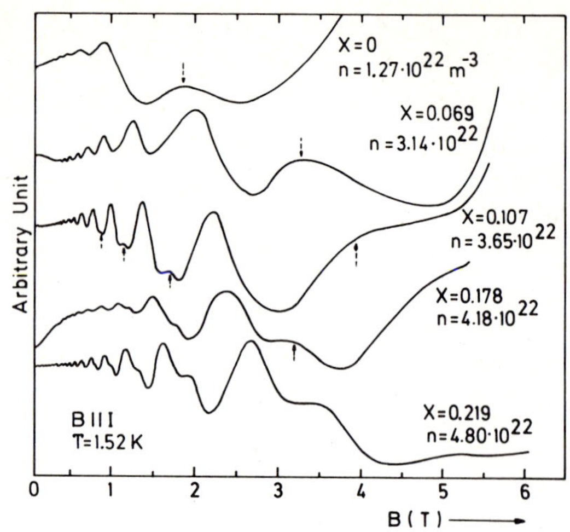

Fig. 75 Fig. 76

Fig. 75. Recorder traces of transverse (upper part) and longitudinal magnetoresistance voltages versus magnetic field for $Hg_{.822}Cd_{.178}Te$ /89/

Fig. 76. Recorder traces of longitudinal magnetoresistance curves versus magnetic field for samples of $Hg_{1-x}Cd_xTe$ with various x values and the carrier concentrations $n \approx 10^{22}m^{-3}$ at 1.52 K /89/

reported analysis of the LM yields a carrier density of $1.33\cdot10^{21}m^{-3}$ whereas Hall measurements indicated $n = 9.4\cdot10^{20}m^{-3}$. In /277/ carrier concentrations as determined from various measurements of $4.5\cdot10^{20} - 1\cdot10^{21}m^{-3}$ have been reported.

The last SdH maximum, the 0^- maximum, is expected to occur approximately at a magnetic field /291/

$$B_o^- = \frac{n^{2/3}}{2.62\cdot10^{14}}\left(\sqrt{\frac{g^*}{2}\frac{m_o^*}{m}} + 0.5361\sqrt{\frac{k_BT}{\hbar\omega_c}}\right)^{-2/3} \quad [T] \qquad (51)$$

dependent on n. However the authors observed an additional peak in the TM at much higher magnetic fields, shifting to lower fields with increasing carrier density. This additional structure has been tentatively ascribed to a resonant state with energy E_R, which modulates the magnetoresistance because of an extra density of states in the conduction band near E_R. This model also explains an anomalous absorption line in the far-infrared spectrum (Sect. 2.3).

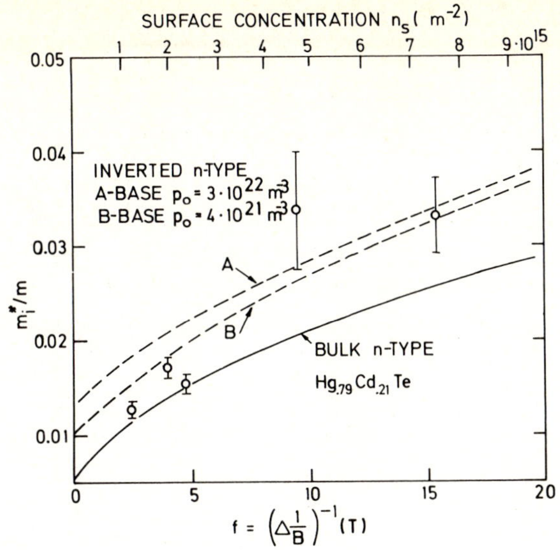

Fig. 77. Experimental values of electron effective mass m_i^* in the inversion layer as a function of the SdH frequency $f = \Delta(1/H)^{-1}$ and surface carrier density $n_s = 4.82\ 10^{10}\ f\ m^{-2}$. Solid curve: bulk $Hg_{.79}Cd_{.21}Te$ material. Dashed line: calculated values for two hole concentrations p_o in the base layer /282/

In measurements on helicons and non-resonant-cyclotron absorption in HgTe GALKIN et al. /280/ observed oscillations in the envelope of the helicon interference structure. These oscillations have been attributed to quantum effects on the imaginary part of the conductivity tensor (microwave SdH effect).

SdH oscillations ascribed to n-type inversion layers have also been observed on thinned p-type $Hg_{.79}Cd_{.21}Te$ samples /282/. In measurements at liquid helium

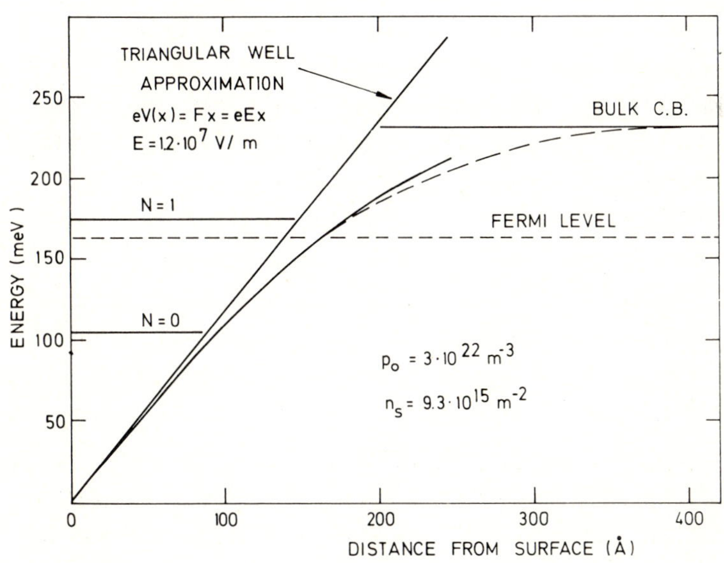

Fig. 78. Results of the computation of the electric sub-bands for a triangular potential well approximation at the inverted surface. An estimate of the potential including the screening effect of carriers in the inversion layer is also shown /282/

temperature in magnetic fields up to 9 T the oscillatory magnetoresistance was found to depend only on the magnetic field component normal to the surface. The 20 - 40 % higher electron effective masses, compared with bulk specimens (Fig. 77), have been explained theoretically as an effect of the surface potential well on the electron distribution (Fig. 78).

SdH oscillations in the longitudinal and transverse magnetoresistance have been studied for $Hg_{.9}Cd_{.1}Te$ with $n = 1.15 \cdot 10^{23}$ m^{-3} in /546/ and for $Hg_{.86}Cd_{.14}Te$ with $n \cong 2 \cdot 10^{21}$ m^{-3} in /514/. ALIEV /546/ was even able to observe SdH oscillations at 77 K. In contrast to the results at 4.2 K, the B_0- peak could not be observed at 77 K even in the transverse magnetoresistance.

The observability of the B_0- maximum in the longitudinal magnetoresistance $\rho_{\parallel}$ has been shown theoretically to depend on the possibility of spin-flip scattering /547,548/. While previous experimental results on the occurrence of this B_0-maximum had been contradictory, the authors of /514/ demonstrated that a B_0-peak is not observed in their n-$Hg_{.86}Cd_{.14}Te$ at 1.7 K, but clearly appears at 10 K. They /514/ argued that in a small-gap semiconductor in the presence of a quantizing magnetic field the ionized impurities do not provide an effective scattering mechanism for spin-flip electron scattering, and that spin-flip processes are instead caused by acoustic phonons through the deformation-potential interaction. This explains the temperature dependence of the B_0- maximum in $\rho_{\parallel}$.

Besides in magnetoresistance, magnetic quantum oscillations have been observed in the photovoltaic effect /549/, in the photomagnetic effect /550,551/, and in the photothermomagnetic effect /552/. Various other investigations of photoelectric properties have also been published /549-562/.

Recently, observations of an oscillatory magnetophotoconductivity effect ("optical SdH effect") have been reported in /553/ for InAs, PbSnTe and PbSe, and interpreted as due to an electron temperature modulation (see also /554/) caused by free carrier absorption in high magnetic fields. Similar observations of very pronounced oscillations in the photoconductivity /534/, the photomagnetic and the photothermomagnetic effects /550-552,561/ on HgCdTe at low temperatures had already been reported, compared with oscillations in the galvanomagnetic properties and sometimes explained as due to electron heating by light. The oscillations are independent of the incident wavelength; in the photoconductivity they are in phase with oscillations in the magnetoresistance, but the photoconductivity modulation is much higher (see Fig. 79) and for a considerable range of incident laser powers independent of this power (as long as the laser does not significantly rise the lattice temperature). With the photomagnetic effect, attempts have been made to determine the electron temperature from these oscillations /561/.

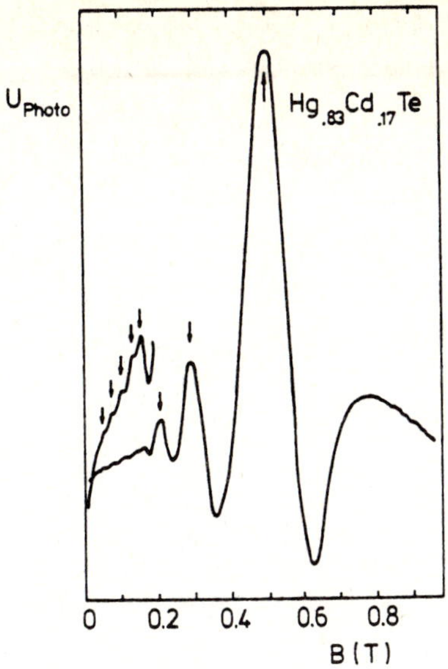

Fig. 79. SdH oscillations in photoconductivity of a $Hg_{.83}Cd_{.17}Te$ sample with $n = 1.5 \cdot 10^{21}$ m^{-3} carriers at 4.2 K. A pulsed GaAs diode laser with a peak power of 100 W and a pulse width of 19 ns has been used for excitation /534/

4.4.2 Magnetophonon Effect

Inelastic scattering of electrons on optical phonons in high magnetic fields has been observed by POMORTSEV et al. in the longitudinal magnetoresistance (OLMR) of HgTe /283/.

KAHLERT and BAUER /110/ have reported the observation of OLMR oscillations (magnetophonon effect) in n-$Hg_{.788}Cd_{.212}Te$ in the range 50 to 130 K in samples with a carrier concentration of $n = 1.6 \cdot 10^{21} m^{-3}$ and $\mu = 8.56\ m^2V^{-1}s^{-1}$ (4.2 K) (Fig. 80). Using $m_0^* = 5 \cdot 10^{-3} m$ and $g_0^* = -172$, they found that the magnetophonon

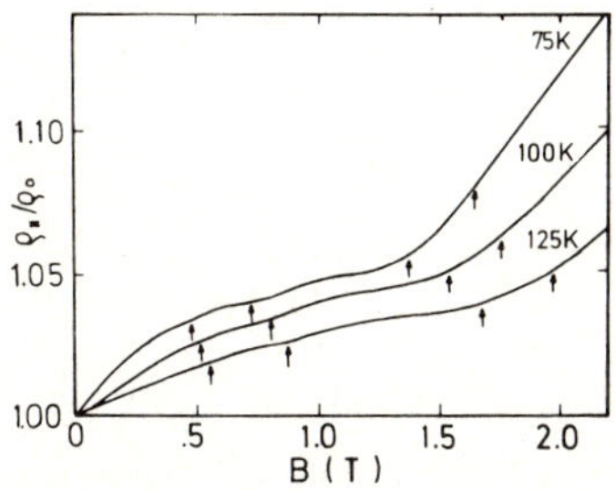

Fig. 80

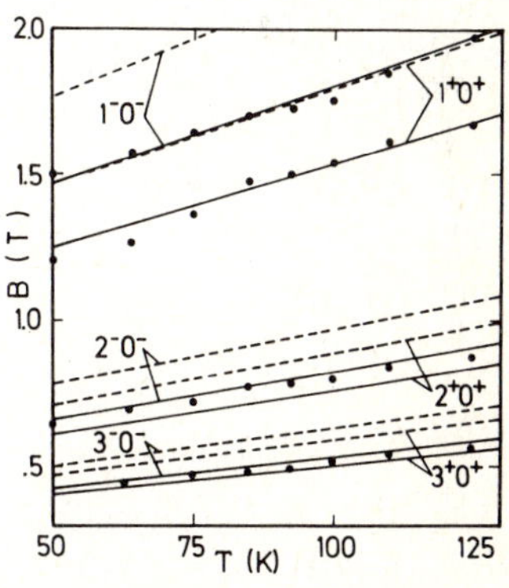

Fig. 81

Fig. 80. Longitudinal magnetoresistance versus B. Parameter is the lattice temperature. The arrows indicate the positions of the magnetophonon resonance as obtained from the second derivative of the curves /110/

Fig. 81. Temperature dependence of the magnetophonon resonances. Solid lines: calculated for $\hbar\omega_{o,1} = 17.1$ meV. Dashed lines: calculated for $\hbar\omega_{o,2} = 19.6$ meV. Dots: experimental data /110/

peaks involved only the LO phonons characteristic of HgTe (17.1 meV). Peaks associated with the other known LO phonon branch characteristic of CdTe (19.6 meV) were not seen. Spin splitting of the N = 1 transition was well resolved (Fig. 81). The peaks were observed to shift to higher fields linearly with temperature. However, the analysis of this shift in terms of the positive temperature coefficient of the energy gap gave $dE_0/dT = 7.6\cdot10^{-4}$ eV/K, which is not in good agreement with other measurements (Sect. 3.4).

While earlier investigations of the magnetophonon effect in the longitudinal magnetoresistance in n-$Hg_{.788}Cd_{.212}Te$ /110/ revealed only structures involving the HgTe-like LO phonons, MOZHAEV et al. /563/ have also observed a contribution of CdTe-like phonons to the transverse magnetoresistance at 77 K.

They studied the oscillating part of the transverse magnetoresistance for a series of HgCdTe samples with compositions in the range between x = 0.18 and x = 0.42. A comparison between calculated and observed peak positions demonstrated that below $x \cong 0.3$ the magnetophonon maxima were governed by the HgTe-like optical mode, while above the observed values shifted gradually to the calculated positions due to the CdTe-like phonon mode. Above x = 0.4 the CdTe-like mode seems dominant. In the case of a very homogeneous sample with x = 0.332, a splitting of the first magnetophonon maximum was observed and attributed to the simultaneous participation of both optical modes.

GELMONT et al. /564/ observed a new type of magnetophonon oscillation in HgTe which, instead of being caused by the inelastic resonant scattering of electrons by optical phonons at higher temperatures, has its origin in electron concentration oscillations at low temperatures. These are due to the resonant character of the phonon-emitted recombination of excess carriers produced by impact ionization in a strong electric field. In ultraquantum magnetic fields this channel of recombination is switched on each time the energy gap between the lowest electron Landau level and some hole level becomes equal to the optical-phonon energy. As a result, the (excess) electron concentration drops sharply and the magnetoresistance has maxima (see also Sect. 6.1).

4.4.3 Cyclotron Resonance

Extensive results of far-infrared magnetotransmission experiments on HgTe have recently been presented in /285/. In experiments on n-type HgTe in magnetic fields up to 5.8 T at 4.2 K in the Faraday and Voigt configurations (wavelength range 2 mm to 100 μm) both intraband transitions in the Γ_8 conduction band (cyclotron resonance) and interband $\Gamma_8 \rightarrow \Gamma_8$ transitions have been observed. From magnetotransmission experiments in the near-infrared region the values of the Luttinger parameters and a complete description of the quantum levels in the Γ_8 bands near the degeneracy point at $\underline{k} = 0$ have been obtained (Sect. 4.4.1):

$$\bar{\gamma} = -8.3 \pm 0.2, \quad \gamma_1 = -12.8 \pm 0.6, \quad \varkappa = -10.5 \pm 0.4.$$

In studying the magnetoplasma cyclotron resonance in the Voigt configuration (an early investigation is described in /122/) the unusual frequency dependence of the dielectric function in zero-gap materials as discussed in /292-295/ has been taken into account.

Investigations on the far-infrared cyclotron resonance in $Hg_{1-x}Cd_xTe$ have been reported in /99/ for x = 0.22 and x = 0.25, in /284/ for x ranging from 0.245 to 0.15, in /119/ for x = 0.176, in /73/ for x = 0.14 and 0.17 and in /97/ for x = 0.204. While results in /99/ are possibly affected by bad experimental conditions (sample thickness ~1 mm) and a somewhat crude analysis, serious differences also occur in the other results as seen in Table 26. The most extensive study has been undertaken by KINCH and BUSS /97/. They observed far-infrared cyclotron resonance in the spectral region 10 cm^{-1} to 330 cm^{-1} in high purity ($n = 5\cdot10^{20}-10^{22}$ m^{-3} at 77 K) n-type $Hg_{.796}Cd_{.204}Te$ samples (thickness 2 - 20 µm) in the temperature range 1.3 - 4.2 K. Measured line shapes at various magnetic field strengths are given in Fig. 82, the variation of the observed transmission minima ν_{CR} with magnetic field in the Faraday

Table 26. Cyclotron resonance effective masses from various references (T = 4.2 K)

x	n ($10^{20}m^{-3}$)	$m_o^*/m\cdot10^3$	Ref.
0.176	8.5	6.2	/119/
0.17	28	0.4	/73/
0.15	70	4	/284/
0.19	4	6	/284/

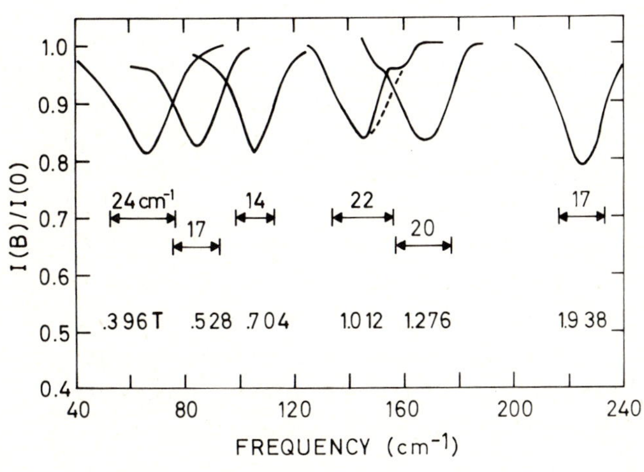

Fig. 82. Normalized transmission I(B)/I(o) in the Faraday configuration at various H ($n \approx 7\cdot10^{20}$ m^{-3}, thickness ≈ 4.3 µm, T = 2.1 K) /97/

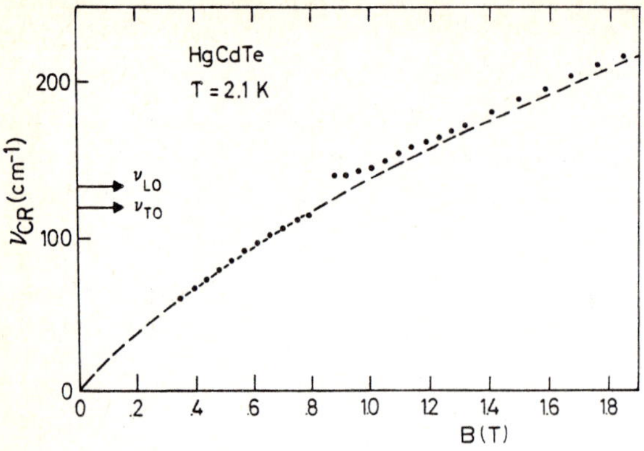

Fig. 83. Cyclotron frequency ν_{CR} versus magnetic field. The dashed line represents a theoretical curve calculated using $m_0^* = 4.66 \cdot 10^{-3} m$, $E_0 = 61.7$ meV, $\Delta = 0.96$ eV /97/

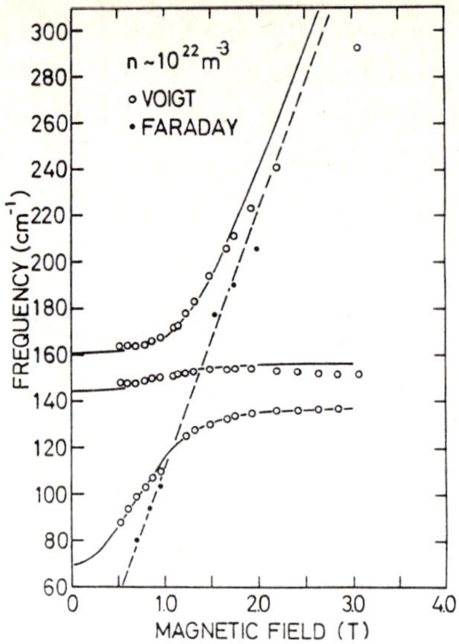

Fig. 84. Observed and calculated transmission minima in Voigt and Faraday configurations versus B for $n \approx 10^{22}$ m^{-3} /97/

configuration in Fig. 83. Comparing the theoretical and experimental data in Fig. 83 for $\nu_{CR} \gtrsim \nu_{LO}$ two features are apparent. At energies below the reststrahl ($\sim$ 120 cm^{-1}) the observed ν_{CR} are depressed below the theoretical curve, and for $\nu_{CR} > \nu_{LO}$ there is a positive shift of ν_{CR}, which for $\nu_{CR} \sim 1.1\ \nu_{LO}$ amounts to 5 ± 1 cm^{-1}. These effects arise from the coupling of the conduction electrons to the LO phonon at $\sim$ 138 cm^{-1} and yield a value of the Fröhlich coupling constant

$$\alpha = \frac{2\pi e^2}{h\varepsilon}\left(\frac{m_0^*}{2h\nu_{LO}}\right)^{1/2}\left(\frac{1}{\varepsilon_\infty} - \frac{1}{\varepsilon_0}\right) \tag{52}$$

of 0.037 ± 0.008, which is in good agreement with calculation. No effects of interaction with the second LO phonon at $\sim$ 157.5 cm^{-1} have been observed in the Faraday configuration, however, in the Voigt configuration the classical coupling of the phonon modes with collective cyclotron excitations is clearly visible /97/ (Fig. 84).

4.4.4 Electron-Spin and Combined Resonance

In addition to modifications of the effective g factor, which is extremely large in narrow gap semiconductors due to interband coupling of orbital angular momentum into the spin states, the spin-orbit interaction in these materials also alters the selection rules for electromagnetic transitions. So transitions which change both Landau (N) and spin (s) quantum numbers become allowed and can be excited by either the electric or magnetic component of the incident radiation. However, the magnetic dipole interaction is much weaker than the electric dipole interaction and so only experimental results on electric dipole excited-electron spin- (EDE-ESR, $\Delta N = 0$, $\Delta s = \pm 1$) /133,286/ and combined resonance ($\Delta N = +1$, $\Delta s = -1$) /133,132,287/ have been reported. EDE-ESR occurs in the cyclotron resonance inactive (CRI) sense of circular polarization. The momentum matrix elements for $E_{N,\pm} \ll E_0$ for EDE-ESR and combined resonance are given by /132,296/:

$$\langle N,-|P_{CRI}|N,+\rangle \simeq \frac{2\sqrt{2}}{3}\pi m_0 |P|^2\cdot\left[1/E_0^2 - 1/(E_0+\Delta)^2\right]k_z\nu_s(N)$$

and (53)

$$\langle N+1,-|P_{\parallel}|N,+\rangle \simeq \frac{2\sqrt{2}}{3}\pi m_0|P|^2\cdot\left[1/E_0^2 - 1/(E_0+\Delta)^2\right]\sqrt{N+1}\,k_H\nu_{c,s}(N)$$

respectively, the transitions and their selection rules are summarized in the schematic energy level diagram of Fig. 85. Intraband magneto-optical investigations utilizing both cyclotron resonance and spin-flip resonance can provide a more precise set of band parameters than measurements of cyclotron resonance alone, since both the separation between Landau levels and between spin levels are measured. Results of such measurements on $Hg_{1-x}Cd_xTe$ have already been mentioned in Sect. 3.3.3.

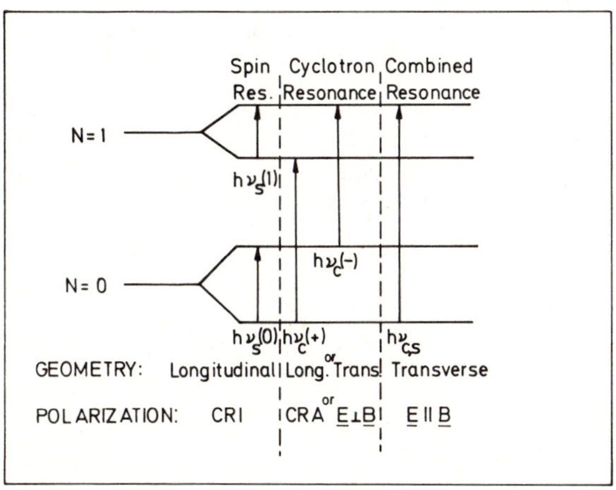

Fig. 85. Schematic energy level diagram for the two lowest Landau levels at $k_z = 0$ including spin. Electric-dipole transitions and polarization selection rules are indicated /132/

As stated in the previous section the interaction between single longitudinal optical phonons and both free and localized carriers in semiconductors produces, at least qualitatively, a splitting of the upper level of an adjacent pair of Landau levels when the energy difference between the two is nearly equal to the LO phonon energy. Detailed studies of this splitting using cyclotron resonance directly are extremely difficult because of the presence of the reststrahlen reflection precisely in the frequency region of interest. These difficulties have been avoided by utilizing the combined resonance transition /287/.

The direct observation of conduction-electron spin resonance in HgCdTe at 4.2 K and ~76 GHz using a superheterodyne spectrometer has been reported in /565/. The n-type samples had compositions near $x \cong 0.4$ and $x \cong 0.22$ and electron concentrations of $\sim 10^{21}\ m^{-3}$. The spin resonance was observed as a single isotropic line, and the resonant magnetic field was directly proportional to the microwave frequency and independent of sample orientation (thus excluding the possibility of magnetoplasma or induced cyclotron resonances).

The g-factors derived are in good agreement with values given in Sect. 3.3.3.3 (13.6 for x = 0.422, ~18 for x = 0.38, 71 for $x \cong 0.22$, 83 for $x \cong 0.21$).

4.4.5 Spin-Flip Raman Scattering

Tunable spin-flip Raman scattering has been observed /120/ in $Hg_{.766}Cd_{.234}Te$ with $n = 5.4 \cdot 10^{20} m^{-3}$ and $\mu = 11\ m^2/Vs$ (at 77 K) using any of the P(10) - P(18) pump laser lines of the (00^01) - (02^00) CO_2 band in magnetic fields between 0.4 and 2.1 T with a maximum radiation intensity at about 1.2 T. The frequency shift and the wavelength of the Stokes radiation obtained using the 1050.4 cm^{-1} CO_2 laser line as a pump are shown in Fig. 86. The maximum frequency shift is about 68 cm^{-1}. The polarization of the scattered radiation has been found to be rotated 90^0 relative to the pump radiation, consistent with the single-particle spin-flip selection rules.

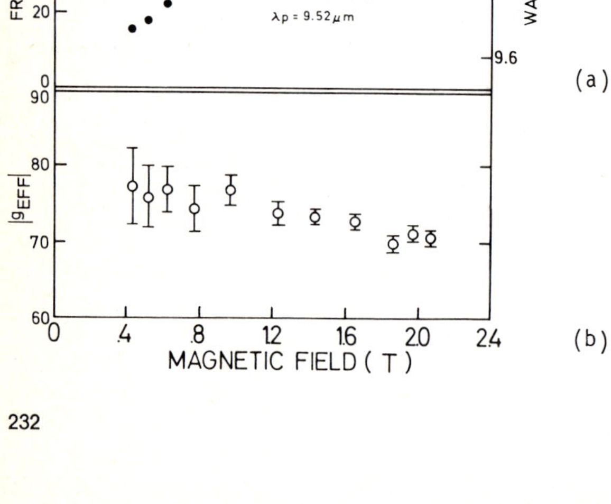

Fig. 86. (a) Frequency shift of $Hg_{.766}Cd_{.234}Te$ spin-flip Raman radiation versus B. (b) Effective g factors for conduction-band electrons calculated from corresponding frequency shifts shown in (a) /120/

Tunable stimulated recombination radiation from $Hg_{.766}Cd_{.234}Te$, resonantly pumped with a chopped-cw CO_2-laser, has also been reported (Sect. 6.4). The polarization of the recombination radiation was measured to be independent of the pump polarization, being perpendicular to the magnetic field between 0 and 0.55 T and parallel to the field from 0.5 to 2.2 T (1077.3 cm^{-1} CO_2 pump). The threshold has been determined to $2 \cdot 10^6$ W/m^2, corresponding to 150 mW of input power, a peak recombination output power of 3 mW has been detected.

The ability to adjust the energy gap of HgCdTe so as to achieve resonance enhancement when employed as the active medium in a spin-flip Raman process pumped by an efficient CO_2 laser has stimulated many experimental investigations in recent years /566-570/. In addition, the large conduction-electron g-factor allows a considerable tuning range with moderate magnetic fields (a basic tunability of 38 cm^{-1}/T for the first Stokes radiation has been measured for $Hg_{.77}Cd_{.23}Te$ /567/). Peak external conversion efficiencies of several percent (up to 9%) have been obtained /568/. WALUKIEWICZ et al. /569,570/ recently considered the role of shallow acceptors as intermediate states in the SFRS process and explained the photon energy and magnetic field dependence of the SFRS cross-section as due to an interference between valence band and acceptor intermediate states.

Besides SFRS, other non-linear effects such as 4-photon and higher order mixing have been observed in HgCdTe /571,572/ and used to determine g-factors at 77 K /571/, to measure spin resonance and study its linewidth below and through the quantum limit /572/, and to determine values for the third-order non-linear optical susceptibilities /573/ (these values, however, scatter considerably in the literature; the sources of non-linearity and their contribution to the observed values are also still a matter of debate /567,574-577/).

The frequency dependence of two-photon absorption coefficients has been calculated for HgCdTe and other zincblende semiconductors based on a three-band, non-parabolic model and shown to depend on $(E_0^3 \, np^2)^{-1}$ in agreement with reported room-temperature coefficients /578,579/.

The high-intensity infrared transmission in HgCdTe using nanosecond and subnanosecond laser pulses at λ = 10.6 /580-583/ and λ = 1.06 µm /584/ has been intensively investigated. In each case a distinct high-intensity transmission limit was observed, beyond which a further increase in incident intensity causes a rapid decrease in the transmission coefficient. This non-linear absorption has been explained by considering a model in which hot-electron generation by intraband absorption leads to an avalanche of excess carriers with a large cross-section for intervalence band absorption for holes /580/.

4.4.6 Various Other Quantum Effects

SdH oscillations in the photomagnetic effect have been observed in n type $Hg_{1-x}Cd_xTe$ ($x = 0.14$ and 0.20) with a carrier concentration of $n = 1.6 \cdot 10^{20} m^{-3}$ at 10 K and $1.6 \cdot 10^{22} m^{-3}$ at 80 K in good agreement with galvanomagnetic measurements /237,288/.

The Auger recombination lifetime varies in an oscillatory manner with increasing magnetic field because the Auger transition probability in the presence of a magnetic field exhibits maxima if the separation between two Landau levels is equal to the band gap energy (Sect. 6.1) /290,297/.

Interband $\Gamma_6 \rightarrow \Gamma_8$ magnetoabsorption in HgTe has been recently investigated by GULDNER et al. /289/ in the spectral region 300 - 400 meV at 4.4 K. The structure of the Γ_8 magnetic levels near Γ has been quantitatively described using the theory of Luttinger and a three-band model including non-parabolicity. A very complex structure for a set of heavy holes has been found.

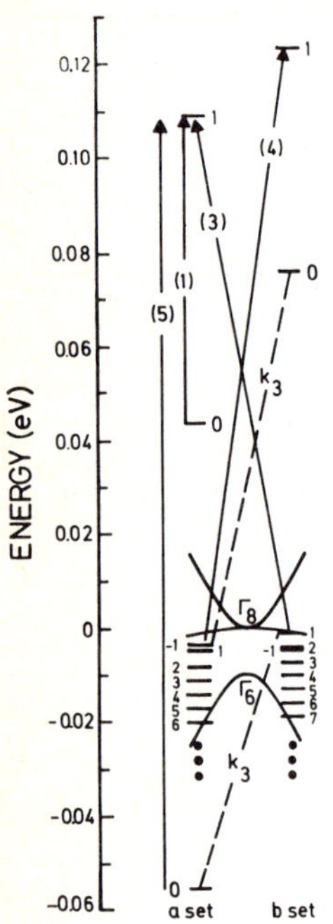

Fig. 87. Magnetic energy levels at 7 T and 25 K and interband transitions observed. A sketch of $E(\underline{k})$ for the Γ_8 and Γ_6 bands at $B = 0$ is also shown /114/

Interband magnetoreflection measurements /114/ will be discussed in Sect. 5.1.2. Experimental results have been analysed based on a magnetic energy level scheme shown in Fig. 87 /159/.

A magnetic field induced energy gap in p-HgTe has been observed in transport measurements in pulsed magnetic fields up to 35 T in the temperature range 77 - 340 K. This suggests a valence and conduction band separation of 0.6 meV/T /80,298/.

4.5 Hot Carrier Properties

Investigations of warm and hot electron effects at different temperatures may yield valuable information on the dominant scattering process in semiconductors. Thus hot electron properties in HgTe and $Hg_{1-x}Cd_xTe$ have been studied using different techniques /44,152,234,236-238,299-301/.

In early investigations on the conductivity in high electric fields acoustic-phonon scattering was believed to dominate the field dependence of the mobility. Impact ionization across the gap, for example generation of electron-hole pairs, has not been observed, only deviations from Ohm's law arising from carrier heating /152/. Above 10 kV/m a marked change in conductivity has been reported in /299/ and attributed to an intervalley transfer process. However, these results seem to be strongly influenced by inhomogeneity effects. KOBYZEV et al. /301/ on the other hand observed a sharp increase in the measured current voltage characteristics above 15 kV/m, apparently the threshold of impact ionization. NIMTZ et al. /238/ have found a non-ohmic behaviour in the microwave transmission of $Hg_{.8}Cd_{.2}Te$ samples as well as in the DC characteristic j(E). Above fields of 9 kV/m deviations from Ohm's law (subohmic behaviour) occur and above 16 kV/m a time-dependent increase of the current density has been observed and interpreted as an excess-carrier generation due to impact ionization (Fig. 88). From the time-dependent j(E) characteristics the generation rate of excess carriers has been calculated according to

$$g = \frac{1}{n}\frac{dn}{dt} = \frac{1}{j}\frac{dj}{dt} \qquad (54)$$

n and j being the instantaneous electron and current densities respectively (Fig. 89). A comparison with InSb shows a much steeper increase of the InSb g(E)-curve, but lower values in the $E < 22$ kV/m range. The non-ohmic behaviour at $9 \leq E \leq 15$ kV/m has been explained by polar-optical scattering (at 77 K).

ELLIOTT et al. /234/ have determined the warm electron coefficient β according to

$$\mu = \mu_o(1 + \beta E^2) \qquad (55)$$

in order to get additional information on the dominant scattering mechanism in the temperature range 77 - 300 K. Their measurements (Fig. 90) resulted in very large positive β values ($0.97\cdot10^{-8}\ m^2V^{-2}$ for x = 0.2 and $0.43\cdot10^{-8}\ m^2V^{-2}$ for x = 0.22 at 295 K), which were about two orders of magnitude greater than the calculated values for cases where polar optical mode scattering controls the energy loss and optical mode or ionized impurity/heavy-hole scattering controls the momentum loss.

Hot carrier experiments at liquid helium temperatures have been carried out by IVANOV-OMSKII et al. /300/ on HgTe and by DORNHAUS et al. /44/ on $Hg_{.8}Cd_{.2}Te$.

Fig. 89

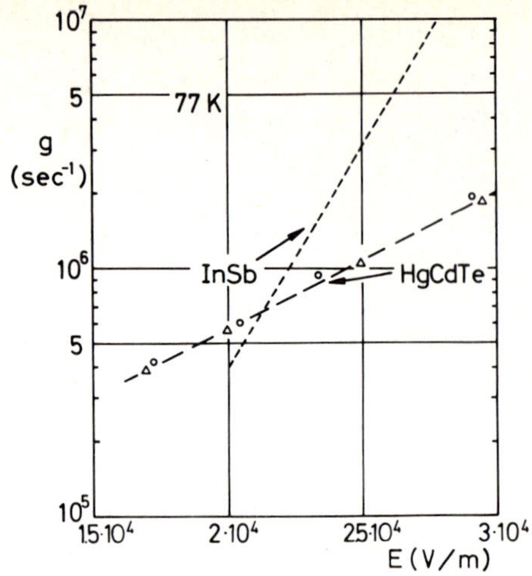

Fig. 88

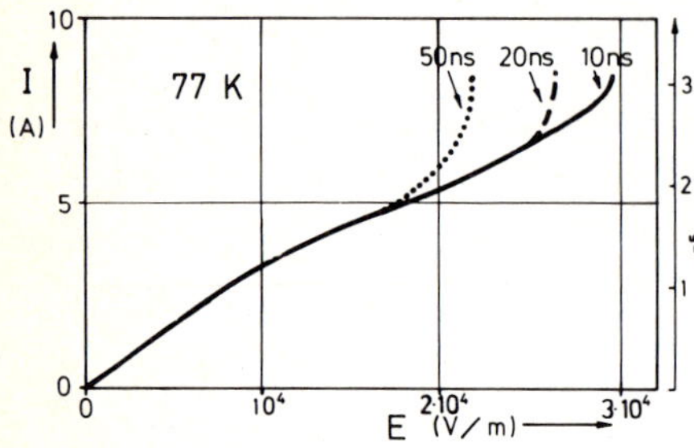

Fig. 88. j(E) characteristic at 77 K. The data were recorded 10, 20, and 50 nsec after the electric field E was applied. The scale of the carrier drift velocity v_d is valid for the exp. data only up to $E \leq 16000$ V/m /238/

Fig. 89. Generation rate g(E) as a function of electric field E /238/

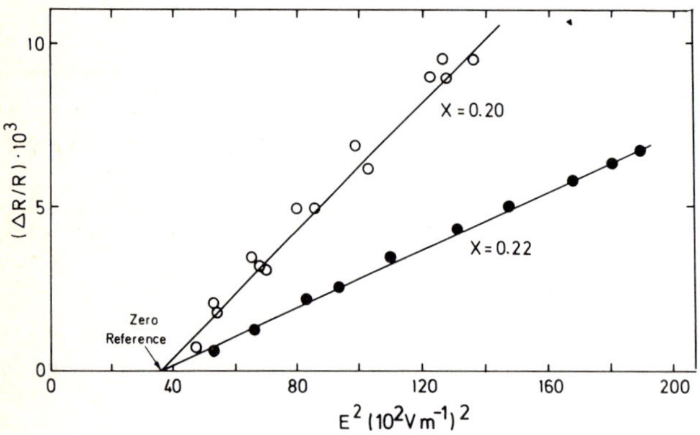

Fig. 90. Variation of sample resistance with square of the electric field /234/

While IVANOV-OMSKII et al. have reported an increase in the carrier density with electric field strength E in HgTe, DORNHAUS et al. have shown from conductivity measurements versus E and B that there is no evidence for ionized impurity scattering nor impurity breakdown in high quality single crystals of $Hg_{.8}Cd_{.2}Te$ (see also Sect.4.2). In contrast to the calculated increase in the σ(E) curve, characteristic for ionized impurity scattering dominating the momentum relaxation, they even have observed a decrease with E (**Fig. 91**). Dips in the d^2E/dj^2 versus E curve have been tentatively ascribed to transitions of electrons from lower to higher Landau-subbands.

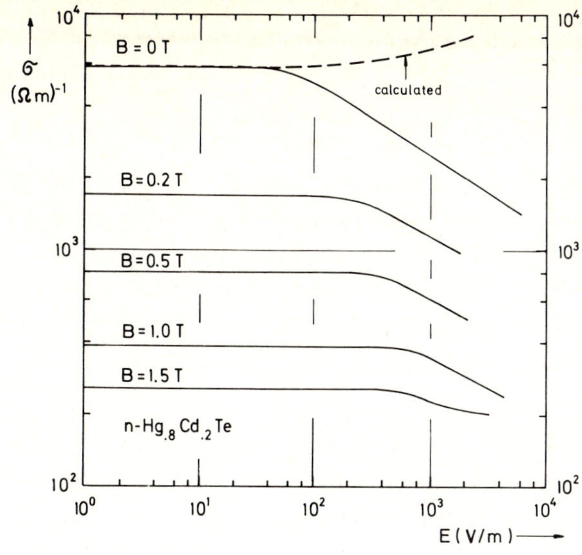

Fig. 91. Conductivity σ versus electric field E at various magnetic fields B. The broken line at B = 0 represents calculated values of the field dependence of the conductivity for ionized impurity scattering dominating the momentum relaxation and polar optical scattering dominating the energy exchange /44/

GEORGITSE et al /236,237/ have studied photoelectric effects in $Hg_{1-x}Cd_xTe$ with $0.14 \le x \le 0.24$ (see Secs. 4.3, 4.4, 6.2) under conditions of electron heating by the incident radiation. They have observed different series of oscillations in the photoconductivity and the photomagnetic effect, which have been attributed to multiphonon processes in the relaxation of hot electrons (Fig. 92). Besides the interaction of electrons with both LO-phonons of the two-mode mixed crystal in higher magnetic field, even interactions with transverse optical phonons have been found.

Above certain threshold values, dependent on the applied magnetic field, current oscillations have been observed at temperatures between 4 and 77 K and with frequencies in the MHz range, increasing with increasing magnetic field strength

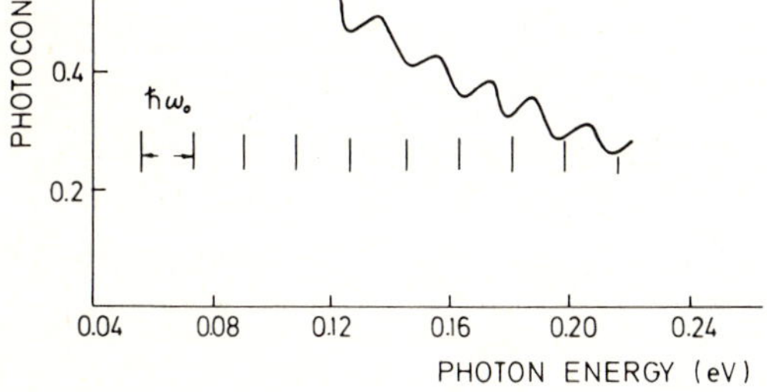

Fig. 92. Photoconductivity spectrum of the alloy $Hg_{.85}Cd_{.15}Te$ at 10 K

/299,301/. A coherent microwave emission has also been obtained at frequencies of 1.5 - 5 GHz dependent on current, magnitude and orientation of the magnetic field, sample orientation and interval from the start of the supply pulse. It seems to be connected with the electron-hole interaction in this material.

The influence of a strong electric field on the galvanomagnetic properties of HgCdTe has been investigated, both experimentally and theoretically, in a number of recent publications /585-590/. The recombination instability of a non-equilibrium electron-hole plasma in crossed electric and magnetic fields /591/ and the magnetoconcentration effect on the Auger recombination in these fields /592/ have also been described. The hot-electron drift velocity in $Hg_{.795}Cd_{.205}Te$ at 77 K was calculated by the Monte Carlo technique in /585/. Band non-parabolicity, admixture of wavefunctions, and the two-mode nature of the polar optical phonons was included as well as the influence of deformation-potential acoustic, alloy and ionized impurity scattering, and good agreement with recent experimental data has been achieved (Fig. 93).

Hot-electron properties in zero-gap semiconductors with non-parabolic band structure have been theoretically investigated in /590/. In addition to the influence of non-parabolic bands and the admixture of p-like components into the

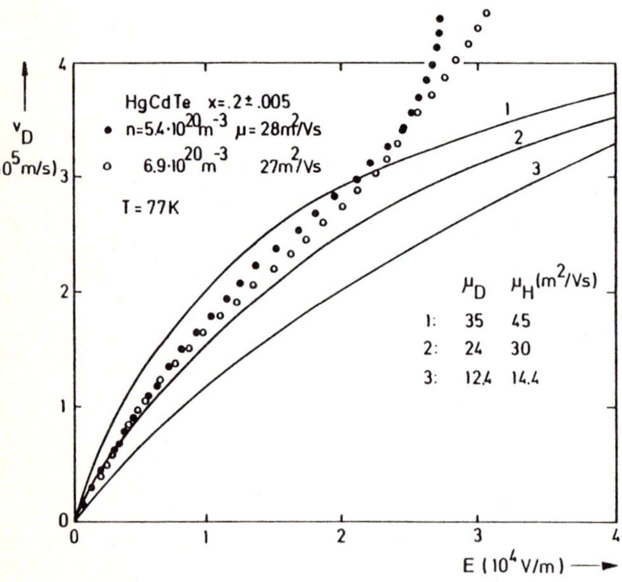

Fig. 93. Drift velocity vs electric field in HgCdTe at 77 K. Solid lines 1-3 are theoretical results for the indicated low-field drift (μ_D) and Hall (μ_H) mobilities, dots are experimental results for two samples. The experimental data are reliable up to fields of about $1.6 \cdot 10^4$ V/m; at higher fields carrier multiplication occurs /585/

conduction-band wavefunctions in zero-gap materials, one is faced with some peculiarities /590/:

- due to the absence of any threshold energy for the thermic generation of charge carriers, one can considerably change their concentration by only weak external perturbations (e.g. electric field in the range of 10 to 100 V/m);
- at low temperatures one has to take into account dielectric anomalies (see Sect. 3.7.1) as the mobility is dominated by scattering via screened Coulomb interaction;
- Coulomb interaction, especially exchange interaction, has an important effect on the spectrum of free carriers in pure zero-gap materials. The semiclassical conception is only valid if all characteristic energies are greater than the Bohr energy. One has to determine carefully whether the adiabatic principle is applicable (see Sect. 3.7).

Calculated values for momentum and energy relaxation times versus electron concentration and for electron temperatures versus electric field are shown in Figs. 94 and 95.

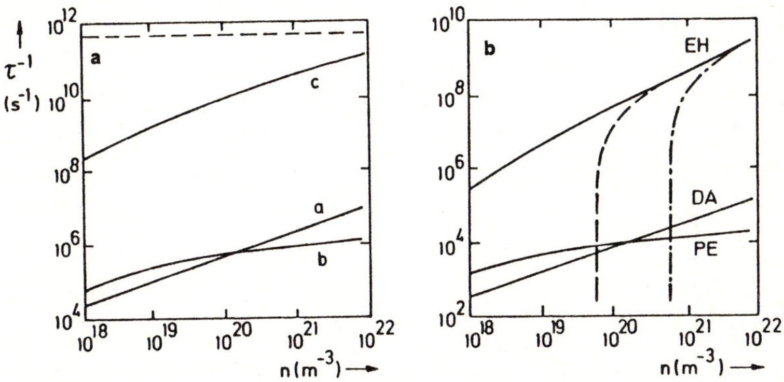

Fig. 94. (a) Calculated momentum and (b) energy relaxation times in dependence on n at T_L = 4.2 K. In (a) the curves are for deformation acoustic (a : DA), piezo-electric (b : PE) and electron-heavy-hole plus ionized impurity (c : EH+I) scattering respectively. In (b) the electron-hole scattering contribution is indicated for intrinsic HgCdTe (———) and $N_D = 6\cdot10^{19}$ m^{-3} (-----) and $N_D = 6\cdot10^{20}$ m^{-3} (-·-·-) /590/

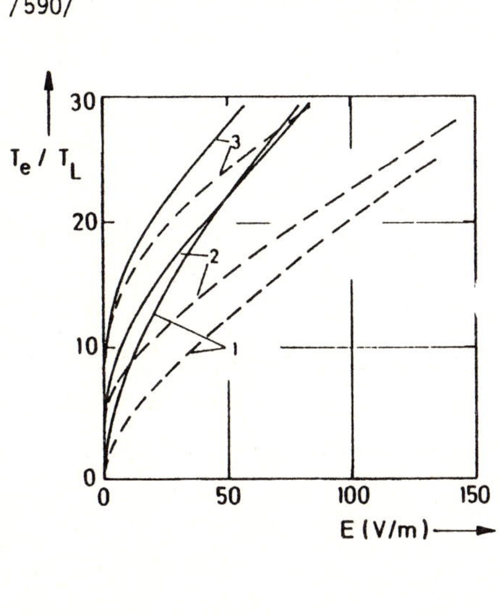

Fig. 95. Electron temperature T_e vs electric field strength for various donor concentrations. (1) Intrinsic, (2) $N_D = 6\cdot10^{20}$ m^{-3}, (3) $N_D = 6\cdot10^{21}$ m^{-3}. Solid curves correspond to a screened Coulomb interaction with concentration-dependent Γ_8-Γ_8 excitations, dashed curves are calculated assuming ε_0 = 20. /590/

4.6 Thermoelectric Effects

The theory of galvano-thermomagnetic effects in semiconductors and semimetals has been derived, amongst others, in /82,246,302-304/. Measurements of the thermoelectric power have been reported by various authors, e.g. on HgTe in /76,123,128/ and on $Hg_{1-x}Cd_xTe$ in /92,94,121,127,95,149/. Measurements of the thermoelectric power as a function of hydrostatic pressure by PIOTRZKOWSKI et al. /76/ played an important role in establishing the Groves-Paul model of band structure in this material. The temperature dependence of the thermoelectric power α was investigated in /94, 127/ and is shown in **Fig. 96**. The magnetic field dependence of α is often used to determine the effective mass value, as in the case of strong degeneracy the high field limit is given by

$$\alpha(\infty) = -\left(\frac{\pi}{3}\right)^{2/3} \frac{k_B^2 T}{e\hbar^2} \frac{m^*}{n^{2/3}} \tag{56}$$

neither depending on the scattering index r nor ℓ /302/.
When trying to find experimental values of $\alpha(\infty)$ by extrapolation, the formula

$$\Delta\alpha(B) = \frac{AB^2}{1 + CB^2} \quad \text{(A, C independent of B)} \tag{57}$$

is very useful /305/. Measurements of $\alpha(B)$ have been performed and reported in /94,92,121,123/ on samples in the concentration range $6.2\cdot10^{22}\ m^{-3} \leq n \leq 4.3\cdot10^{24}\ m^{-3}$. The influence of the inelasticity and non-parabolicity on the scattering of electrons in HgTe has been calculated and the results have been compared with experimental results on the carrier density dependences of α and $\alpha(\infty)$ in /246/.

For pure HgTe measurements of the temperature dependence of the thermoelectric power, the Nernst coefficient, the phonon drag, the Righi-Leduc coefficient and the thermal conductivity have been undertaken /248,306/. The temperature and magnetic field dependence of the Nernst photothermomagnetic effect is given in /307/.

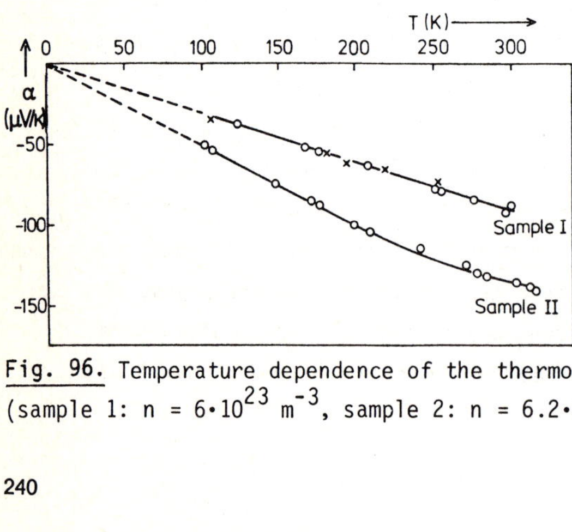

Fig. 96. Temperature dependence of the thermoelectric power α of $Hg_{.9}Cd_{.1}Te$ (sample 1: $n = 6\cdot10^{23}\ m^{-3}$, sample 2: $n = 6.2\cdot10^{22}\ m^{-3}$) /94/

Recent investigations on thermomagnetic properties of n- and p-type HgTe and HgCdTe have been reported in /593-598/. In principle, thermomagnetic effects provide a powerful tool to study the band structure of semiconductors. For example, the thermoelectric power for degenerate statistics in the high field limit of a transverse magnetic field is directly related to the bandshape [see (56)]. A quantitative analysis of the temperature dependence of the thermoelectric power in HgTe, however, has only been successful above 60 K when the hole contribution can be neglected /509,520,593/. At low temperatures such an analysis is difficult since the one-carrier approximation breaks down. This occurs because the phonon drag of holes is very large and can even cause a change of sign of the thermoelectric power (Fig. 97).

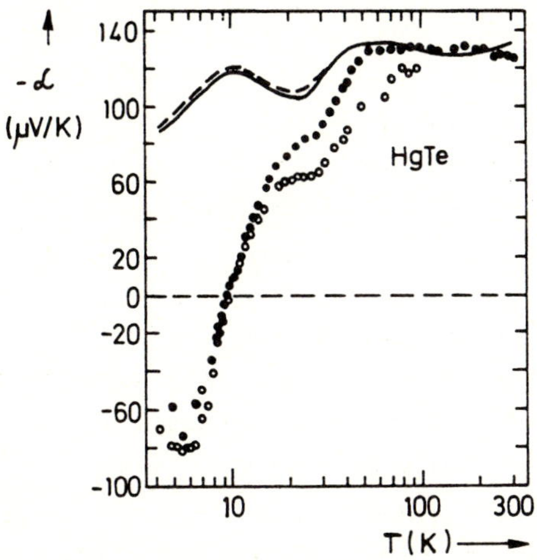

Fig. 97. Temperature dependence of thermoelectric power /520/. Dots represent experimental results; lines were theoretically obtained neglecting holes and phonon drag (———), and with phonon drag but neglecting holes (-----)

For non-degenerate statistics the thermoelectric power is very sensitive to the contributing types of scattering. Thus a more quantitative comparison of calculated and measured values will require a better knowledge of the phonon and hole scattering mechanisms and should be based upon simultaneous solution of the triply coupled Boltzmann equations for electrons, holes and phonons /520/.

In /593/ thermoelectric power and longitudinal and transverse Nernst-Ettingshausen effects have been measured for heavily doped ($\sim 10^{24}$ m^{-3}) n- and p-type HgTe. The main purpose of the n-HgTe measurements was a systematic study of the temperature dependence of the effective mass at the Fermi level. These data provide a description of the temperature dependence of band-structure parameters. The results could be reasonably explained assuming that the temperature dependence of the effective mass results from the temperature dependence of the band gap E_0 alone, giving a value of $dE_0/dT = 5 \cdot 10^{-4}$ eV/K.

This is in contradiction to results from the magneto-Seebeck effect described in /595/, where the authors claimed that the deduced temperature dependence of the effective mass in HgTe was in close correspondence to the lattice dilatation contribution to the variation of the energy gap versus temperature alone. This contribution, however, is the smaller part of the total change and has the wrong sign (see Sect. 3.4).

The thermomagnetic measurements of p-HgTe in /593/ aimed at the determination of the heavy-hole mass. Due to the strong coupling between carriers and acoustic phonons in p-HgTe a strong phonon-drag contribution to the thermoelectric power has been observed. To analyse the experimental data this phonon-drag contribution was calculated and one of the deformation potential constants was obtained by fitting the temperature dependence of the thermoelectric power in p-type HgTe. A value of $a = 2.7 \pm 0.3$ eV was found (see also Table 24).

4.7 Magnetic Susceptibility

In the semimetal HgTe the magnetic susceptibility can be satisfactorily accounted for by two components, a diamagnetic susceptibility of the crystal lattice χ_ℓ and a paramagnetic susceptibility of the electron gas χ_c /308/. In the semiconducting CdTe the main components turned out to be the lattice susceptibility χ_ℓ and a paramagnetic susceptibility of complex lattice defects of the type of donor-acceptor complexes /309/.

On the mixed crystal $Hg_{1-x}Cd_xTe$ there are only two experimental investigations available:

IVANOV-OMSKII et al. /310/ reported measurements of the magnetic susceptibility versus composition x, temperature and growth conditions (annealing in mercury vapour), SCHLABITZ et al. /311/ investigated the dependence of susceptibility on temperature and carrier concentration of n- and p-type samples with x = 0.2.

The molar lattice susceptibility χ_ℓ (SI units [2]) is assumed to be independent of temperature and to increase linearly from $\chi_\ell^{HgTe} = -1420\ m^3\ mol^{-1}$ to $\chi_\ell^{CdTe} = -1060\ m^3\ mol^{-1}$. The lattice susceptibility represents the sum of two parts, a dia-

---

[2] χ is the rationalized molar susceptibility defined by

$$\chi = \mu_o \frac{M}{B} \cdot \frac{\mathcal{M}}{\zeta}\ (m^3 mol^{-1})$$

where $\mathcal{M}$ is the mass of one mole in kg and ζ the density in kgm^{-3}. It is related to the traditional unrationalized cgs susceptibility $\chi_{trad.}$ (unit: $cm^3\ mol^{-1}$) by

$$\chi = 4\pi \cdot 10^6\ \chi_{trad.}$$

magnetic part arising from the Larmor precession of electrons in a magnetic field and polarization paramagnetism as described by van Vleck, increasing with decreasing energy gap /312/.

In **Fig. 98** the temperature dependence of the susceptibility χ of a semimetallic ($x = 0.03$) and a semiconducting ($x = 0.28$) sample before and after the annealing procedure is shown. Annealing always brings the values of χ closer to the straight line connecting the lattice susceptibilities of HgTe and CdTe (**Fig. 99**).

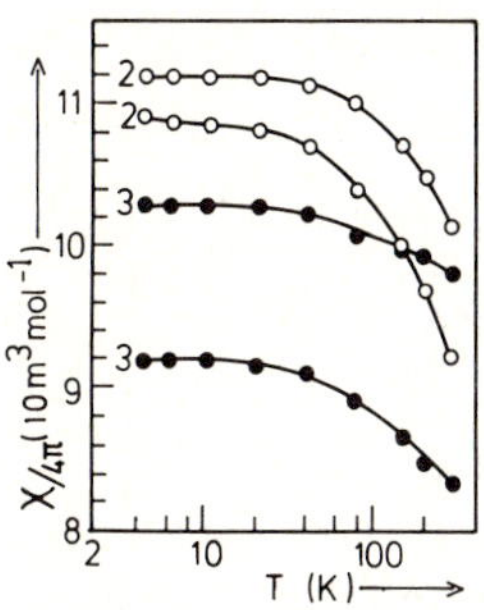

Fig. 98

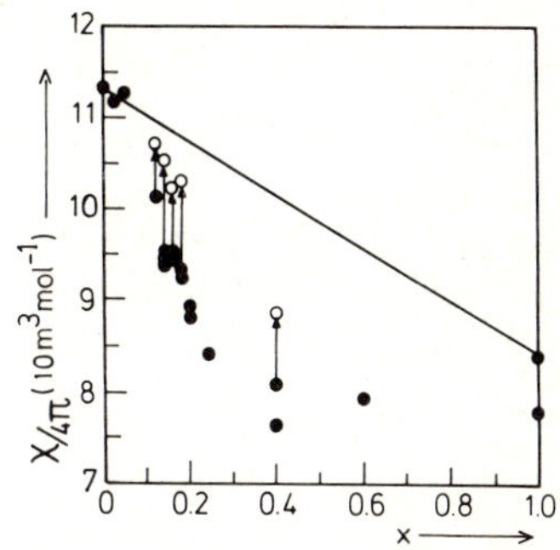

Fig. 99

Fig. 98. Magnetic susceptibility versus temperature curves for semimetals ($x = 0.03$, curves 2 and 2') and semiconductors ($x = 0.18$, curves 3 and 3') before (2,3) and after (2',3') annealing /310/

Fig. 99. Magnetic susceptibility of $Hg_{1-x}Cd_xTe$ versus composition x at 4.2 K before (•) and after (o) annealing /310/

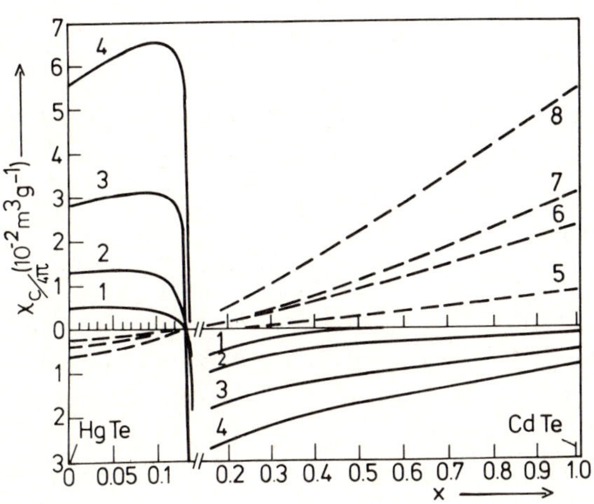

Fig.100. Theoretical charge carrier susceptibility versus alloy composition for 4.2 K

n =	10^{21} m^{-3}	(1)	
	10^{22} "	(2)	
	10^{23} "	(3)	
	$5 \cdot 10^{23}$ "	(4)	
p =	10^{22} m^{-3}	(5)	
	$5 \cdot 10^{22}$ "	(6)	
	10^{23} "	(7)	
	$5 \cdot 10^{23}$ "	(8)	/310/

In the region of intrinsic conduction ($T > 60$ K) the diamagnetic susceptibility of the electron gas is shown to give a vanishing contribution. So the observed strong decrease of the susceptibility with temperature is tentatively attributed to an increasing paramagnetic component associated either with complex defects, e.g. donor-acceptor associates, or deep donors or acceptors. In the low-temperature region SCHLABITZ et al. ascribed the slight increase in the susceptibility to the contribution of charge carriers in n- and p-type samples.

At the semimetal-semiconductor transition the susceptibility of the electron gas shows an abrupt reversal of its sign and an increase in its magnitude /141/ (Fig. 100).

5. Optical Properties

In this chapter on optical properties we present data on reflectivity and transmission spectra which are related to three physical processes. Spectra arising from electronic interband transitions are described in Sect. 5.1, divided into three parts according to the various experimental methods applied. Thus Sect. 5.1 deals with the properties of different band edges in the mixed crystal dependent on composition and temperature. Studies of the optical phonons in the $Hg_{1-x}Cd_xTe$ alloy are discussed in Sect. 5.2. Most of the data have been obtained from the analysis of infrared reflectance and absorption spectra, besides a few other physical effects caused by the optical phonons. The most interesting data such as phonon frequencies and their contributions to the dielectric constant will be presented in this section. The last part is concerned with optical effects of plasma resonance and infrared conductivity related to intraband transitions of free carriers.

In Sect. 3.7 the general aspects of the dielectric properties of narrow-gap semiconductors were discussed. The results of various magnetooptical investigations were also presented there.

5.1 Fundamental Reflectivity and Absorption

5.1.1 Reflectance and Absorption

Measurements of optical absorption and reflectance in solid solutions of $Hg_{1-x}Cd_xTe$ have been carried out mainly at energies near the band gap. They have provided valuable information on the dependence of the band gap on composition and temperature /26,30,100,101,104,105,313/. Data deduced from these experiments have been discussed already in Chap. 3.4 and led to analytical expressions for the x and T dependence of E_g and m_0^*. Typical results of the absorption coefficient as a function of photon energy for various compositions are shown in Fig. 101. The observed steepness of the absorption edge is characteristic for the direct energy gap in $Hg_{1-x}Cd_xTe$ alloys and does not vary noticeably in the temperature range between 10 and 300 K /100/. The absorption experiments shown in Fig. 101 refer to

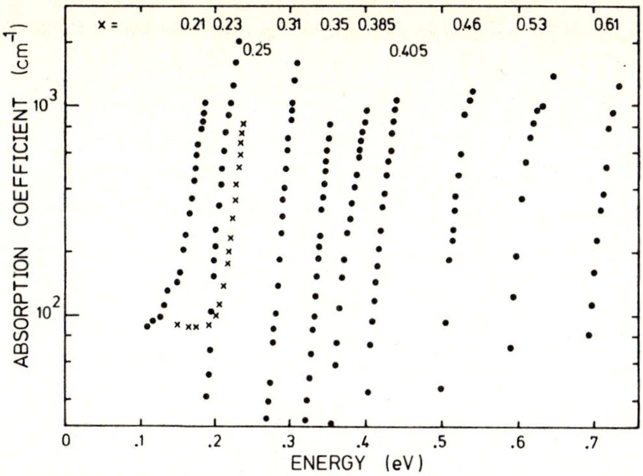

Fig.101. Room-temperature absorption coefficients versus photon energy for mixed crystals $Hg_{1-x}Cd_xTe$ /100/

samples cut from single crystals grown from a melt by a modified Bridgman method. Similar results have been obtained with epitaxial films /26,30/.

Reflectance measurements up to 4 eV with various compositions of $Hg_{1-x}Cd_xTe$ at photon energies much greater than the energy of the band gap have been carried out by GALAZKA and KISIEL /141/ and by KISIEL et al. /139/. They measured the reflectance in order to get information about higher interband gaps and their dependence on x. As shown in Fig.102a, two reflectance peaks are resolved within the investigated energy range. The observed structures, which shift non-linearly with x to higher energies, were related to the transitions $E_1(\Lambda_{4.5} - \Lambda_6)$ and $E_1 + \Delta_1$ $(\Lambda_6 - \Lambda_6)$ at the Λ-point of the Brillouin zone. In Sect. 3.6 these results are compared with calculated band structure data and with data obtained in several other experiments. In contrast to the reflectance spectrum in these experiments with single crystal samples the transmission and reflectivity spectra of epitaxial films show a linear dependence of the transition energies on x /138,314/. For $x < 0.1$ the energy functions plotted in Fig.102b seem to have a non-monotonic dependence on x. As discussed in Sect. 5.1.3 such an effect has not been observed in electroreflectance experiments.

Recently the temperature dependence of E_1 and $(E_1 + \Delta_1)$ transitions has been measured for samples with $0 \leq x \leq 0.168$ by KISIEL et al. /140/ in the temperature range between 100 and 315 K. A very interesting observation concerns the temperature dependence of E_1 with a sample of $x = 0.125$ where near 180 K the E_1-T relation deviates markedly from a linear one. Since this temperature corresponds to $E_0 = 0$ at the Γ point the authors suggest that the anomalous behaviour may be caused by an interaction between the Γ point and the Λ point at which the E_1 transition takes

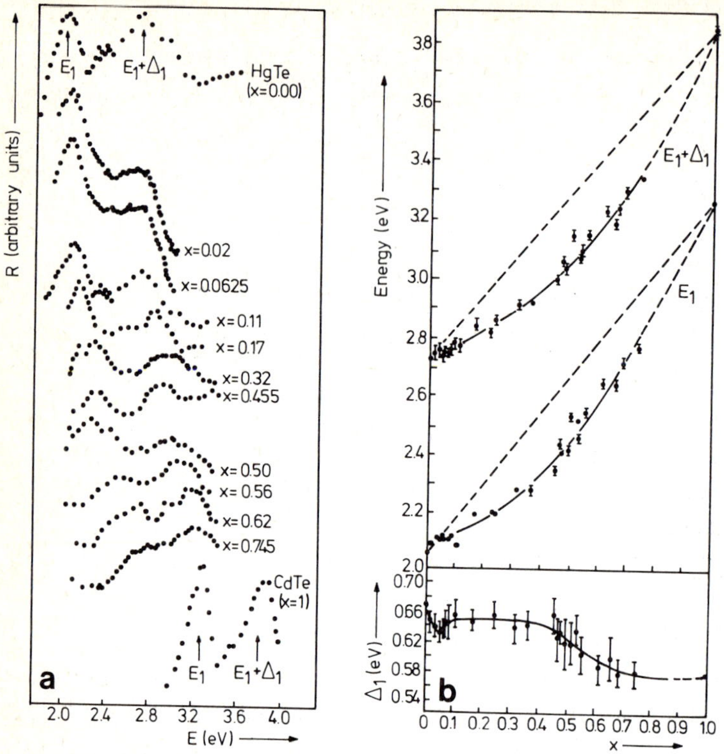

Fig.102. (a) Reflectivity versus photon energy for mixed crystals $Hg_{1-x}Cd_xTe$ at T = 315 K /141/. (b) Plot of energy of the reflectivity peaks versus composition x /141/

place in the Brillouin zone. This question is more extensively discussed in Sect. 3.6 and it seems that there is a link between the non-monotonic temperature dependence /140/ and the dependence on x described above /139/.

5.1.2 Magnetoreflectance

Magnetooptical experiments providing more detailed information than the corresponding zero field ones have made valuable contributions to the understanding of the electronic energy band structure. Magnetoreflection in $Hg_{1-x}Cd_xTe$ has been studied in fields up to 10 T for compositions near the Γ_6 - Γ_8 band crossover /73,96,114/. Figure 103 shows the reflected light intensity from n type $Hg_{.389}Cd_{.161}Te$ as a function of magnetic field as observed by GROVES et al. /114/. The traces were recorded during an up and down sweep of the magnetic field for two polarizations of the reflected light at a photon energy of 146 meV. The structure of the magnetic field-induced changes in reflectance correspond to various transitions between magnetic levels of conduction and valence bands. The analysis of the experimental results

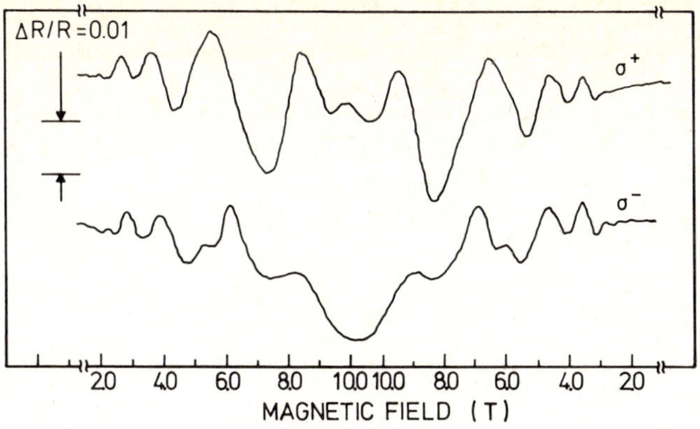

Fig.103. Traces of magnetoreflectance during an up and down sweep of magnetic field at a photon energy of 0.146 eV /114/

yields effective masses of valence and conduction bands /73,96,114/, which are discussed and presented in Chap. 3.

5.1.3 Electroreflectance

The introduction of modulation spectroscopy has considerably widened the scope of experiments to study transitions above the band gap. Electroreflectance is the most frequently used version of this technique. Recently MORITANI et al. /135,136/ have published experimental electroreflectance data for various $Hg_{1-x}Cd_xTe$ alloys associated with transition energies corresponding to the E_0, $E_0 + \Delta_0$, $E_1 + \Delta_1$, E_2, and $E_2 + \delta$ band transitions. Their work includes the first measurement of the spin-orbit splitting Δ_0 of the valence band in its dependence on x. As shown in Fig.102b the results of the Δ_1 energy as measured by GALAZKA and KISIEL /141/ show a minimum near $x = 0.07$. This non-monotonic behaviour was neither observed in the electroreflectance experiments /135,136/ nor in transmission experiments with epitaxial films /138/. For comparison the results of the different experiments aimed at the clarification of the band structure are presented in Fig. 36. However, KISIEL et al. /139/ have quite recently discussed further experimental evidence of an abnormal behaviour of Δ_1 and E_1 near $x = 0.07$. It has been suggested /139,140/ that the non-monotonic behaviour might be related to the transition from the inverted energy band structure to the InSb like structure. So far it is not known why the values of Δ_1 and E_1 at compositions $x < 0.1$ as obtained in different experiments differ so markedly.

5.2 Optical Phonons

Most of the knowledge on the optical phonon spectra of the $Hg_{1-x}Cd_xTe$ alloy was obtained from spectroscopic investigations in the infrared. Reflectance and transmission experiments at wavelengths above 20 μm have been carried out by different authors /315-318/ to study the properties of the optical phonons in their dependence on composition and temperature. It is of great interest to determine whether the mixed crystal will show a one- or two-mode behaviour and to ascertain the composition dependence of frequency and oscillator strength. Apart from their basic interest these data are necessary for describing and understanding various transport properties as discussed in Chap. 4. Early theoretical predictions about the one- or two-mode behaviour of the phonon spectrum have not been definite /317/. From local mode data LUCOVSKY et al. /319/ predicted a two-mode behaviour, which was experimentally proved for the first time by CARTER et al. /315/ by measurements of infrared transmission. The two-mode behaviour was confirmed later in a number of other experiments for various compositions /236,316-318,320/. The results on the dependence of the phonon frequency on x as deduced from reflectance experiments by BAARS and SORGER /317/ are presented in Fig.104 together with data from other experiments /316,318,320/. The two-mode behaviour is evident by inspecting Fig. 104 but the results obtained by different authors differ significantly (by amounts up to a few cm^{-1}). This cannot be ascribed to the fact that the experiments were carried out at different temperatures, since BURKHARD et al. /318/ have shown that at least at a composition with x = 0.20 the temperature shift of the phonon frequencies between 4 and 350 K is less than 2 cm^{-1}. Typical results of the real and imaginary part of the dielectric constant as calculated from the reflectance spectrum by a Kramers-Kronig analysis are shown for two temperatures in Fig. 105. The carrier density at 4.2 K of the investigated sample is $7.7 \cdot 10^{22}$ electrons/m^3.

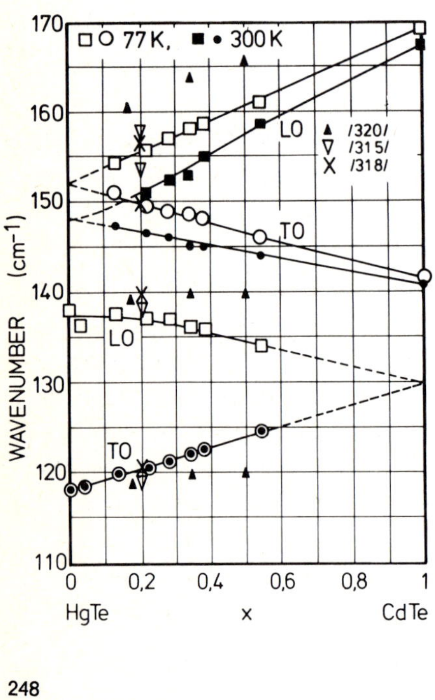

Fig.104. Longitudinal and transverse phonon frequencies for mixed crystals $Hg_{1-x}Cd_xTe$ /317/. In the figure also data are presented from /315,318,320/

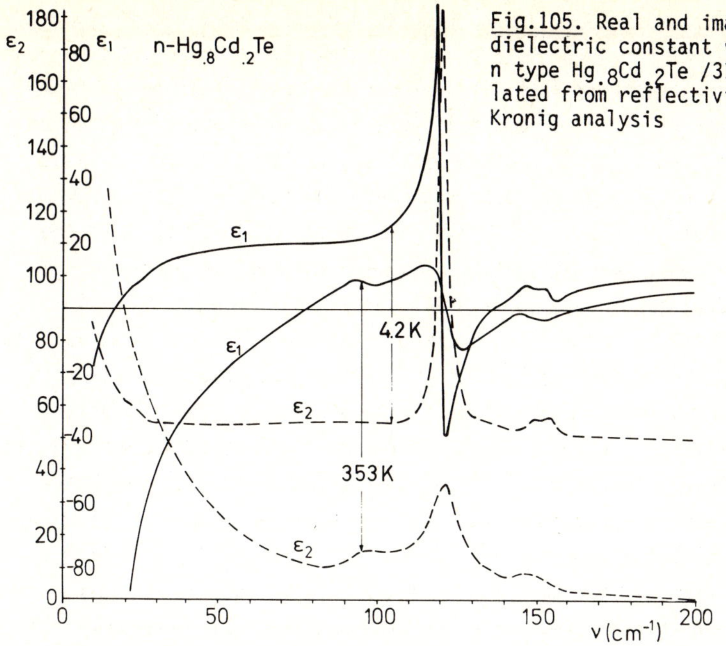

Fig.105. Real and imaginary part of the dielectric constant versus frequency for n type $Hg_{.8}Cd_{.2}Te$ /318/. The data calculated from reflectivity according a Kramers-Kronig analysis

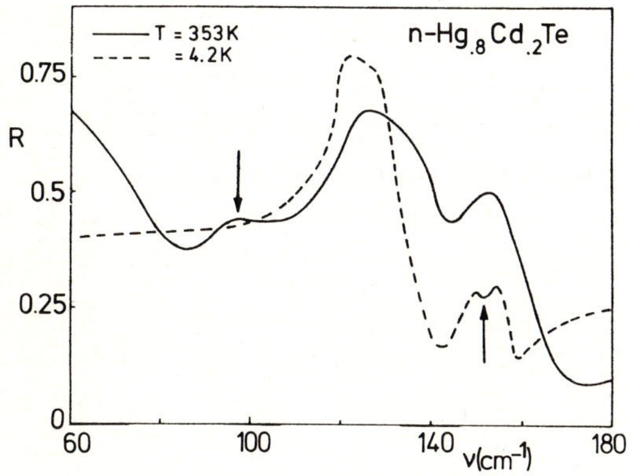

Fig.106. Reflectivity versus frequency for n type $Hg_{.8}Cd_{.2}Te$, the data correspond to Fig. 83 /318/

In the infrared reflectance measurements further structures besides the phonon bands have been observed. BAARS and SORGER /317/ have found a reflectance peak near the wave number of 100 cm^{-1} which increases with temperature, and suggested a subtractive phonon process of the type $\omega_{LO} - \omega_{TA}$. Their experimental result has been confirmed recently by BURKHARD et al. /318/ as shown in Fig.106, who, in addition, have observed a structure even in the reflectance band of the CdTe-like phonons at temperatures below 77 K. We think that this structure might be caused by different normal modes of which the two bands consist. In some other two-mode mixed crystals as for instance the system $GaAs_yP_{1-y}$ considerable fine structure in the two main bands was observed. This structure was described by VERLEUR and

BARKER /321/ using an isodisplacement model with thirteen coordinates. As shown in Fig. 1a for the $Hg_{1-x}Cd_xTe$ mixed crystal they consider five basic units of nearest-neighbour ions around a Ga-ion. Their model yields twelve optic modes which form the two main bands and can describe the experimental data of $GaAs_yP_{1-y}$ by an appropriate set of parameters. It is assumed that like-negative ions (As, P) have a tendency to cluster around positive ions (Ga) within a range of some lattice parameters. The deviation from the random distribution is described by a cluster parameter β, which is defined bv the relations for the probability of finding an As ion next to another As ion (P_{AA}) and vice versa for the P ions (P_{PP})

$$P_{AA} = y + \beta (1 - y)$$
$$P_{PP} = (1 - y) + \beta y \quad . \tag{58}$$

Thus when $\beta = 0$ the distribution of the like-charged ions is random depending on the average concentration y only, whereas for $\beta = 1$ the crystal consists of the basic units 1 and 5 (Fig. 1a). VERLEUR and BARKER /321/ obtained a good fit of the experimental data for $GaAs_yP_{1-y}$ by assuming a rather large amount of clustering. This interesting feature on the number of normal modes and on the clustering of the $Hg_{1-x}Cd_xTe$ alloy has not been investigated so far. A study of this problem might even enlighten some of the obscure discrepancies between experimental and theoretical data mentioned in previous sections. For instance studies on Raman scattering should give valuable information about additional modes and cluster formation. Unfortunately the only data published on Raman scattering in $Hg_{1-x}Cd_xTe$ by MOORADIAN and HARMAN /320/ do not contribute anything to this problem. A review of the vibrational properties of disordered solids including a critical discussion of the main theoretical models was given by SIEVERS and BARKER /322/ quite recently.

Experimental evidence for the LO phonons through their electron-phonon coupling was obtained in cyclotron resonance absorption experiments by KINCH et al. /97/. They deduced from the shift of the cyclotron resonance absorption line near the LO_{HgTe} phonon frequency a polaron coupling constant, discussed in Sect. 4.4.3. The electron-LO phonon interaction of both phonon modes was observed by GEORGITSE et al. /236/ through the oscillatory behaviour of the photoconductivity as a function of the energy of the incident radiation. In the photomagnetic effect they could detect additional oscillations which they interpreted in terms of TO phonon interaction with the photoionized carriers.

Figure 107 shows the static and high frequency dielectric constants as a function of x. The data given in /317,318,322/ are calculated from reflectance measurements (values obtained from samples with high carrier density corrected for plasma effects) while those given in /315/ are deduced from geometrical resonances in the transmission of thin samples. The main results are that the dielectric constant is not a linear function of x and that a temperature dependence was not observed

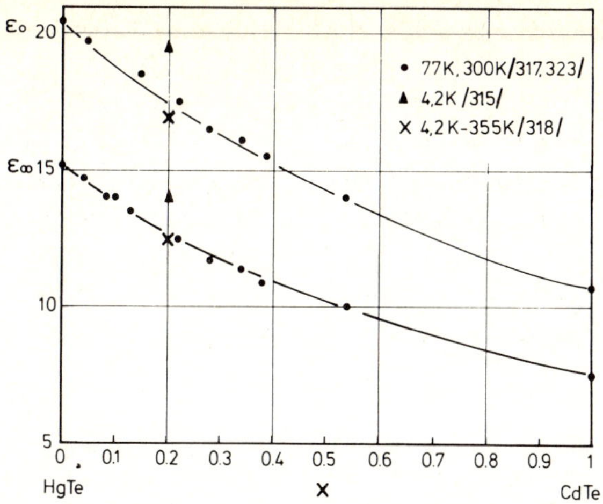

Fig.107. Static ε_0 and high frequency ε_∞ dielectric constants for the mixed crystal $Hg_{1-x}Cd_xTe$ /317,323/

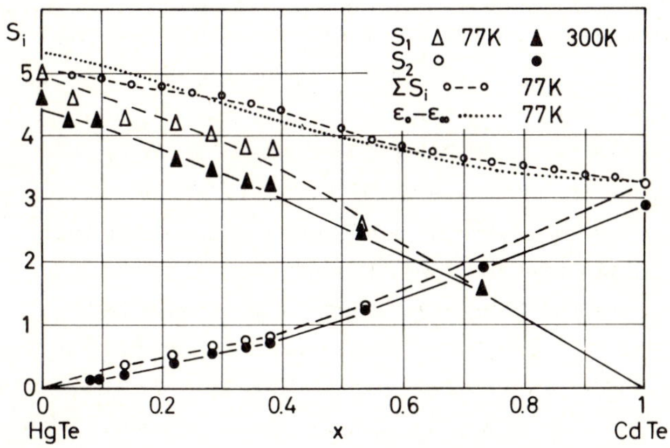

Fig.108. Oscillator strength versus composition x /323/

within the experimental resolution. Oscillator strengths of the two modes are given in Fig.108 and correspond to the analysis of BAARS and SORGER /317/. The pronounced difference between the strengths of the HgTe- and CdTe-like modes is evident, also the non-linear variation of the strengths with composition. The oscillator strengths were deduced from an oscillator fit and are compared with the results of a Kramers-Kronig analysis with $\sum_j S_j = \varepsilon_0 - \varepsilon_\infty$ /323/. Investigations of the phonon dispersion of the mixed crystal have so far not been published.

Both acoustical- and optical-phonon dispersion relations were not available at the time of the first edition. In the meantime results for HgTe from a neutron study were published by KEPA et al. /599,600/. The results are presented in Fig. 109. The solid curves of the optical-phonon dispersion represent a calculated fit (based on the rigid-ion model) between the measured zone-centre frequencies. The parameters were taken from the experimental data of the acoustical-phonon dispersion relation.

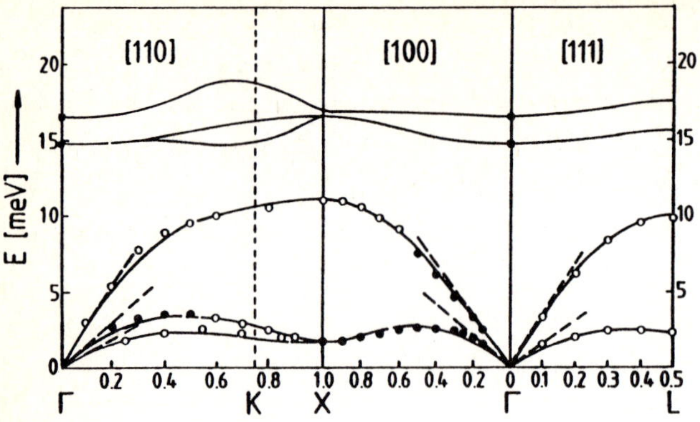

Fig. 109. Experimental results of phonon dispersion relations for HgTe obtained by neutron scattering; full circles represent data obtained at Swierk, open circles represent data obtained at Risø. Full squares are zone-centre optical frequencies. The solid curves represent a rigid-ion model fit to the experimental data. The initial slopes indicated by the broken lines were calculated from elastic constants /599,600/

Optical-phonon frequencies for Cd-rich compositions were investigated by HOCLET et al. /465/. Their results confirm earlier data by BAARS and SORGER /317/. In addition, the vibrational-mode frequency of the isolated Hg atom replacing Cd in the binary CdTe compound was determined to be at 131 cm^{-1}. The vibrational-mode frequencies of various other isoelectronic impurities in substitution for Cd in CdTe were also discussed.

Reflectivity measurements with $Hg_{.87}Cd_{.13}Te$ have been carried out to investigate the different far-infrared contributions to the dielectric function near the semimetal-semiconductor transition. Three contributions habe been considered - that from the phonons, the electronic-interband transitions, and plasmons. The best fit to the experimental result was obtained with a frequency-dependent contribution of the electronic-interband transition (see Sect. 3.7).

5.3 Free Carrier Absorption and Reflectance

The infrared absorption and reflectance spectra, especially near the plasma frequency, were investigated by various authors, mostly in n-type material /96,97, 104,317,318,323/. Such an analysis yields data on the effective mass as a function of carrier density and x, and values for the high frequency conductivity $\sigma(\omega)$. Qualitatively the data on the effective masses are in agreement with those determined by magnetic quantum effects. However, with the plasma frequency depending on parameters such as n and ε_o according

$$\omega_p^2 = n\, e^2/(\varepsilon\varepsilon_0\, m^*) \quad , \tag{59}$$

the data on the effective mass in this strongly non-parabolic material are not as reliable as those obtained from magnetic quantum effects. In Sect. 3.3.3 the results are compared with those determined by other experiments.

In the contribution of the plasma oscillation to the dielectric constant the damping can be described by a relaxation time

$$\Delta\varepsilon_p(\omega) = -\,\omega_p^2/(\omega(\omega + i\tau_p^{-1})) \tag{60}$$

thus we get for the conductivity

$$\sigma(\omega_p) = \varepsilon\, \omega_p^2\, \tau_p . \tag{61}$$

SORGER /323/ and BURKHARD et al. /318/ determined $\sigma(\omega_p)$ by their fit-parameters of the reflectance. The comparison with the dc conductivity σ shows /318/ that for $x = 0.2$ in the temperature range between 4 and 77 K $\sigma(\omega_p)/\sigma \approx 0.3$. Above 77 K the ratio increases with temperature and becomes ≈ 1 above 200 K. This results is in quantitative agreement with the observations of SORGER /323/ at 77 K and 300 K for several x values. The behaviour is plausible since at low temperatures we expect a considerable amount of carrier heating due to the absorption of the infrared radiation and relaxation effects since $\omega\tau \approx 1$. Both, the relative carrier heating and the relaxation effects decrease with increasing lattice temperature.

In the frequency range between the plasma edge and the HgTe phonon band near 30 cm^{-1} BURKHARD et al. /318/ observed an absorption peak which is shown in Fig. 9. As was mentioned in Sect. 2.3 it was suggested by DORNHAUS et al. /45/ to relate this absorption line to the transition between the conduction band and an acceptor level.

5.4 Photoluminescence and Deep Level Transient Spectroscopy

Photoluminescence studies in HgCdTe alloys have been reported in /406,407,601-604/.

In /601/ photoluminescence from $Hg_{.715}Cd_{.285}Te$ at 2 K was obtained using a CO laser with an excitation energy of 244.6 meV. The spectrum consisted of one line, which was interpreted as a band-acceptor line. The deduced acceptor energy was 14 ± 2 meV. The line width of 3.2 meV has been tentatively attributed to composition fluctuations on the scale of the Bohr radius.

In addition this study revealed that the radiative recombination is spin dependent. This information was obtained from measuring the rate of circular polarization of the fluorescence radiation. The analysis of the intensities of circularly and linearly polarized radiation in magnetic fields yielded an electron lifetime of 4 ns and a spin relaxation time of 2 ns at a temperature of 2 K.

In n-$Hg_{.7}Cd_{.3}Te$ /602/ and in n- and p-$Hg_{.52}Cd_{.48}Te$ three lines have recently been observed in the photoluminescence spectrum at low temperatures and two lines in Au-doped p-$Hg_{.68}Cd_{.32}Te$ at temperatures up to 30 K /406,407/. The higher-energy line has been interpreted as due to band-to-band transitions whereas the lower-energy line was attributed to donor-to-acceptor luminescence at low and to acceptor (bound-hole) to-free-electron luminescence at higher temperatures. Donor ionization energies of 2 meV /406,407/ and 3 ± 1 meV /602/ have been derived as well as acceptor energies of 14.0 ± 1.5 meV for $x = 0.32$ and 15.5 ± 2 meV for $x = 0.48$ in /406,407/ and 20 ± 2 meV in /602/. The third observed line at intermediate energies was ascribed to bound-exciton recombination in both cases.

POLLA and JONES /603-605/ have recently used deep level transient spectroscopy to characterize traps in Hg^+-vacancy doped HgCdTe samples. They were able to measure trap densities, energies, and capture cross-sections for electrons and holes. Two defect levels (Shockley-Read recombination centres) were observed for p-type non-stoichiometric (Hg lean) samples with compositions between $x \cong 0.2$ and $x \cong 0.4$, one slightly below $(1/2)E_0$ and one at $(3/4)E_0$. For most of the samples the ratio of the $(1/2)E_0$ trap density to shallow acceptor concentration was approximately 0.1 to 0.2. Capture cross-sections for electrons and holes on the $(1/2)E_0$ level are 10^{-19}-10^{-20} m^2 and 10^{-21}-10^{-22} m^2, for the $(3/4)E_0$ centre 10^{-20} m^2 and 10^{-21}-10^{-24} m^2. It was possible to correlate the trapping parameters with the observed excess carrier lifetimes in the samples (see Sect. 6.1). Data have also been presented for copper-doped samples /605b/ suggesting two levels pinned to the valence band at $E_v + 0.06$ eV and $E_v + 0.15$ eV.

6. Infrared Devices

As mentioned in the introduction the early work on the development of crystals of $Hg_{1-x}Cd_xTe$ alloys was directed towards their potential application as an intrinsic infrared photoconductor. Compositions with a maximum response at wavelengths near 10 μm were of particular interest. Since carrier lifetime governs both the responsivity and the rise time of photoconductive detectors we begin with a discussion of the dominant recombination mechanisms and the resulting carrier lifetime. In the subsequent sections photoconductivity and photovoltaic effects and their application as photodetectors are discussed. The last section deals with the generation of infrared radiation and the underlying mechanisms.

6.1 Recombination Mechanisms and Carrier Lifetime

In semiconductors one can distinguish three recombination mechanisms by which the thermodynamic equilibrium is established /324/. If the recombination takes place via recombination centres, the process is called Shockley-Read mechanism. These centres might be connected with lattice defects or impurities and have energy levels in the band gap. In the case of dominant recombination through Shockley-Read centres the carrier lifetime is proportional to the number of centres and thus might be controlled by crystal preparation. This contrasts with the so-called intrinsic mechanisms governed by radiative or Auger recombination, where the lifetime cannot be influenced by techniques of crystal preparation.

In the radiative process the recombination energy is converted into a photon and the lifetime is proportional to the number of carriers. The Auger mechanism represents the inverse process of impact ionization and proceeds via electron-electron interaction. An electron and a hole recombine while a third carrier takes up the recombination energy. Auger recombination shows the strongest dependence on carrier density since statistically this process represents a three particle collision. In semiconductors the carrier lifetime is usually determined by Shockley-Read and radiative recombination processes. The Auger process is expected to be of importance only in semiconductors with a narrow band gap and a pronounced difference in conduction band and valence band effective masses.

Up to now the recombination behaviour of the $Hg_{1-x}Cd_xTe$ alloy has been studied only in n-type material with compositions near $x = 0.2$. The results of the early investigations are discussed by LONG and SCHMIT /2/ and REYNOLDS et al. /325/. More recently definite results on the dominant intrinsic recombination process were obtained in three different types of experiments /238,326-329/.

PETERSEN /330/ investigated theoretically the carrier lifetime for Auger recombination in n type $Hg_{1-x}Cd_xTe$ taking into account the strong non-parabolic bands in the narrow band gap alloys. He presented numerical values for $x = 0.20$ and 0.27 in the temperature range between 77 and 300 K, which show that the dominant intrinsic recombination process for these compositions is expected to be the Auger process. In Fig.110 experimental results from KINCH et al. /326,329/ are compared with theoretical data of the intrinsic carrier lifetime τ_{Ai} for dominant Auger recombination and τ_{Ri} for dominant radiative recombination. The comparison of the experimental and calculated data shows that for temperatures above 120 K the lifetime is determined by Auger recombination. Below 120 K the samples are in the extrinsic range and in the case of Auger-dominated lifetime τ_A the following relation should hold /324/

$$\tau_A \approx \tau_{Ai} \, (n_i/n_o)^2 \qquad (62)$$

where n_i and n_o are the intrinsic and equilibrium carrier densities, respectively.

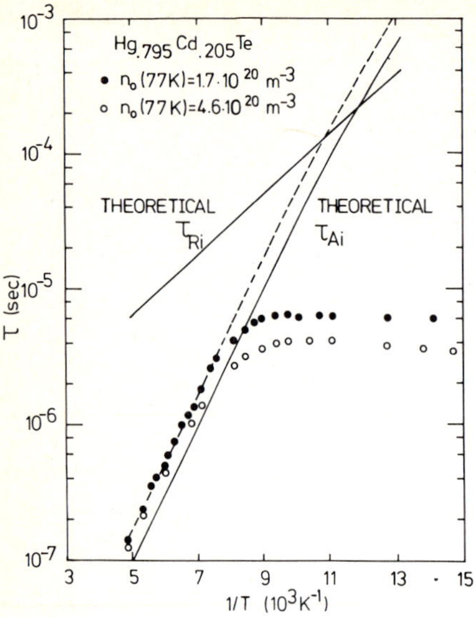

Fig.110. Carrier lifetime versus reciprocal temperature. Here τ_{Ri} and τ_{Ai} (solid lines) represent calculated values for the intrinsic radiative and Auger lifetime, respectively /326,329/

This yields for the lifetimes of the two samples (Fig.110) according to their carrier densities a ratio of 7.3 whereas the experimental values in Fig.110 correspond to a ratio of only 1.5 at 77 K. Thus we conclude that the observed lifetimes are also influenced by a Shockley-Read recombination at low temperatures in this experiment.

Equation (62) holds for excess carrier densities which are small compared to the equilibrium density. At large deviations from thermal equilibrium the lifetime changes with excess carrier density n_e according /324,238/

$$\tau_A \approx 2\, \tau_{Ai}\, (n_i/(n_e + n_o))^2 \qquad (63)$$

Since only the Auger recombination mechanism shows such a strong dependence on n_e /324/ the measurement of the transient decay of large excess carrier densities gives a statistical criterion for the dominant recombination mechanism. Such a transient carrier decay was measured by NIMTZ et al. /238,327/ using a microwave technique and excess carriers generated by impact ionization in electric fields (Fig.111). The comparison with calculated decays for Auger and radiative recombination shows the steep decay as expected for Auger. BARTOLI et al. /328/ measured the photoconductivity of n-$Hg_{.8}Cd_{.2}Te$ as a function of a 10.6 μm CO_2-laser flux at 80 K. Their experimental data show at carrier densities above $1.5 \cdot 10^{21}$ m^{-3} an Auger-dominated lifetime. Thus it can be concluded that at temperatures $\geq$ 77 K the lifetime of high mobility bulk $Hg_{1-x}Cd_xTe$ ($x \approx 0.2$) is determined by Auger recombination. The lifetime depending on n_o can be calculated according (63) with τ_{Ai} and n_i as given in /330/ and /199/, respectively. In early work lifetimes were

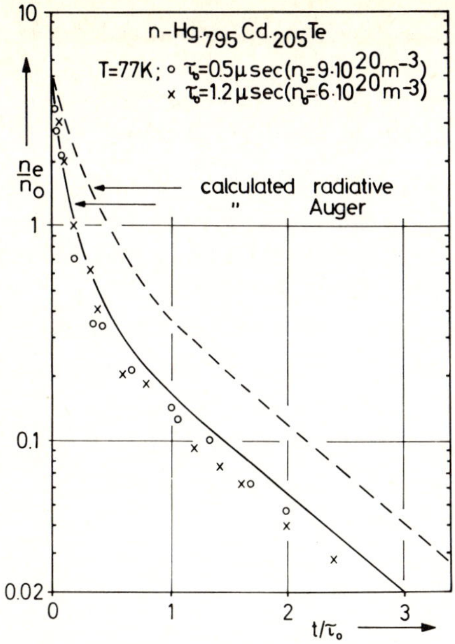

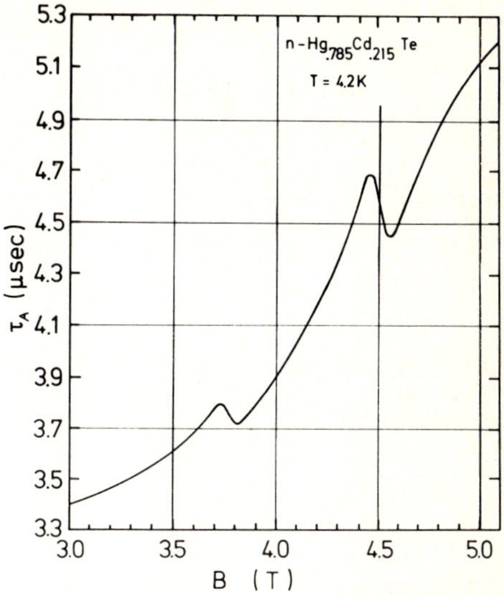

Fig.111. Relative excess carrier density versus decay time t /7/. τ_0 is the lifetime for $n_e << n_0$. The solid and the broken lines represent calculated data for Auger and radiative recombination dominated transient decay, respectively /238/

Fig.112. Auger governed lifetime versus magnetic field ($n = 5\cdot 10^{20}$ m^{-3}) /290/

often observed which are shorter or even longer than the Auger lifetime /105,325, 331/. This is assumed to arise from competing Shockley-Read mechanisms in the bulk of less pure crystals or in thin samples with a high rate of surface recombination /238,326,327/. An enhanced recombination is met especially in photovoltaic detectors caused by Shockley-Read recombination or minority carrier sweep-out. The effect of minority carrier sweep-out in photoconductors was investigated by WILLIAMS et al. /332/.

Recently TAKESHIMA /297/ calculated the Auger recombination in quantizing magnetic fields (Sect. 4.4.6). An oscillatory behaviour of the lifetime was predicted as an effect of Landau quantization on the density of states. On account of its dominant Auger recombination even at low carrier densities n type $Hg_{1-x}Cd_xTe$ with $x \leq 0.25$ should be an excellent material to test this prediction.

Quite recently this new effect was observed by DORNHAUS et al. /290/ in n type $Hg_{.785}Cd_{.215}Te$. Preliminary results are presented in Fig.112. In magnetic fields up to 5 T two structures were found in the Auger-governed lifetime of excess carriers which are assumed to be caused by the magnetic quantization of the conduction band. As expected the observed structures increase with magnetic field, however, contrary to the numerical data of TAKESHIMA /297/ the amplitudes are much smaller and the lifetime increases with magnetic field /290/.

During the last few years various papers have been published on recombination which consolidate the importance of the Auger effect as intrinsic recombination mechanism in the narrow-gap compositions of the $Hg_{1-x}Cd_xTe$ alloy /606-612,626/. The studies include low-temperature data and degenerate statistics, but also p-type material. An introduction to the special features of interband recombination in narrow-gap semiconductors was presented recently by NIMTZ /613/.

In earlier calculations of the transition rate for Auger collision processes, different estimates of overlap integrals were used. Besides taking into consideration a degenerate electron gas, which becomes important at low temperatures, GELMONT /607,608/ and GERHARDTS et al. /606/ calculated the overlap integrals of conduction- and valence-band Bloch functions from a k•p calculation. Numerical data of the Auger lifetime in n-type crystals are plotted in Fig. 113. In the intrinsic temperature range ($T \gtrsim 100$ K) the lifetime is governed by

$$\tau \propto n^{-2} \propto \exp(E_0/kT) \quad . \tag{97}$$

At lower temperatures the material becomes extrinsic, and the lifetime is essentially dependent on the density of states in the valence band

$$\tau \propto N_v(T) \propto T^{3/2} \quad , \tag{98}$$

i.e. it decreases in the intermediate temperature interval with decreasing T. At very low temperatures the lifetime becomes dominated by an exponential with an activation energy of the order m_0^*/kT /606/. In Fig. 113 some experimental data

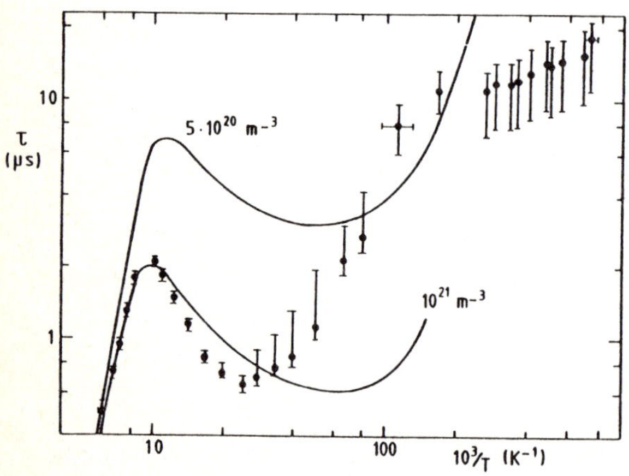

Fig. 113. Experimental lifetime data τ vs temperature T of n-type $Hg_{.8}Cd_{.2}Te$ [n_0 (77 K) = $4\cdot10^{20}$ m^{-3}]. Solid lines give theoretical values for comparison /606/

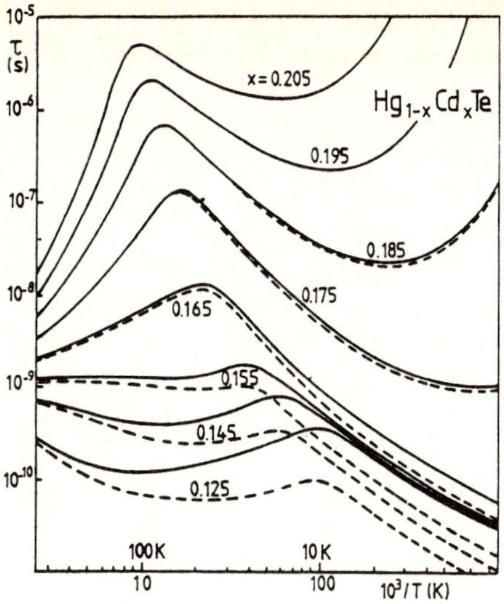

Fig. 114. Auger lifetime τ vs temperature T for several x values [n_0 (77 K) = 10^{21} m^{-3}, E_p = 19 eV], (———) with screening, (-----) without screening /606/

are presented for comparison with the theoretical data. The agreement is good in the intrinsic and extrinsic range at temperatures above 50 K. At lower temperatures the predicted temperature dependence is found only qualitatively. The lifetime saturation observed at low temperatures is not due to a volume recombination mechanism but caused by an experimental limitation.

The variation of the lifetime with band gap, i.e. with composition of the alloy, is shown in Fig. 114. According to the theoretical data, the low-temperature exponential increase of lifetime disappears in semimetallic crystals. In the same figure, the influence of the screening of the Coulomb interaction due to free carriers is demonstrated. The broken lines represent unscreened data calculated with λ = 0 with the screening parameter given by

$$\lambda^2 = 4\pi e^2 \varepsilon^{-1}\, dn/d\xi \tag{99}$$

where ξ represents the chemical potential. Obviously, the free carrier screening becomes more and more important with decreasing energy gap.

Recently some investigations have been devoted to elucidating the recombination processes relevant to p-type crystals /609,611,614/. This research on p-type material has become necessary in order to achieve np-photodiode operation at the theoretical performance limits. CASSELMAN and PETERSON /611/ showed that in p-type crystals the Auger hole-hole collision process, involving a transition from the light- to the heavy-hole band, is important as long as the excess carrier density is small compared with the equilibrium hole density. This transition involving three bands is sketched in Fig. 115 and compared with the elec-

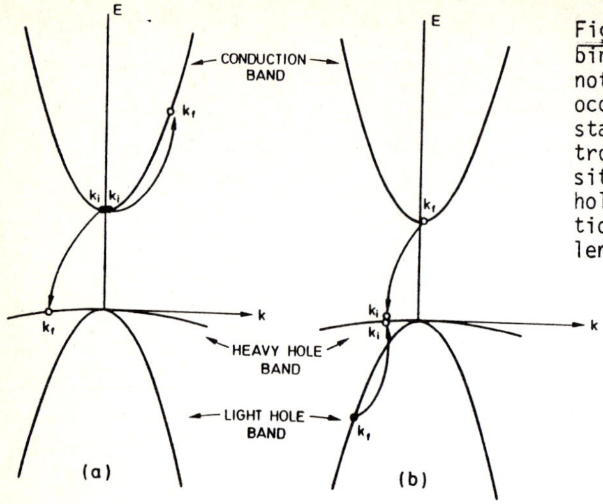

Fig. 115a,b. Principle Auger recombination transitions; i and f denote initial and final states, (•) occupied states, and (o) unoccupied states. (a) The most important electron-electron collision Auger transition, and (b) the most important hole-hole collision Auger transition; the latter includes both valence bands /611/

tron-electron collision process, which generally represents the most important one. Numerical lifetime data for p-type $Hg_{.8}Cd_{.2}Te$ are shown in Fig. 116; both Auger transitions of Fig. 115 were taken into consideration and the data are compared with theoretical radiative lifetimes /609/. It follows from these data that the Auger effect represents the most important intrinsic recombination mechanism in p-type material. Experimental data obtained in p-type $Hg_{.8}Cd_{.2}Te$ crystals have been published in /405,615,616/. From an experimental investiga-

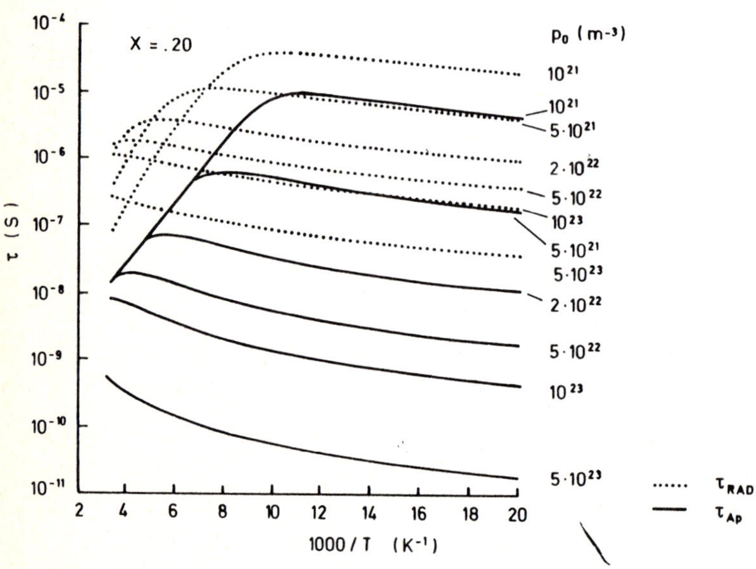

Fig. 116. Calculated radiative (τ_R) and Auger (τ_A) lifetimes as a function of reciprocal temperature for x = 0.20 and various hole densities p_0 /609/

tion of the recombination mechanisms in n^+-on-p $Hg_{1-x}Cd_xTe$ ($0.2 < x < 0.4$), POLLA et al. /614/ concluded that a Shockley-Read process governs the lifetime of excess carriers. From the lifetime dependence on temperature they deduced an energy of the composite trap level 0.14 eV above the valence band edge in x = 0.39 material, 0.12 eV in x = 0.32 material, and 0.06 eV in x = 0.22 material.

Interband recombination assisted by single elementary excitations, for instance, optical phonons or plasmons, has been the subject of various theoretical and experimental investigations during the last decade /613/. It was predicted that such recombination channels are very efficient if band-gap energy and single-excitation energy are of the same value. In searching for the importance of such recombination channels DORNHAUS and NIMTZ /617,618/ varied the band gap by varying a magnetic field to establish resonance condition (Fig. 117).

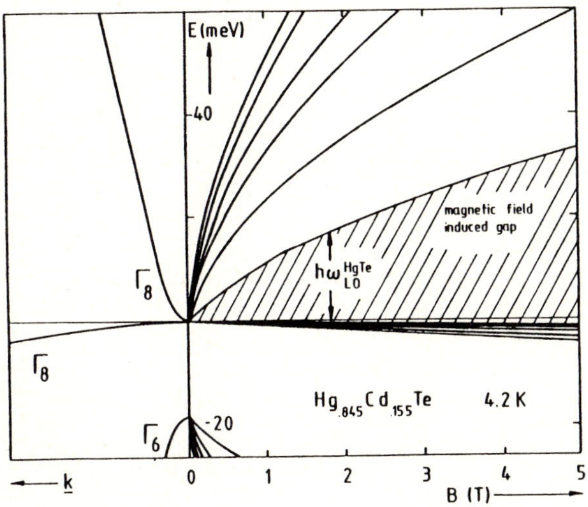

Fig. 117. Band structure of $Hg_{.845}Cd_{.155}Te$. Left side shows the energy-momentum relation without magnetic field. Right side shows the magnetic field dependence of the band extrema at the Γ point /617-618/

Starting with semimetallic compositions of n-type $Hg_{1-x}Cd_xTe$, with x in the range between 0.14 and 0.165, the magnetically induced band gap can easily be adjusted to the relevant energies of elementary excitations by applying moderate magnetic fields, and in addition, the resonance can easily be switched on and off. The results of photoconductivity measurements show that these recombination channels dominate the recombination behaviour around the energy resonance between the band gap and single LO phonon or plasmon /617,618/. Accordingly, these channels severely limit photoconductive detectors at wavelengths above 50 µm. On the other hand, the channels may be used to generate high densities of LO phonons or plasmons. Theoretical problems involved in a plasmon-assisted recombination in a magnetic field were investigated by WOLFF et al. /619a,b/.

A new magnetic quantum effect was predicted recently by TAKESHIMA /297/. According to this prediction the Auger transition rate should oscillate in quantizing magnetic fields. Generally the Auger transition is handicapped by the simultaneous conservation of both energy and momentum. This results in an effective band gap up to 50% higher than the intrinsic energy gap as shown by BEATTIE and LANDSBERG /620/. However, in a magnetic field vertical transitions between conduction- and valence-band extrema become allowed if the band gap equals the energy separation between two levels of the conduction-band Landau spectrum. The quantum effect was discovered by DORNHAUS et al. /613,621/ in various narrow-gap alloys of the $Hg_{1-x}Cd_xTe$ system. The experimental data in **Fig. 118** demonstrate the oscillatory behaviour of the lifetime in magnetic fields; lifetime minima correspond to maxima in the Auger transition rate. The experimental data show that non-parabolicity of the conduction band and the spin polarization in the magnetic quantum limit play an important role in this quantum effect /613/. The influence of both properties were considered in the theoretical papers by GERHARDTS /622,623/.

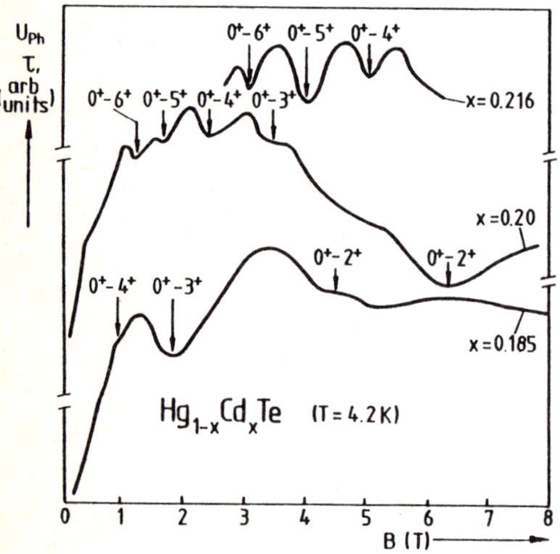

Fig. 118. Experimental photoconductivity and lifetime vs magnetic field for three different n-type $Hg_{1-x}Cd_xTe$ compositions. The experimental lifetime minima were identified by calculated maxima of the Auger transition rate and indicated by arrows; e.g. at the minimum 0^+-3^+ at a field of 1.8 T for x = 0.185, the band-gap energy equals the energy between the 0^+ and the 3^+ conduction-band Landau level /613,621/

6.2 Photoconductivity

The basic phenomenon underlying the working of intrinsic photodetectors is the photoionization of electron-hole pairs. There are two ways in which it can manifest itself, namely as photoconductivity or as photovoltaic effect. The usual experimental set-up of a photoconductive detector is sketched in **Fig.119**. The photoconductive crystal with resistance R_D is biased with voltage V and, by virtue of $R_D \ll R_L$, operated under constant current conditions. If Φ_S represents an incident photon flux density the signal voltage in n-type material is /329/

$$V_s = \eta\, V\, \Phi_s\, \tau/(n_o t) \qquad (64)$$

where η represents the quantum efficiency and τ the carrier lifetime. This relation reveals the conditions necessary for a high photoconductive response to a given radiation: a low thermal equilibrium carrier concentration and a long excess carrier lifetime. The first experimental results on photoconductivity and related effects in $Hg_{1-x}Cd_xTe$ alloys were reported by LAWSON et al. /1/. Since then this topic has formed the subject of a large number of investigations /1,2,4,26,30,49,105,236,237, 288,313,325,329,332-348/. Figure 120 shows typical results, obtained by KRUSE et al. /105/, on the wavelength dependence of the photoconductive response. The maximum response is shifted with increasing x and increasing temperature to smaller wavelengths. This is explained by the dependence of the band gap on composition and temperature and demonstrates the possibility of adjusting the maximum response to users' requirements. For low signal intensities the photoconductive response has to compete with noise: the usefulness of an infrared detector is given by the spectral detectivity D^*, which is proportional to the ratio of the photoconductive responsivity and the total noise voltage. The main noise sources are the Johnson or thermal noise and the generation-recombination noise /2,329/. As demonstrated by WILLIAMS et al. /332,338/ the ultimate detectivity might be limited also by the sweep-out of minority carriers under the action of the bias voltage.

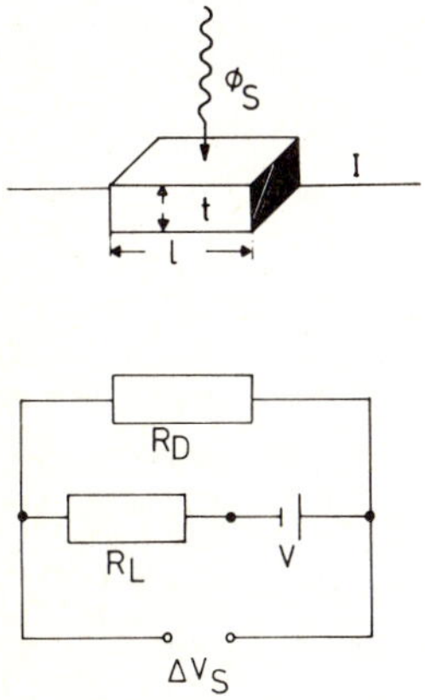

Fig. 119

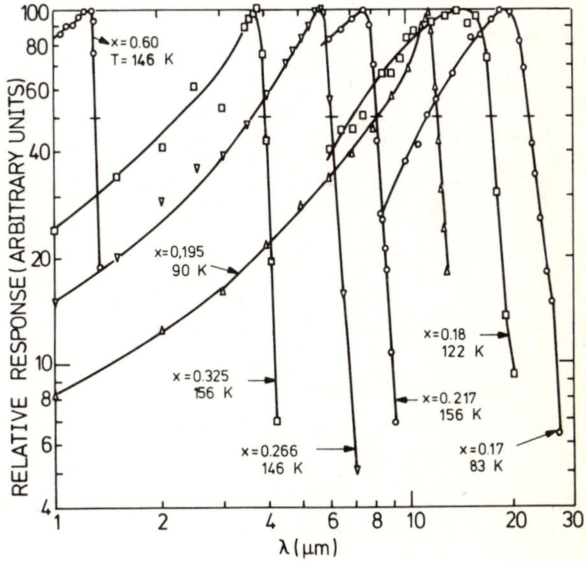

Fig. 120

Fig.119. Photoconductor geometry and bias circuit

Fig.120. Relative photoconductive response vs wavelength for $Hg_{1-x}Cd_xTe$ samples having the alloy compositions x indicated /105/

Usually the detector geometry is of the same order of magnitude as the carrier diffusion length L. Then the surface recombination might dominate the bulk recombination and the effective carrier lifetime might be more or less influenced by the surface recombination. Experimental results of the surface recombination velocity in n-type material (x = 0.195 and 0.205) at 77 K are reported by KINCH et al. /326/ and are of the order of 1 m/sec. Taking into account the effect of surface recombination and assuming Auger limited bulk lifetime, the temperature dependence of the ultimate detectivity for n type $Hg_{.795}Cd_{.205}Te$ was calculated by KINCH and BORELLO /321/. Their results, presented in Fig.121, show a detectivity of $4 \cdot 10^{10}$ m $Hz^{1/2}$ W^{-1} at a temperature below 80 K. The constant detectivity at low temperature arises from the sweep-out of minority carriers, while the decrease with increasing temperature above 80 K is accounted for by the temperature dependence of the carrier concentration.

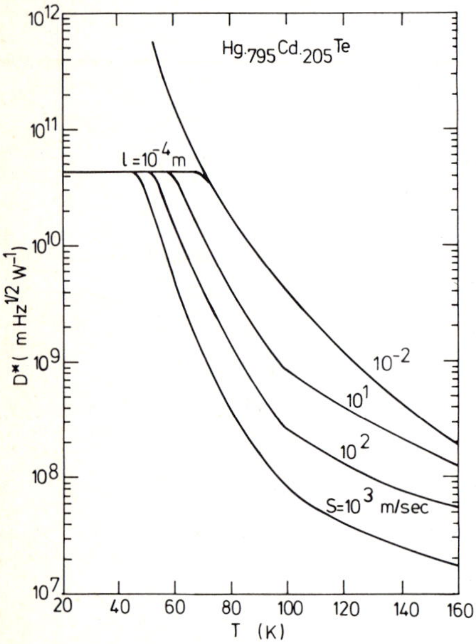

Fig.121. Calculated detectivity vs T for the n-type $Hg_{.795}Cd_{.205}Te$ photoconductor with $n_0 = 4 \cdot 10^{20}$ m^{-3}; t = 10 µm with surface recombination velocity s as parameter /321/

High detectivities between 10^8 and 10^9 m $Hz^{1/2}$ W^{-1} were observed in the 8 - 14 µm region operating at 77 K /340/, much higher than those in $Pb_{1-x}Sn_xTe$ alloys /4/ (10^6 - 10^7 m $Hz^{1/2}$ W^{-1}). Thus, it seems that in this infrared region the $Hg_{1-x}Cd_xTe$ mixed crystal is the most favourable photoconductive detector material available at present. The main problem encountered in large scale preparation of photoconductors based on $Hg_{1-x}Cd_xTe$ is the evaporation of Hg atoms from uncoated surfaces causing defects, as discussed in Sect. 2.3.1.

The optical immersion of $Hg_{1-x}Cd_xTe$ photoconductive detectors was studied by SLAWEK and RANDALL /348/. The immersion of a 2.5 x 2.5 mm^2 detector element to a germanium hemispherical lens improved markedly the voltage responsivity and the effective detector area was approximately 10 x 10 mm^2.

$Hg_{.8}Cd_{.2}Te$ detectors are widely used to measure CO_2 laser radiation. Thereby the detectors are often thermally loaded due to intense laser radiation. The thermal dependence of detector properties and the thresholds for permanent damage were studied by BARTOLI et al. /345,346/ and by ALLEN et al. /349/.

The variation of spectral response with mole fraction of Cd to Hg was applied to fabricate multispectral photodetector arrays or n-colour photodetectors /350, 351/. The description of a multispectral scanner built for a skylab for earth scanning was given by ALDRICH and BECK /350/.

Photo effects in the presence of a magnetic field were investigated in $Hg_{1-x}Cd_xTe$ by GEORGITSE et al. /236,237/. The results yield information on electron-phonon interaction and on the optical heating of carriers. Some aspects are discussed in Sects. 4.4,5.

6.2.1 Photoconductive Infrared and Microwave Devices

Recently various investigations were devoted to improving photoconductive detectors and to calculating the ultimate performance /612,624-629/. The influence of background radiation on detector responsivity and noise level is discussed in /630-632/. Properties of photoconductive detectors operated at room temperature were investigated in /633/, and the application of graded-gap material for broadband photoconductive detectors was studied by PAWLIKOWSKI /634/.

In addition to the established qualification of HgCdTe crystals for infrared applications, it was recently shown that this material may also be used profitably for millimeter-wave sensors /635/. WEBER and KULPA /636/ investigated the photoconductivity from free-carrier microwave absorption in n-type $Hg_{.8}Cd_{.2}Te$ at temperatures between 4.2 K and 40 K. The responsivity of this device was about 0.01 times that of an InSb bolometer; however, the observed temporal response of HgCdTe was at least 10 times faster.

It was shown by NIMTZ et al. /635,637/ that n-type $Hg_{.8}Cd_{.2}Te$ performes an excellent low power millimeter-wave superheterodyne detection if operated at temperatures below 3 K and magnetic fields of about 7 T. This hot carrier bolometer effect is based on the recently discovered Coulomb correlation effect discussed in Sect. 4.3. If compared with the classical InSb hot carrier frequency mixer, the new device offers two advantages: 1) a higher responsivity up to a factor of 10 and 2) the IF band width is more than doubled.

6.3 Photovoltaic Effects

Photovoltaic effects are usually observed in structures with built-in potential barriers, when excess carriers are injected optically into the vicinity of such a barrier. In the last few years they were widely investigated in various structures of the $Hg_{1-x}Cd_xTe$ alloy. Nowadays photovoltaic detectors from these alloys are commercially available at least with maximum response in the wavelength range near 10 μm. Structures with p-n junctions are prepared usually by one of three techniques, namely diffusion of Hg atoms into p-type single crystals prepared from the melt or into p-type epitaxial grown layers /2,4,29,333,334,344,352-358/, proton or electron bombardment /4,37,49,50,359/, or ion implantation /360-363/.

In Fig.122 the sketch of a mesa photo diode is shown, the p-n junction was formed by diffusion of Hg atoms into the p-type substrate which produces an n-type layer on the surface /4/. This technique seems to be the most simple and up to now the most successful one. Typical spectral response curves of photovoltaic detectors made by a diffusion technique are shown in Fig.123 /353/. Diodes made by diffusion of Hg atoms into epitaxial layers exhibit a spectral response which is flatter over a wavelength range as seen in Fig.124 /29/. This broader response is assumed to be connected with the composition gradient perpendicular to the surface, which is a characteristic of epitaxial layers. The formation of p-n junctions in $Hg_{1-x}Cd_xTe$ by proton bombardment was reported by FOYT et al. /359/, and discussed by HARMAN and MELNGAILIS /4/. Instead of the diffusion process the type conversion is achieved by the proton bombardment of the surface. As discussed in Sect. 2.3.1 the nature

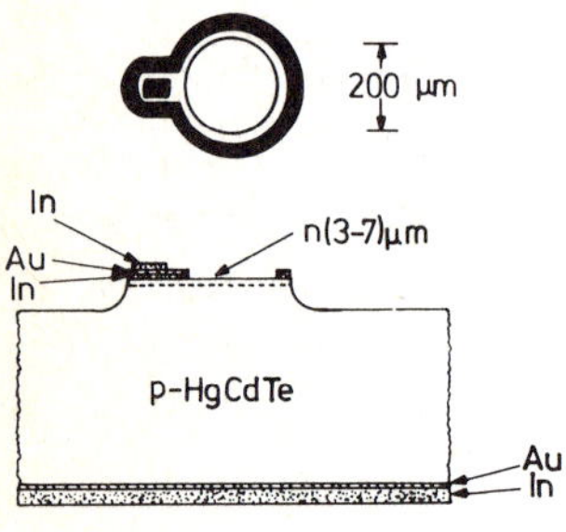

Fig. 122

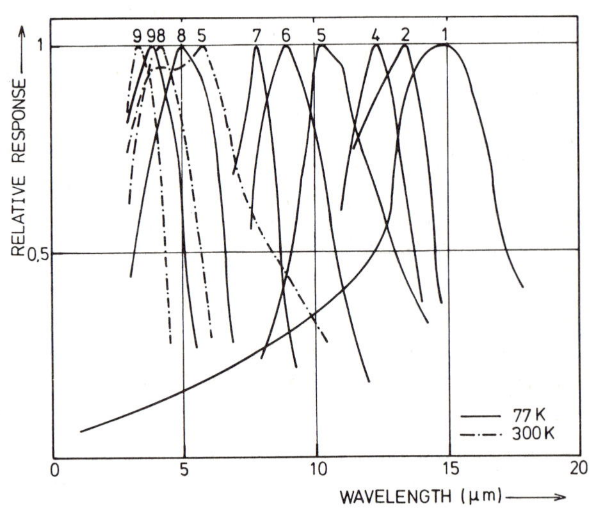

Fig. 123

Fig.122. Structure of a $Hg_{1-x}Cd_xTe$ mesa photodiode with a 200 μm diameter active area /4/

Fig.123. Relative photovoltaic response of $Hg_{1-x}Cd_xTe$ vs wavelength. (9) $x = 0.28$; (8) $x = 0.23$; (7) $x = 0.21$; (6) $x = 0.20$; (5) $x = 0.19$; (4), (2), and (1) $0.19 < x < 0.15$ /353/

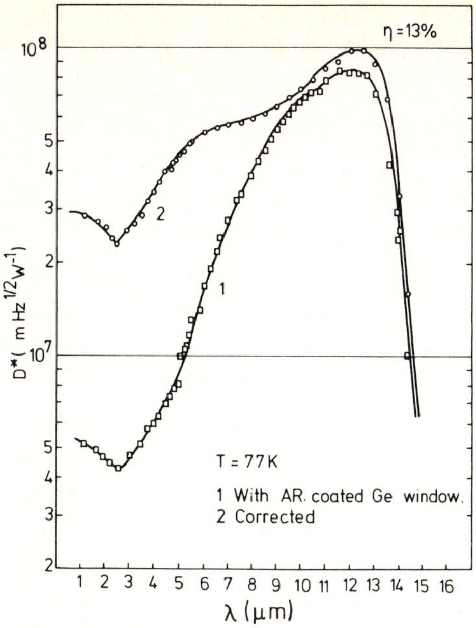

Fig.124. Detectivity vs wavelength for an epitaxial $Hg_{1-x}Cd_xTe$ diode. Curve 1 is obtained with an AR-coated Ge window, curve 2 is calculated from curve 1, η is the peak quantum efficiency /29/

of the defects generated by the bombardment and causing this conversion is as yet unknown. The measured data such as detectivity are similar to those obtained by the diffusion technique. Also by the implantation of Hg- and Al-ions p-type material is converted into n type at the surface. The preparation of high quality photovoltaic elements by this method has been reported in /360-363/.

Photovoltaic effects in graded band gap structures were also investigated in HgCdTe /24,334,364,365/. These structures, where the alloy composition and thus the band gap vary with position, are usually prepared by an epitaxial deposition of HgCdTe on CdTe substrates.

The speed of response of fast photovoltaic detectors is capacitance limited and was found to be $\leq$ 1 nsec. /4,24,344,355/. The fast response of HgCdTe alloys is helped by the relatively small dielectric constant between 10 and 20 compared with, for example, a value of 400 in PnSnTe, which is another narrow band gap material for intrinsic infrared photodetectors. Signals at frequencies up to 60 GHz have been detected using varactor mixing in fast HgCdTe mesa photodiodes /356/.

The investigations on photoeffects have shown the HgCdTe alloy to be a very useful material for photodetector devices. Both, photoconductive and photovoltaic diodes are applied in the detection of infrared radiation. KINCH and BORELLO /329/ have studied the ultimate performance of both modes. Their conclusion is that the highest detectivity D^* can be achieved with PC diodes at temperatures above 70 K, while at temperatures below 70 K the higher D^* can be achieved with the PV mode. The limiting factors determining the ultimate D^* in the case of photoconductors are the surface recombination velocity and the degree of purity of material available at present. The D^* of photovoltaic diodes is limited by the noise contribution of generation-recombination currents in the depletion region and possible by perturbing effects at the HgCdTe surfaces /329/. Calculated data of D^* for diffusion and for generation-recombination current limited cases are shown in Fig.125./329/.

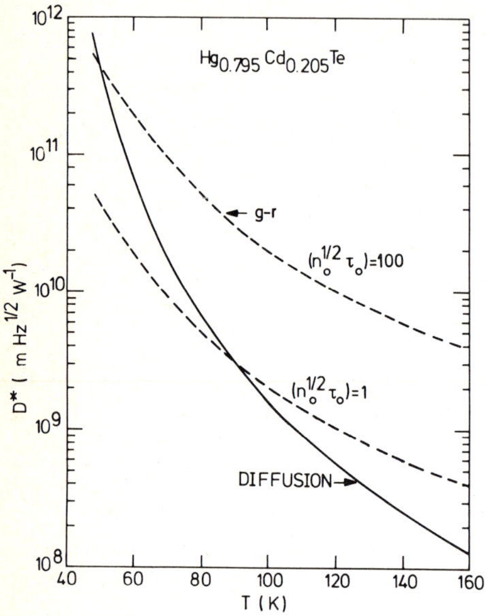

Fig.125 Calculated detectivity vs temperature for generation-recombination current and for diffusion limited cases /329/

Since the first edition the photovoltaic effect of ion-implanted junctions and the performance of photodetectors have been studied in various papers /638-642/. The effect of trap tunneling on the performance of photodiodes was investigated by WONG /643/, and the 1/f noise was studied in $Hg_{.7}Cd_{.3}Te$ photodiodes by TOBIN et al. /644/. The photovoltaic properties of metal-narrow-gap junctions /645/ and of HgCdTe-CdTe junctions /646,647/ were investigated. In HgCdTe-CdTe photodiodes an improvement of the R_0A product was observed compared with an np-HgCdTe homojunction diode.

Photodiodes from the HgCdTe alloy for the wavelength range of 1-1.3 μm were investigated by FIORITO et al. /648/ and compared with those from other semiconductor material.

Arrays of GHz response HgCdTe photodiodes for application in CO_2 laser radar have been studied by SPEARS /649,650/. The n-on p-type photodiodes have an over 50% quantum efficiency, being uniform over the 280-μm-diameter quadrantal array and independent of wavelength from 4 to 12 μm.

6.4 Infrared Radiation Sources

Infrared radiation both spontaneous and coherent was observed in different types of experiments with $Hg_{1-x}Cd_xTe$. Injection luminescence from p-n $Hg_{.65}Cd_{.33}Te$ junctions was first reported by VERIE and GRANGER /352/. An emission peak at 3.7 µm was observed at liquid nitrogen temperature. Spontaneous and coherent emission was also observed by MELNGAILIS and STRAUSS /107/ when $Hg_{.68}Cd_{.32}Te$ was irradiated by the beam of a GaAs laser. As shown in Fig.126 coherent emission evidenced by an abrupt increase in emission intensity as well as by the appearance of mode structure was observed above a threshold intensity of the pump radiation.

In spite of the proved qualification of $Hg_{1-x}Cd_xTe$ alloys as laser material for the 3 to 15 µm range, most investigations have been carried out with lead salt materials which appear to be more stable /4/. However, for the generation of infrared radiation, where magnetic quantum levels are involved, narrow band gap $Hg_{1-x}Cd_xTe$ alloys have an important advantage over lead salt materials by virtue of their smaller effective mass and related large g factor. As discussed in Sect. 4.4 these properties increase the magnetic tunability of spin-flip Raman scattering and of radiative transitions between Landau levels. SATTLER et al. and WEBER et al. /120/ have observed stimulated spin-flip Raman scattering emission with $Hg_{.766}Cd_{.234}Te$ at 12 K using a CO_2-laser pump (Sect. 4.4.5, Fig. 86). The average tuning rate was found to be as high as 34 cm^{-1}/T. Peak power of over 1 mW of this tunable spin-flip Raman laser emission with a conversion efficiency of about 0.3% was reported /120/.

Magnetically tunable far-infrared radiation was observed corresponding to radiative transitions between Landau levels in $Hg_{.8}Cd_{.2}Te$ by DORNHAUS et al. /44/. The population of higher Landau levels, necessary for a non-zero radiative transition

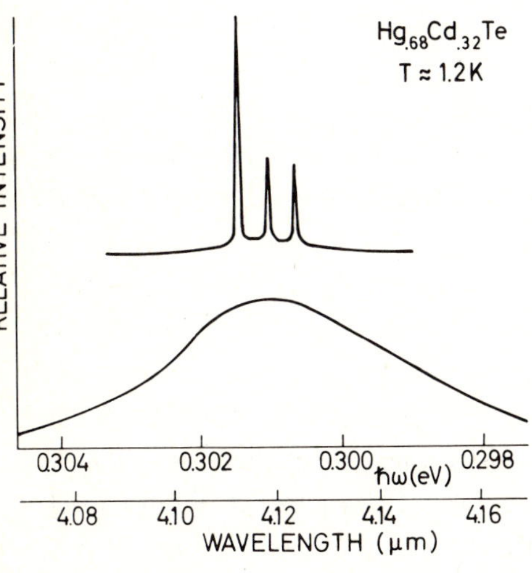

Fig.126. Emission spectra of optically excited $Hg_{.68}Cd_{.32}Te$ at 12 K. Lower signal represents spontaneous emission, the upper signal shows coherent emission with resolution-limited modes of the 135 µm long cavity /107/

rate, is obtained by the application of an ac or pulsed voltage. This cyclotron resonance emission covers the region of the far-infrared radiation. However, it should be stated that for the application of this magnetically tunable radiation InSb seems to be a more suitable material /366/, since $Hg_{1-x}Cd_xTe$ alloys have two reststrahlen bands which reduce the effective emission compared with InSb.

A new type of far-infrared radiation was observed recently by TYSSEN et al. /43/ with $Hg_{.8}Cd_{.2}Te$ samples. On bulk n-type material a p-type surface layer was obtained by evaporation of Hg atoms. As was mentioned already in Sect. 2.3.1, the application of an electric field of about 100 V/m yields a radiation as shown in Fig.127. The radiation intensity increases with increasing electric field, however, vanishes in higher fields when sample heating occurs. The experiment was carried out at 4.2 K and the radiation was analysed by Fourier spectroscopy. The emission is observed in the frequency range between 70 and 350 cm^{-1} and shows a pronounced minimum at 150 cm^{-1}. Comparing this emission spectrum with the reflectivity of Fig.106 it is evident that the emission maximum has the same frequency as the maximum of the reflectivity which corresponds to the HgTe reststrahl, whereas at the frequency of the second reflectivity maximum corresponding to the CdTe reststrahl, the minimum of the emission occurs. This coincidence which was found with the emission of all five investigated samples made us assume that most of the Hg atoms are evaporated from the surface. The radiation seems to be excited in the p-n junction near the surface by hot carriers. Recently, however, experiments have shown that the radiation may be due to reflection of background radiation.

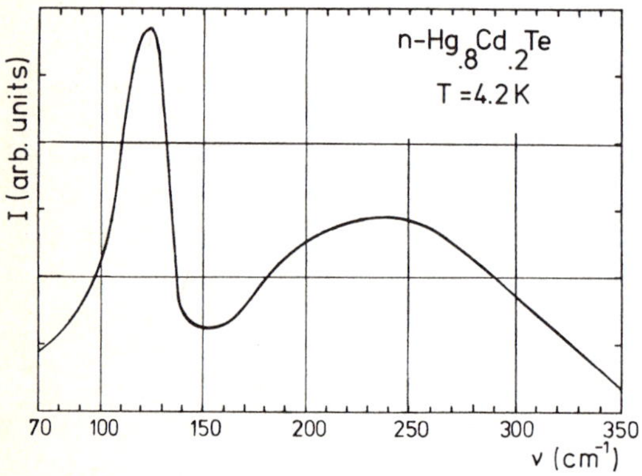

Fig.127. Hot carrier infrared emission spectrum from $Hg_{.8}Cd_{.2}Te$

6.5 Charge Transfer Devices and Field-Effect Transistors

In order to obtain a substantial performance improvement of present-day infrared systems the number of detectors employed in the focal plane has to be increased significantly /651,652/. This requirement demands a high degree of integration, which can only be achieved by constructing charge-coupled or charge-injected devices (CCD, CID). CCD and CID arrays of HgCdTe were developed successfully by various groups /651-655/, and the results are promising for the planned large-scale integration of intrinsic photodetectors.

The metal-insulator-semiconductor (MIS) properties of $Hg_{.79}Cd_{.21}Te$ and its native oxide were studied by NEMIROVSKY and KIDRON /656/ (see Sect. 7.2). The construction and characteristics of an n-channel field-effect transistor were quite recently presented by NEMIROVSKY et al. /657/. The cross-section of this transistor and an integrated capacitor is shown in Fig. 128. This device can be operated at a temperature of 77 K and is designed for switching or amplifying requirements in monolithic HgCdTe structures.

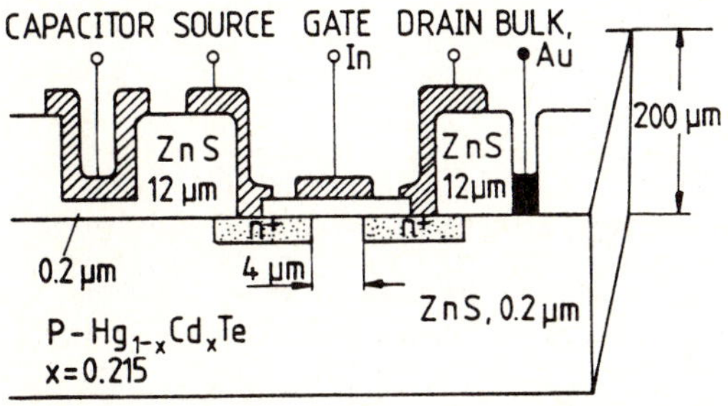

Fig. 128. Cross-section of ion-implanted MIS transistors and capacitor. Channel length 4 µm; channel width 200 µm; gate insulator thickness 0.2 µm; field insulator thickness 1.2 µm; n^+ region junction depth 0.3 µm; effective carrier concentration of p-type substrate $N_A = 2\cdot10^{22}$ m^{-3} at 77 K; capacitor area $7.28\cdot10^{-8}$ m^2 /657/

7. Surfaces, Interfaces, Graded-Gap Materials

In 1976, when the first issue of this review was prepared, not much work on HgCdTe surface and interface phenomena had been published. It is the aim of this new chapter to present a short summary of this field. Section 7.1 deals with free HgCdTe surfaces and the interface between HgCdTe and its native oxide. Section 7.2 reviews investigations on MOS and MIS devices, thus complementing Sect. 6.5. Section 7.3 summarizes transport and optical studies of inversion layers, and Sect. 7.4 presents recent calculations on CdTe/HgTe superlattices. Section 7.5 deals with tunneling experiments, and Sect. 7.6 compiles data on metal-HgCdTe contacts. Section 7.7, finally, briefly reviews the field of homogeneous and graded-gap epitaxial layers.

7.1 HgCdTe Surfaces

Using the Kane model, wavefunctions and dispersion laws were obtained recently for Tamm surface states of electrons and holes arising on the ideal surface of HgCdTe crystals in the parabolic and strong non-parabolic limits /658/.

Real surfaces of HgCdTe mixed crystals as well as interfaces between HgCdTe and its native oxide or other insulating layers have been the subject of a number of investigations /659-673/.

7.1.1 Free HgCdTe Surfaces

Strong fundamental studies of the surface and interfaces of HgCdTe have been started only recently. SPICER et al. /659a/, attempting to find similarities and differences between the much better understood 3 - 5 compound surfaces and the HgCdTe ones, reached the following conclusions:

- There are indications that, similar to GaAs, intrinsic surface states on HgCdTe surfaces have been moved out of the bandgap by surface reconstruction. The reconstructed surfaces are expected to be strongly stressed.
- The most striking difference between 3 - 5 compounds and HgCdTe is the relative instability of Hg in the HgCdTe lattice. The Hg-Te bond energy is at least 30% less than the Cd-Te bond energy (see also Sect. 3.6). Although this is a bulk property, it is also critical to the understanding of surfaces. It will be reflected by a difference in the Hg and Cd bonding at the surface and in the production of defects at the surface or in the bulk which may migrate easily to the surface.

The importance of bonding of Hg in the HgCdTe lattice is illustrated by the results of several studies on diffusion and Hg surface evaporation /424,427,665-667/ (see also the discussion in Sect. 2.3.3). NITZ et al. /424/ measured the

Table 27. Diffusion constants and surface penetration velocities of Hg at various temperatures /424/

Temperatures T [K]	Velocity K [m/s]	Diffusion constants D [m^2/s]
470	$6.9 \cdot 10^{-9}$	$3.0 \cdot 10^{-15}$
420	$1.8 \cdot 10^{-9}$	$1.5 \cdot 10^{-15}$
360	$1.8 \cdot 10^{-10}$	$2.5 \cdot 10^{-15}$
330	$4.1 \cdot 10^{-11}$	$8.0 \cdot 10^{-17}$
300		$2.8 \cdot 10^{-17}$
300	$D_{Cd} = 1.1 \cdot 10^{-12}$	

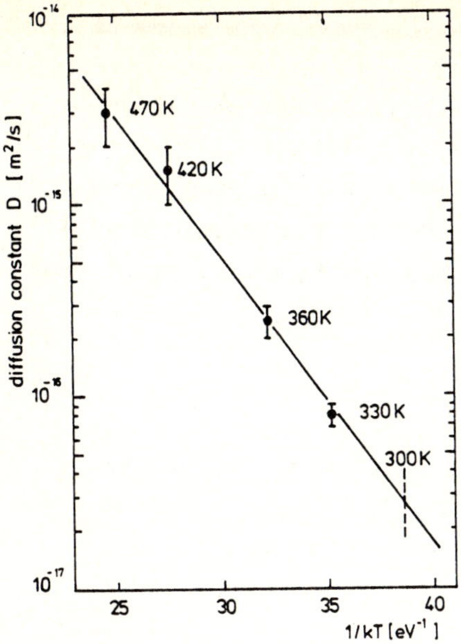

Fig. 129. Arrhenius plot of Hg diffusion constant D as a function of reciprocal temperature /424/

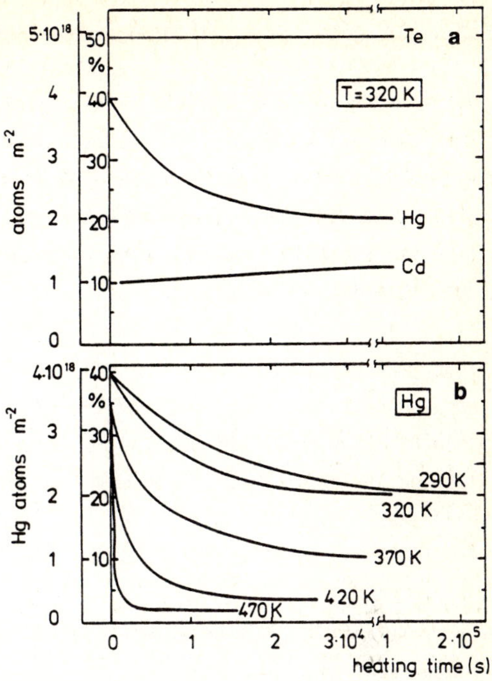

Fig. 130a,b. Surface concentration of elements as a function of heating time. (a) Cd, Hg, and Te at 320 K; (b) Hg at various temperatures /424/

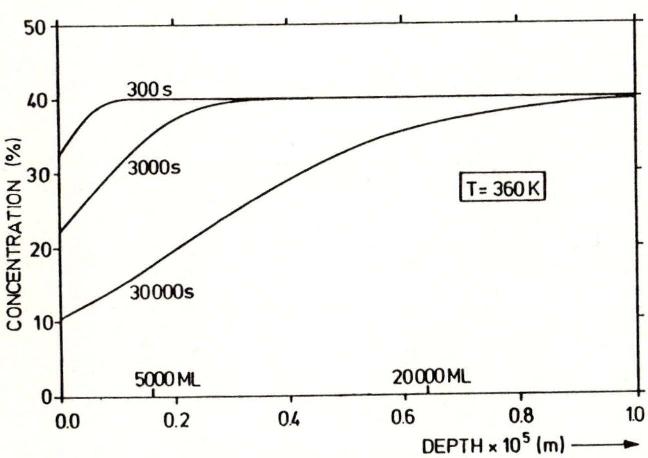

Fig. 131. Computed concentration profiles for HgCdTe samples with an initial Hg concentration of 40% for 360 K and with storage time as parameter /424/

change in surface composition with time at room and elevated temperatures of $Hg_{.8}Cd_{.2}Te$ samples prepared by scraping the surface in vacuum or by ion bombardment. Diffusion constants as determined in this study, are shown in Fig. 129 and surface penetration velocities at various temperatures are given in Table 27. The value D_{Hg} (300 K) is in good agreement with the value estimated by NIMTZ et al. /421/. Data on the surface concentration of elements as a function of heating time are shown in Fig. 130. These data have been used to calculate depth distribution profiles of HgCdTe samples at various times and temperatures (Fig. 131).

SILBERMANN et al. /427/, on the other hand, found no evidence for loss of Hg at room temperature for periods of tens of hours from p-type crystal surfaces prepared by cleaving in vacuum. They suggest that structural imperfections in the lattice in the surface region may have an important influence on the compositional stability of the alloy. This is substanciated by the observation of a preferential loss of Hg produced by ion sputtering or exposition to excited oxygen.

The preferential ejection of Hg from the surface of HgCdTe alloy crystals by ion bombardment techniques has been firmly established /424,427,660,664/. In addition it has been reported that even low doses of electrons induce desorption

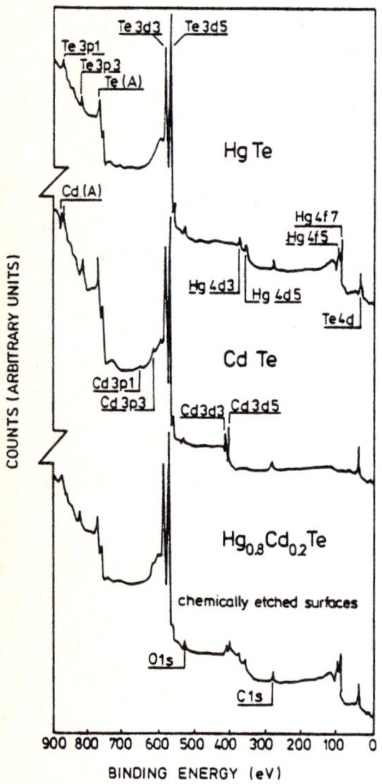

Fig. 132. XPS spectra of HgTe, CdTe and $Hg_{.8}Cd_{.2}Te$ surfaces after etch-polishing in a methanol solution containing 2% bromine /659/

of Hg from HgCdTe surface regions /424,664/. Thus the derivation of details of concentration profiles for free or oxide covered HgCdTe surfaces after cleaning or during depth profiling by ion beam bombardment could be hazardous without a basal knowledge of the possible preferential sputtering effects [sputtering yields for the individual elements Cd, Hg and Te have, e.g., been discussed in /424/ as well as corrected elemental sensitivity factors for XPS (X-ray photoemission) and AES (Auger electron spectroscopy)].

The influence of different etching procedures on the surface composition of HgCdTe samples has been investigated in various studies using XPS /660a,662b,663, 664d/ and electroreflectance spectroscopy /664f/. Agreement has been obtained in that the standard bromine-methanol chemical etching procedure damages (Hg annealed) materials to a considerable depth depleting them in Cd and leaving their surfaces covered with a thin Te-rich layer (Fig. 132). These layers seem to contain activated Te, rich in dangling bonds /664f/. LEED examinations did not produce any evidence for a crystalline nature of the layers. Exposed to laboratory atmosphere the surfaces are quickly contaminated with oxygen and carbon, finally resulting in a thin TeO_2 layer.

There are indications for differences in the mercury-to-tellurium content near the surface depending on the detailed etch procedure (Br_2 spray, Br_2 polish etc. /664d/).

Treatments with other surface etches (lactic or citric acid) seem to produce surfaces with little or no TeO_2, they may result in a polyester-type protective coating, however /663/.

A clean and defect-free bulk surface may be generated by a succession of anodization-dissolution steps after applying a chemomechanical (Br_2 methanol) polish /664f/.

7.1.2 Native Oxides, Production and Composition

The native oxide on HgCdTe is known to electrically stabilize the surface of n-type material by forming an accumulated interface layer /660,664g/. Consequently, native oxides, grown by several different methods, have been widely used as passivating and insulating films on HgCdTe. In spite of the technological importance especially of the anodic oxide for device fabrication /664j/, the details of the oxidation process are still not fully understood.

The most common method for the production of oxide layers is anodization in a solution of 0.1 M KOH in a mixture of 90% ethylene glycol and 10% water /664j/ at constant current density (~0.3 mA/cm^2), the voltage is allowed to increase with time. The thickness of the anodic oxide is proportional to the voltage and can also be determined by examining the color of the oxide film. Finally, the samples are "voltage annealed" at a lower voltage to obtain a denser oxide.

Although anodic oxides grown by this or slightly modified procedures have been extensively investigated, considerable discrepancies emerged in the literature concerning the molecular composition of the oxide layers. Results have usually been obtained by combining sputter profiling with XPS or AES, in some cases additional information from ellipsometry data has been used. There seem to be two main reasons for the controversial results:

- The methods employed are not truly molecular specific, e.g., in XPS the chemical shifts are rather indicative for a certain oxidation stage. Differences between, e.g., the chemical shifts of the 3d or 4d levels of Te in TeO_2 and $CdTeO_3$ are quite small and it seems very difficult to differentiate them in a mixture of oxides, which might not even be quilibrium reaction products. Better reliability might only be obtained in matching complete XPS signature data.
- As already mentioned in the previous section ion sputtering might seriously affect oxide compositions. Indications have been found /660b/ that, e.g., Ar-ion sputtering strongly depleted the surface of $HgTeO_3$ (and of anodized HgTe) from mercury, whereas TeO_2 was found to be accumulated as a sputter artifact.

Taking these results into consideration the most probable composition of the anodic oxides of HgCdTe seems to ly between the stoichiometric compounds (Hg, Cd) TeO_3 and (Hg, Cd) Te_2O_5. Whether the mercury deficiency indicated, e.g., by the elemental composition of 2.5 - 4% Hg, 13 - 14% Cd, 24 - 25% Te and 58 - 60% O typically found in XPS results for the anodic oxides on $Hg_{.8}Cd_{.2}Te$ /660b, 664a,b/ is solely an artifact of the sputter treatment /660b/ remains to become firmly established.

As anodic oxides on HgCdTe are plagued by excessive fixed charge and by long-term instabilities (possibly due to the instability of $HgTeO_3$), alternate methods of growing a native oxide or improve its quality by additional ZnS – or SiO_2 overlayers have been persued.

Photochemical oxidation may be a promising technique for passivating HgCdTe surfaces with relatively low fixed charge and interface state density /664c/. While the oxidation of HgCdTe surfaces in an O_2 atmosphere is an extremely slow process /660a/ (due to the very small sticking coefficient of O_2 in the ground state on HgCdTe surfaces /659a/) additional exposure to ultraviolet light allows the growth of even very thick oxides (> 2000 Å) /664c/.

Plasma oxidation might also result in improved interface quality and lower interface state density /664d,k/.

Overcoating native oxides with SiO_2 layers by chemical vapour deposition has been shown to result in an interfacial reaction reducing part of the native oxide /664e/.

7.1.3 Electronic Properties of the Native Oxide-HgCdTe Interface

Electronic properties of the native oxide-HgCdTe interface have been derived from galvanomagnetic measurements and capacitance-voltage studies /656,660b, 662,664a,g,h/. We only quote some characteristic values here:

- the fixed surface (oxide) charge density, which is always positive for both n- and p-type HgCdTe, has been found in the range between $5 - 10 \cdot 10^{15}$ m^{-2},
- in favourable cases a nearly flat distribution of fast surface states over the band gap has been obtained with values of $\sim 2 \cdot 10^{16}$ m^{-2} $(eV)^{-1}$ near the band edges and $\sim 4 \cdot 10^{15}$ m^{-2} $(eV)^{-1}$ at mid-gap, thus resulting in an integrated density of $\sim 5 \cdot 10^{15}$ m^{-2},
- the slow surface state density has been determined to be approximately $1 \cdot 10^{14}$ m^{-2},
- a typical value for the surface mobility is $\mu_s \cong 3$ $m^2V^{-1}s^{-1}$, the resistivity of anodic films is in the range between 10^8-10^{11} Ωm and film breakdown occurs at field strengths of 10^7-10^8 Vm^{-1}, all values obtained at 77 K.
- the oxide energy gap has been determined to 3.4 eV /664g/, the oxide dielectric constant to $\varepsilon_{ox} \cong 14.8$ /662a/.

Although an unambiguous correlation of the chemical and electrical properties of the HgCdTe oxide system has not been obtained up to now, first steps towards this ultimate goal may be found in /660-664/.

7.2 MOS and MIS Devices

Metal-insulator-semiconductor (MIS) [and especially metal-oxide-semiconductor (MOS)] devices are a powerful tool for the investigation of surface and bulk properties of semiconducturs. The proper use of MIS techniques requires an accurate theoretical model with which experimental data can be compared. In the case of HgCdTe there are at least three additional factors which are not normally addressed in the general treatises on MIS device physics /664h/:

- conduction band degeneracy
- interband tunneling
- inversion layer quantization.

In addition, short wavelength visible light ($\hbar\omega \geq 2.2$ eV) is known /664g/ to affect the fixed charge density at the anodic oxide HgCdTe interface; the effect is believed to be the result of internal photoemission of electrons in HgCdTe into the oxide followed by trapping. The influence of the surface electric field on MOS/MIS capacitance of inversion layers in semiconductors with non-parabolic bands has been calculated in /668,669/. Surface potential relaxation due to charge exchange via tunneling between oxide traps and the semiconductor surface has been investigated both experimentally and theoretically in /664g,669b/.

Low-frequency capacitance characteristics have been experimentally obtained and reported for MOS /656,657,660b,662,664c,670/ and MIS /671,672/ structures with ZnS as insulator. The latter devices exhibited an appreciably higher density of surface states /672/.

The workfunction for $Hg_{.8}Cd_{.2}Te$ has been estimated to be 4.23 ± 0.67 eV /672/.

Based on the results described above a series of devices such as capacitors, transistors /657/ and charge-coupled devices /651-655/ have been developed (see Sect. 6.5). Space does not permit a detailed discussion of design characteristics of such devices in this review; the reader is referred, e.g. to /673-675/.

7.3 Transport and Optical Studies of Inversion Layers

While the first observation of Shubnikov-de Haas oscillations from electrons in the free surface inversion layer of p-HgCdTe by ANTCLIFFE et al. /282/ was already mentioned in Sect. 4.4.1, the influence of Landau levels in the two-dimensional electron gas on the differential capacitance voltage characteristics of MIS structures on p-$Hg_{.8}Cd_{.2}Te$ has been reported for the first time in /676/.

Antcliffe et al. concluded that two series of oscillations observed in their investigations were both associated with the ground subbands, but that they were signals from different surface regions. OHKAWA and UEMURA have proposed a different explanation for the two series /677/. The main points of their work are:

- While the effective mass approximation for a single band can usually be successfully applied to calculate the properties of the quasi-two-dimensional electron system in the insulator-semiconductor interface, this is no longer valid when the perturbed energy due to the surface electric field is larger than the band gap. Then the surface states are mixtures of the conduction bands and the valence bands.
- When the band gap is small and the spin-orbit interaction is large, the inversion asymmetry of the surface potential causes a large k-linear term in the two-dimensional dispersion relation, which removes the spin degeneracy.
- The g factor becomes so large that a reversal of the ordering of the Landau levels of different subbands is expected.

Ohkawa and Uemura thus interpreted the observed two series of oscillations in Antcliffe's data as due to the ground and first excited subband and a beat pattern around 1 T as due to the inversion in Landau level order.

NARITA et al. /678-680/ have recently studied surface inversion layers on p-HgCdTe by means of far-infrared cyclotron resonance absorption (Fig. 133) and microwave non-resonant absorption. Although they have deduced surface electron

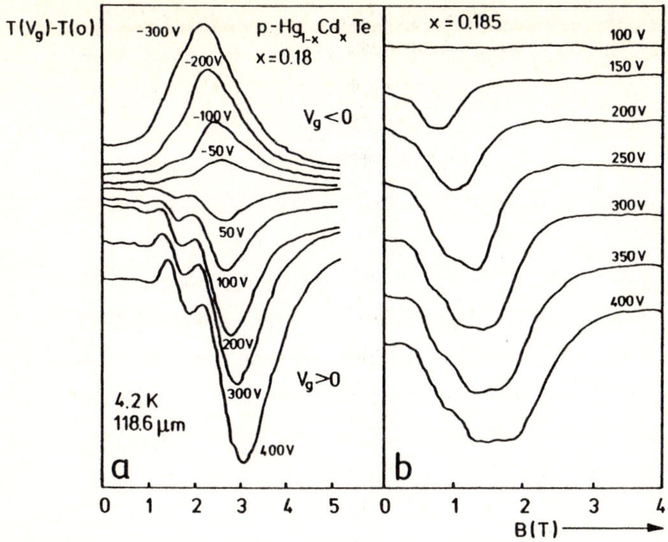

Fig. 133a,b. Gate voltage dependence of the cyclotron resonance absorption of electrons in the inversion layer on p-HgCdTe /678/. (a) Sample with naturally formed inversion layer; (b) sample without natural inversion layer

cyclotron masses for the ground and excited subbands depending on surface carrier density, appreciable discrepancies between the various experimental results /678-680/ as well as between theory /681/ and experiment remain unexplained up to now.

7.4 CdTe-HgTe Superlattices

The study of superlattice systems made from alternating layers of two semiconductors has been a fruitful source of new knowledge about interfaces, band structures and novel quantum effects on a submicron scale. Recently a superlattice system consisting of alternating layers of CdTe and HgTe has been proposed as a new infrared material /682,683/ to be fabricated by molecular beam epitaxy. The electronic properties have been calculated using the tight-binding approach. The energy gap is found to decrease monotonically with increasing HgTe-layer thickness for a fixed ratio of CdTe-to-HgTe layer thicknesses, thus allowing additional flexibility over the random alloy in adjusting the band gap for a given Cd-to-Hg ratio. The symmetry of the valence-band maximum state is found to change at certain HgTe-layer thicknesses. The electronic states of the CdTe-HgTe-(001) superlattice on CdTe layers have been analysed in terms of bulk CdTe evanescent states /684/. The states at the conduction-band minimum have been found to be relatively delocalized across the superlattice layers, which should lead to substantial electronic transport perpendicular to the layers, i.e. relatively large electron mobilities.

7.5 Tunneling Spectroscopy of HgCdTe Crystals

Electron tunneling in metal-insulator-semiconductor junctions has been applied to the study of HgTe and HgCdTe crystals using either ZnS, SiO_2 /685/ or Al_2O_3 barriers /686,687/ and Pb counter electrodes. The dU/dI-versus-U characteristics of the junctions were asymmetric, parabolic in shape with maxima in the negative bias range. Structures due to phonon-assisted tunneling gave phonon energies in good agreement with optical measurements.

7.6 Metal-HgCdTe Contacts

Metal-HgCdTe contacts have been investigated in considerable detail in a series of recent papers /688-693/. While there have been reports /694/ that the properties of metal contacts on degenerate PbSnTe can be explained with a simple M-S contact model requiring neither surface /695/ states nor an interfacial layer, studies on metal-HgCdTe contacts indicate the effect of the surface on the transport properties of contacts. The observed weak dependence of contact barrier height on the metal workfunction differed remarkably from a simple Schottky model.

For n-HgCdTe ($N_D \gtrsim N_A$) and compensated samples ($N_D < N_A$) ohmic contacts have been obtained which show a weakly increasing contact resistivity with increasing metal workfunction (In, Zn, Sn, Au; only Pt gives a distinctly higher contact resistance). p-type samples exhibited rectifying contact properties (except for Pt with ohmic contacts to p-type material) with increasing barrier heights Φ_{Bp} by decreasing workfunction Φ_m from which experimentally followed (at 300 K) /691/

$$\Phi_{Bp} = -0.1\ \Phi_m + 0.81\ \text{eV} \quad . \tag{100}$$

Transport mechanisms and photovoltaic properties of metal-HgCdTe contacts have been discussed in /690,693/ respectively.

7.7 Homogeneous and Graded-Gap Epitaxial HgCdTe Layers

Epitaxial growth of HgTe on CdTe single crystal substrates under certain experimental conditions results in "graded-gap" HgCdTe layers. These layers exhibit a slowly varying composition across a depth profile. Hence the band parameters, such as the energy gap, effective mass as well as the lifetime of carriers, depend on the position in the layer. While for some applications of epitaxial HgCdTe these gradients are highly undesirable, they also permit the investigation of several interesting physical effects and have some useful applications.

Transport, photoconductivity, PEM effect and non-linear electrical effects in graded-gap layers have been investigated /696-702/. Most of the results together with a review of the older literature on this field have been recently summarized in /696/. The most interesting new physical effect with graded-gap materials is the occurrence of a non-linear electrical effect which stems from the gradient in the effective mass across the sample /703/. A double-frequency internal electric field is created which is parallel to the mass gradient and perpendicular to the variable applied field /699/.

Recently the conditions for the growth of epitaxial layers of uniform composition have been investigated, and successful preparation by different techniques (LPE /704-708/, RF-triode sputtering /709/, and close space vapour transport /710/) has been reported (Sect. 2.2). Especially the liquid-phase epitaxy technique seems well suited for the fabrication of HgCdTe epitaxial layers on CdTe substrates from a Te-rich melt. Consequently this technique has been used for the production of high-performance photodiodes and multilayer structures (Sect. 6.3).

Acknowledgements

We would like to express our gratitude to those who have helped with the production of this article. In particular we thank Professor E.E. Schneider for his labour in critically reading the manuscript and suggesting a number of modifications, and Professor P. Grosse for his encouragement to write this article. We would also like to thank Michael Schifferdecker, Egon Tyssen, Karl-Heinz Müller, and Donald Eigler for their valuable help with the preparation of the paper and we are indebted to Fräulein Elles and Frau Pitzler for their technical services.

References

1 W.D. Lawson, S. Nielsen, E.H. Putley, A.S. Young: J. Phys. Chem. Sol. 9, 325 (1959).
2 D. Long, J.L. Schmit: in Semicond. and Semimetals, ed. by R.K. Willardson, A.C. Beer (Academic, N.Y. 1970), Vol. 5, p. 175.
3 T.C. Harman: in Physics and Chemistry of II-VI Compounds, ed. by M. Aven, J.S. Prener (North-Holland Publ., Amsterdam 1967), p. 767.
4 T.C. Harman, I. Melngailis: in Applied Solid State Science, ed. by R. Wolfe (Academic, N.Y. 1974), Vol. 4, p. 1.
5 J.C. Woolley, B. Ray: J. Phys. Chem. Sol. 13, 151 (1969).
6 J. Blair, R. Newnham: in "Metallurgy of Elemental and Compound Semiconductors" (Wiley, Interscience, 1967), Vol. 12, p. 393.
7 A.D. Shneider, I.V. Gavrishchak: Sov. Phys. Sol. State 2, 1865 (1960).
8 J.G. Mavroides, D.F. Kolesar: Solid State Commun. 2, 363 (1964).
9 T. Alper, G.A. Saunders: J. Phys. Chem. Sol. 28, 1637 (1967).
10 L. Sniadower, M. Psoda, R.R. Galazka: Phys. Stat. Sol. 28, K 121 (1968).
11 S.I. Novikova, N.K. Abrikosov: Sov. Phys. Sol. State 5, 1558 (1964).
12 D.G. Thomas: J. Appl. Phys. 32, 2298 (1961); H.J. McSkimin, D.G. Thomas: J. Appl. Phys. 33, 56 (1961).
13 R.P. Chasmar, E.W. Durham, A.D. Stuckes: in Proc. Int. Conf. Semicond. Physics, Prague 1960 (Czech. Acad. Sci., Prague, and Academic Press, N.Y. 1961), p. 1018.
14 Thermophysical properties of Matter (The TPRC Data Series), Vol. 1, ed. by Y.S. Touloukian, R.W. Powell, C.Y. Ho, P.G. Klemens (Plenum, N.Y. - Washington 1970), p. 1267, 1321, 1407; G.A. Slack, S. Galginaitis: Phys. Rev. 133, A 253 (1964).
15 C.S. Guenzer, A. Bienenstock: Phys. Rev. B8, 4655 (1973); D.N. Talwar, B.K. Agrawal: J. Phys. C7, 2981 (1974).
16 D. Berlincourt, H. Jaffe, L.R. Shiozawa: Phys. Rev. 129, 1009 (1963).
17 J.L. Schmit, C.J. Speerschneider: Infrared Physics 8, 247 (1968).
18 B. Ray, P.M. Spencer: Phys. Stat. Sol. 22, 371 (1967).
19 B.E. Bartlett, J. Deans, P.C. Ellen: J. Material Sciences 4, 266 (1969).
20 R.R. Galazka: Acta Physica Polonica 24, 791 (1963).
21 E.Z. Dziuba: J. Electrochem. Soc. 116, 104 (1969).
22 R. Ueda, D. Ohtsuki, K. Shinobara, Y. Ueda: J. Cryst. Growth 13/14, 668 (1972).
23 H. Rodot, J. Henoc: Compt. Rend. Acad. Sci. 256, 1954 (1963).
24 G.S. Almasi, A.C. Smith: J. Appl. Phys. 39, 233 (1968).
25 Y. Marfaing, G. Cohen-Solal, F. Bailly: Int. Conf. Cryst. Growth, Boston 1966, ed. by H.S. Peiser (Pergamon Press, Oxford 1967), p. 549.
26 L.A. Bovina, V.P. Meshcheryakova, V.I. Stafeev, E.S. Banin: Sov. Phys. Semicond. 7, 26 (1973).
27 G. Cohen-Solal, Y. Marfaing, F. Bailly: Revue de Physique Appliquée 1, 11 (1966).
28 C. Sella, G. Cohen-Solal, F. Bailly: Compt. Rend. Acad. Sci. B264, 179 (1967).
29 G. Cohen-Solal, Y. Riaud: Appl. Phys. Lett. 19, 436 (1971).
30 O.N. Tufte, E.L. Stelzer: J. Appl. Phys. 40, 4559 (1969).
31 F. Bailly, G. Cohen-Solal, L. Svob, Y. Marfaing: Compt. Rend. Acad. Sci. 269, 456 (1969).
32 F. Bailly, L. Svob, G. Cohen-Solal, R. Triboulet: J. Appl. Phys. 46, 4244 (1975).
33 L. Svob, Y. Marfaing, R. Triboulet, F. Bailly, G. Cohen-Solal: J. Appl. Phys. 46, 4251 (1975).
34 V. Leute, W. Stratmann: Z. Physik. Chemie 90, 172 (1974).
35 D.B. Holt: J. Phys. Chem. Sol. 23, 1353 (1962).
36 F. Buch, C.N. Ahlquist: J. Appl. Phys. 45, 1756 (1974).
37 J. Melngailis, J.O. Ryan, T.C. Harman: J. Appl. Phys. 44, 2647 (1973).
38 C. Finck, S. Otmezguine, G. Weill, C. Vérié: in Proc. XIth Int. Conf. Phys. Semicond., Warsaw 1972 (PWN-Polish Scientific Publishers, Warsaw 1972), p. 944.
39 A. Mauger, J. Friedel: Phys. Rev. B12, 2412 (1975).

40 C.T. Elliott, I. Melngailis, T.C. Harman, A.G. Foyt: J. Phys. Chem. Sol. 33, 1527 (1972).
41 G. Bastard, D. Nozières: Phys. Rev. B13, 2560 (1976).
42 C.T. Elliott, J. Melngailis, T.C. Harman, J.A. Kafalas, W.C. Kernan: Phys. Rev. B5, 2985 (1972).
43 G. Haider, E. Tyssen, R. Dornhaus, H. Happ, K.H. Müller, G. Nimtz: Verhandl. DPG, HL 32, 7/1976.
44 R. Dornhaus, H. Happ, K.H. Müller, G. Nimtz, W. Schlabitz, P. Zaplinski, G. Bauer: in Proc. XIIth Int. Conf. Phys. Semicond., Stuttgart 1974, ed. by M.H. Pilkuhn (Teubner, Stuttgart 1974), p. 1157.
45 R. Dornhaus, G. Nimtz, W. Schlabitz, H. Burkhard: Solid State Commun. 17, 837 (1975).
46 N.H. Parada: Phys. Rev. B3, 2042 (1971).
47 L.A. Hemstreet: Phys. Rev. B12, 1212 (1975).
48 M. Jaros: J. Phys. C8, L 550 (1975).
49 C.E. Mallon, J.A. Naber, J.F. Colwell, A.B. Green: IEEE Trans. Nucl. Sci. 20, 214 (1973).
50 R.E. Leadon, C.E. Mallon: Infrared Physics 15, 259 (1975).
51 K. Seeger: Semiconductor Physics (Springer, Wien, N.Y. 1973), p. 49.
52 B.L. Gel'mont, M.I. D'yakonov: Sov. Phys. JETP 35, 377 (1972).
53 N.B. Brandt, O.N. Belousova, L.A. Bovina, V.I. Stafeev, Ya.G. Ponomarev: Sov. Phys. JETP 39, 157 (1974).
54 A. Mauger, S. Otmezguine, C. Vérié, J. Friedel: in /44/, p. 1166.
55 D.J. Chadi, J.P. Walter, M.L. Cohen, Y. Pétroff, M. Balkanski: Phys. Rev. B5, 3058 (1972).
56 M.L. Cohen, T.K. Bergstresser: Phys. Rev. 141, 783 (1966).
57 S. Bloom, T.K. Bergstresser: Phys. Stat. Sol. 42, 191 (1970).
58 H. Overhof: Phys. Stat. Sol. (b) 43, 221 (1971).
59 H. Overhof: Phys. Stat. Sol. (b) 45, 315 (1971).
60 M. Balkanski: in Elementary Excitations in Solids, Molecules and Atoms, Part B, ed. by J.T. Devreese, A.B. Kunz, T.C. Collins (Plenum Press, London and N.Y. 1974), p. 213.
61 H. Overhof: in Computational Methods in Band Theory (Plenum Publ. Corp, N.Y. 1970), p. 218.
62 F. Herman, S. Skillman: Atomic Structure Calculations (Prentice Hall Inc., Englewood Cliffs (N.J.) 1963).
63 S. Katsuki, M. Kunimune: J. Phys. Soc. Jap. 31, 415 (1971).
64 D.J. Chadi, M.L. Cohen: Phys. Rev. B7, 692 (1973).
65 A.O.E. Animalu, V. Heine: Phil. Mag. 12, 1249 (1965).
66 A.O.E. Animalu: Phil. Mag. 13, 53 (1966).
67 D.R. Penn: Phys. Rev. 128, 2093 (1962).
68 G. Weisz: Phys. Rev. 149, 504 (1966).
69 S. Bloom, T.K. Bergstresser: Solid State Commun. 6, 465 (1968).
70 A. Kisiel, P.M. Lee: J. Phys. F2, 395 (1972).
71 F. Herman, O.D. Kuglin, K.F. Cuff, R.L. Kortum: Phys. Rev. Lett. 11, 541 (1963).
72 C. Vérié: in Festkörperprobleme X (Pergamon-Vieweg, Braunschweig 1970), p. 3.
73 A.J. Strauss, T.H. Harman, J.G. Mavroides, D.H. Dickey, M.S. Dresselhaus: in Proc. Int. Conf. Phys. Semicond., Exeter 1962 (The Institute of Physics and the Physical Society, London SW 7 1962), p. 703.
74 C. Vérié, E. Decamps: Phys. Stat. Sol. 9, 797 (1965).
75 M. Rodot, R. Triboulet: Compt. Rend. Acad. Sci. (France) 254, 825 (1962).
76 R. Piotrzkowski, S. Porowski, Z. Dziuba, J. Ginter, W. Giriat, L. Sosnowski: Phys. Stat. Sol. 8, K 135 (1965).
77 G. Weill, C. Vérié: Compt. Rend. Acad. Sci. B263, 463 (1966).
78 C. Vérié: Phys. Stat. Sol. 17, 889 (1966).
79 S. Otmezguine, F. Raymond, G. Weill, C. Vérié: in Proc. Xth Int. Conf. Phys. Semicond., Cambridge 1970 (USAEC Div. of Techn. Inf., Oak Ridge 1970), p. 536.
80 P. Byszewski, E.Z. Dziuba, R.R. Galazka, K. Szlenk, K. Szymborski, W. Walukiewicz: Phys. Stat. Sol. (b) 71, 117 (1975).
81 C. Vérié: in II-VI Semicond. Compounds, Intern. Conf., Providence 1967, ed. by D.G. Thomas (W.A. Benjamin Inc., N.Y. 1968), p. 1124.

82 W. Zawadzki: Adv. Phys. 23, 435 (1974).
83 E.O. Kane: J. Phys. Chem. Sol. 1, 249 (1957).
84 R. Bowers, Y. Yafet: Phys. Rev. 115, 1165 (1959); Y. Yafet: Phys. Rev. 115, 1172 (1959).
85 W. Zawadzki: in New Developments in Semiconductors, ed. by Wallace, Harris, Zuckerman (Noordhoff Intern. Publ., Leyden 1973), p. 441.
86 B. Lax, J.G. Mavroides, H.J. Zeiger, R.J. Keyes: Phys. Rev. 122, 31 (1961).
87 W. Dobrowolski: Phys. Stat. Sol. (b) 70, K 63 (1975).
88 G.A. Antcliffe: Phys. Rev. B2, 345 (1970).
89 K. Suizu, S. Narita: Rep. of the Departm. of Mat. Physics in the Faculty of Eng. Science of Osaka University Toyonaka, Osaka, Japan 1973; S. Narita, K. Suizu: Progr. Theor. Phys. Suppl. No. 57, 187 (1975).
90 W. Giriat: in /81/, p. 1058.
91 W. Giriat: Phys. Lett. 24A, 515 (1967).
92 L. Sosnowski, R.R. Galazka: in /81/, p. 888.
93 A.D. Shneider, I.V. Gavrishchak: Sov. Phys. Sol. State 5, 881 (1963).
94 R.R. Galazka, L. Sosnowski: Phys. Stat. Sol. 20, 113 (1967).
95 E.M. Aliev, S.A. Aliev, T.G. Gadzhiev, M.I. Aliev: Phys. Stat. Sol. 40, K 41 (1970).
96 T.C. Harman, A.J. Strauss, D.H. Dickey, M.S. Dresselhaus, G.B. Wright, J.G. Mavroides: Phys. Rev. Lett. 7, 403 (1961).
97 M.A. Kinch, D.D. Buss: in The Physics of Semimetals and Narrow Gap Semiconductors, Dallas 1970, ed. by D.L. Carter, R.T. Bate, J. Phys. Chem. Sol. 32, 461 (1971), Suppl. 1.
98 J.D. Wiley, R.N. Dexter: Phys. Rev. 181, 1181 (1969).
99 B. Ellis, T.S. Moss: in Proc. III. Int. Conf. on Photoconductivity, Stanford, Calif. 1969, ed. by E.M. Pell (Pergamon, N.Y. 1971), p. 211.
100 M.W. Scott: J. Appl. Phys. 40, 4077 (1969).
101 M.D. Blue: Phys. Rev. 134, A 226 (1964).
102 J.L. Schmit, E.L. Stelzer: J. Appl. Phys. 40, 4865 (1969).
103 M.V. Kot, V.G. Tyrzin, A.V. Simashkevich, Yu. E. Maronchuk, V.A. Mshenskii: Sov. Phys. Sol. State 4, 1128 (1962).
104 I.M. Nesmelova, N.S. Baryshev, F.P. Volkova, A.P. Cherkasov: Sov. Phys. Semicond. 6, 822 (1972).
105 P.W. Kruse, D. Long, O.N. Tufte: in /99/, p. 223.
106 W. Saur: Infrared Physics 8, 255 (1968).
107 I. Melngailis, A.J. Strauss: Appl. Phys. Lett. 8, 179 (1966).
108 A.D. Shneider, I.S. Zhmurko: Ukrainian J. Phys. 9, 32 (1964).
109 C.T. Elliott, J. Melngailis, T.C. Harman, J.A. Kafalas, W.C. Kernan: Phys. Rev. B5, 2985 (1972).
110 H. Kahlert, G. Bauer: Phys. Rev. Lett. 30, 1211 (1973).
111 D. Long: Phys. Rev. 176, 923 (1968).
112 J. Stankiewicz, W. Giriat: Phys. Stat. Sol. (b) 49, 387 (1972).
113 C. Vérié, G. Martinez: Compt. Rend. Acad. Sci. B266, 720 (1968).
114 S.H. Groves, T.C. Harman, C.R. Pidgeon: Solid State Commun. 9, 451 (1971).
115 J.A. Van Vechten, T.K. Bergstresser: Bull. Amer. Phys. Soc. Ser. II 15, 254 (1970).
116 J. Stankiewicz, W. Giriat: Phys. Stat. Sol. (b) 48, 467 (1971).
117 T.C. Harman: J. Phys. Chem. Sol. 25, 931 (1964).
118 K. Suizu, S. Narita: Solid State Commun. 10, 627 (1972).
119 S. Narita, R.S. Kim, O. Ohtsuki, R. Ueda: Phys. Lett. 35A, 203 (1971).
120 J.P. Sattler, B.A. Weber, J. Nemarich: Appl. Phys. Lett. 25, 431 (1974); B.A. Weber, J.P. Sattler, J. Nemarich: Appl. Phys. Lett. 27, 93 (1975).
121 M. Mali: Phys. Stat. Sol. 13, 215 (1966).
122 A. Mycielski, R.R. Galazka: in Proc. Int. Conf. Phys. Semicond., Moscow 1968 (Publishing House Nauka, Leningrad 1968), p. 875.
123 W. Szymanska, L. Sniadower, W. Giriat: Phys. Stat. Sol. 10, K 11 (1965).
124 L. Sniadower, V.I. Ivanov-Omskii, Z. Dziuba: Phys. Stat. Sol. 8, K 43 (1965).
125 L. Sniadower, R. Kowalczyk: Phys. Stat. Sol. 25, 233 (1968).
126 S.H. Groves, R.N. Brown, C.R. Pidgeon: Phys. Rev. 161, 779 (1967).
127 R.R. Galazka, T. Zakrzewski: Phys. Stat. Sol. 23, K 39 (1967).

128 I. Ivanov-Omskii, F.P. Kesamanly, B.T. Kolomiets, A.S. Mekhtiev, V.A. Skripkin: Phys. Stat. Sol. 27, K 169 (1968).
129 V.I. Ivanov-Omskii, B.T. Kolomiets, A.A. Mal'kova, Y.F. Markov, A.S. Mekhtiev: Sov. Phys. Semicond. 2, 122 (1969).
130 V.I. Ivanov-Omskii, B.T. Kolomiets, A.A. Mal'kova, A.S. Mekhtiev: Phys. Stat. Sol. 32, K 83 (1969);
V.I. Ivanov-Omskii, B.T. Kolomiets, Y.F. Markov, A.S. Mekhtiev, K.P. Smekalova: Sov. Phys. Semicond. 1, 1203 (1967).
131 G.A. Shneider: Phys. Stat. Sol. (a) 31, K 53 (1975).
132 B.D. McCombe: Report on Electric Dipole Allowed Spin-Flip Resonances in Narrow-Gap Semiconductors, Max-Planck-Institut für Festkörperforschung, Stuttgart 1974.
133 B.D. McCombe, R.J. Wagner, G.A. Prinz: Phys. Rev. Lett. 25, 87 (1970).
134 N.N. Berchenko, V.E. Krevs, M.V. Pashkovskii: Sov. Phys. Sol. State 14, 2447 (1973).
135 A. Moritani, K. Taniguchi, C. Hamaguchi, J. Nakai: J. Phys. Soc. Jap. 34, 79 (1973).
136 A. Moritani, K. Taniguchi, C. Hamaguchi, J. Nakai, R. Ueda, D. Ohtsuki, Y. Ueda: J. Phys. Soc. Jap. 31, 945 (1971).
137 A. Moritani, C. Hamaguchi, J. Nakai: Surf. Science 37, 769 (1973).
138 R. Ludeke, W. Paul: J. Appl. Phys. 37, 3499 (1966).
139 A. Kisiel, M. Podgórny, A. Rodzik, W. Giriat: Phys. Stat. Sol. (b) 71, 457 (1975).
140 A. Kisiel, M. Podgórny, A. Rodzik, W. Giriat: Phys. Stat. Sol. (b) 70, 767 (1975).
141 R.R. Galazka, A. Kisiel: Phys. Stat. Sol. 34, 63 (1969).
142 M. Cardona: Solid State Phys. Suppl. 11, 70 (1969).
143 V.V. Sobolev: Optics and Spectroscopy 18, 456 (1965).
144 M. Cardona, D.L. Greenaway: Phys. Rev. 131, 98 (1963).
145 M. Cardona, K.L. Shaklee, F.H. Pollak: Phys. Rev. 154, 696 (1967).
146 C.S. Guenzer, A. Bienenstock: Phys. Lett. A34, 172 (1971).
147 H.Y. Fan: Phys. Rev. 82, 900 (1951);
M.I. Cohen: Phys. Rev. 128, 131 (1962);
M.E. Mostoller: Ph. D. thesis (Harvard University, 1964).
148 S.C. Yu, H. Brooks (unpublished);
S.C. Yu: Ph. D. thesis (Harvard University, 1964);
R.V. Kasowski: Phys. Rev. 187, 891 (1969).
149 R. Piotrzkowski, S. Porowski: in /81/, p. 1030.
150 S. Porokowski, T. Zakrzewski: Phys. Stat. Sol. 11, K 39 (1965).
151 M. Lallemand, G. Weill, A. Lacam, M. Rodot: Compt. Rend. Acad. Sci. 260, 4469 (1965).
152 M. Rodot: in /122/, p. 609.
153 J. Stankiewicz, W. Giriat, A. Bienenstock: Phys. Rev. B4, 4465 (1971).
154 M.I. Daunov, E.L. Broyda: Phys. Stat. Sol. (b) 55, K 155 (1973).
155 G.L. Bir, G.E. Pikus: Sov. Phys. Sol. State 3, 2221 (1962).
156 P.S. Kireev, V.V. Ptashinskii, A.M. Sokolov, V.G. Kovalev: Sov. Phys. Semicond. 7, 266 (1973).
157 O.N. Tufte, E.L. Stelzer: in /81/, p. 1110.
158 K. Takita, N. Tanimura, S. Tanaka: in /44/, p. 1152.
159 J.M. Luttinger: Phys. Rev. 102, 1030 (1956).
160 D.L. Camphausen, G.A.N. Connell, W. Paul: Phys. Rev. Lett. 26, 184 (1971).
161 R.J. Elliott, J.A. Krumhansl, P.L. Leath: Rev. Mod. Phys. 46, 465 (1974).
162 P. Dean: Rev. Mod. Phys. 44, 127 (1972).
163 R. Bell: Rep. Prog. Phys. 35, 1315 (1972).
164 F. Yonezawa, K. Morigaki: Prog. Theor. Phys. Suppl. 53, 1 (1973).
165 L. Nordheim: Ann. Physik 9, 607 and 641 (1931).
166 R.H. Parmenter: Phys. Rev. 97, 587 (1955).
167 J.C. Phillips: Bonds and Bands in Semiconductors (Academic Press, N.Y. and London 1973), p. 212.
168 D. Richardson: J. Phys. C5, L 27 (1972).
169 D. Richardson: J. Phys. C4, L 289 (1971).
170 D. Stroud: Phys. Rev. B5, 3366 (1972).

171 J.A. Van Vechten, T.K. Bergstresser: Phys. Rev. B1, 3351 (1970).
172 J.C. Phillips: Comments Sol. State Phys. 4, 9 (1971).
173 J.A. Van Vechten: Solid State Commun. 11, 7 (1972).
174 N.A. Semikolenova, E.N. Khabarov: Sov. Phys. Semicond. 8, 1459 (1975).
175 E.D. Siggia: Phys. Rev. B10, 5147 (1974).
176 J.A. Van Vechten, T.K. Bergstresser: Bull. Amer. Phys. Soc. Ser. II 15, 254 (1970).
177 J.A. Van Vechten: Phys. Rev. 187, 1007 (1969).
178 J.A. Van Vechten, O. Berolo, J.C. Woolley: Phys. Rev. Lett. 29, 1400 (1972).
179 O. Berolo, J.C. Woolley, J.A. Van Vechten: Phys. Rev. B8, 3794 (1973).
180 P. Soven: Phys. Rev. 156, 809 (1967).
181 B. Velický, S. Kirkpatrick, H. Ehrenreich: Phys. Rev. 175, 747 (1968).
182 D.W. Taylor: Phys. Rev. 156, 1017 (1967).
183 Y. Onodera, Y. Toyozawa: J. Phys. Soc. Jap. 24, 341 (1968).
184 D. Stroud, H. Ehrenreich: Phys. Rev. B2, 3197 (1970).
185 V. Kumar, D. Kumar, S.K. Joshi: Phys. Rev. B11, 2831 (1975).
186 T. Morita, T. Horiguchi, C.C. Chen: J. Phys. Soc. Jap. 38, 981 (1975).
187 A. Chen, G. Weisz, A. Sher: Phys. Rev. B5, 2897 (1972).
188 P. Středa, L. Smrčka: Phys. Stat. Sol. (b) 70, 537 (1975).
189 H. Braeter, V.B. Priezzhev: Phys. Stat. Sol. (b) 72, 353 (1975).
190 A.Yu. Zakharov: Sov. Phys. Sol. State 17, 823 (1975).
191 R. Friedberg, J.M. Luttinger: Phys. Rev. B12, 4460 (1975).
192 C.E. Carroll: Phys. Rev. B12, 4142 (1975).
193 K.S. Song: Solid State Commun. 13, 1397 (1973).
194 P.N. Sen: Phys. Rev. B8, 5613 (1975).
195 L.M. Falicov, F. Yndurain: Phys. Rev. B12, 5664 (1975).
196 R. Alben, M. Blume, H. Krakauer, L. Schwartz: Phys. Rev. B12, 4090 (1975).
197 R.R. Galazka: Phys. Lett. 32A, 101 (1970).
198 C.T. Elliott: J. Phys. D4, 697 (1971).
199 J.L. Schmit: J. Appl. Phys. 41, 2876 (1970).
200 W.F. Leonard, M.E. Michael: J. Appl. Phys. 45, 958 (1974).
201 H. Rodot, M.R. Triboulet: Compt. Rend. Acad. Sci. 254, 852 (1962).
202 W. Giriat, Z. Dziuba, R.R. Galazka, L. Sosnowski, T. Zakrzewski: in Proc. 7th Int. Conf. Phys. Semicond., Paris 1964, ed. by M. Hulin (Dunod, Paris 1964), p. 1251.
203 M. Rodot, H. Rodot, C. Vérié: in /202/, p. 1237.
204 C.T. Elliott, I.L. Spain: Solid State Commun. 8, 2063 (1970).
205 R.A. Reynolds, M.J. Brau, H. Kraus, R.T. Bate: in /97/, p. 511.
206 V.V. Sologub, V.I. Ivanov-Omskii, V.M. Muzhdaba, S.S. Shalyt: Sov. Phys. Semicond. 13, 1452 (1971).
207 T.C. Harman, J.M. Honig, P. Trent: J. Phys. Chem. Sol. 28, 1995 (1967).
208 E.Z. Dziuba: Phys. Stat. Sol. (b) 62, 307 (1974).
209 T. Zakrzewski, E.Z. Dziuba: Phys. Stat. Sol. (b) 52, 665 (1972).
210 W. Scott, R.J. Hager: J. Appl. Phys. 42, 803 (1971).
211 O. Ohtsuki, R. Ueda, K. Shinobara, Y. Ueda, S. Narita: Proc. Int. Conf. Physics on Semicond. and Semimetals, Nice 1973, ed. by C. Vérié.
212 O. Ohtsuki, R. Ueda, K. Shinobara, Y. Ueda: Jap. J. Appl. Phys. 10, 1288 (1971).
213 Z. Dziuba: Acta Physica Polonica 25, 757 (1964).
214 T. Okazaki, K. Shogenji: J. Phys. Chem. Sol. 36, 439 (1975).
215 V.V. Ptashinskii, P.S. Kireev: Sov. Phys. Semicond. 6, 1398 (1973).
216 Y. Yafet, R.W. Keyes, E.N. Adams: J. Phys. Chem. Sol. 1, 137 (1956).
217 N.J. Parada, G.W. Pratt: Phys. Rev. Lett. 22, 180 (1969).
218 G.W. Pratt: J. Nonmetals 1, 103 (1973).
219 P. Nozières, C. Lewiner: Journ. de Phys. 34, 901 (1973).
220 J.N. Chazalviel: Phys. Rev. B11, 3918 (1975).
221 R. Dornhaus, G. Nimtz, P. Gibart: in Verh. DPG 8, 661 (1974); R. Dornhaus, G. Nimtz: in Verh. DPG, Freudenstadt (1976).
222 W. Zawadzki, W. Szymanska: Phys. Stat. Sol. (b) 45, 415 (1971).
223 W. Zawadzki, W. Szymanska: J. Phys. Chem. Sol. 32, 1151 (1971).
224 L.L. Korenblit, V.E. Sherstobitow: Sov. Phys. Semicond. 2, 573 (1968).
225 L.L. Korenblit, V.E. Sherstobitow: Sov. Phys. Semicond. 2, 564 (1968).

226 B.L. Gel'mont, V.I. Ivanov-Omskii, V.K. Ogorodnikov: Phys. Stat. Sol. (b) 69, 717 (1975).
227 D.L. Rode: Phys. Rev. B2, 4036 (1970);
D.L. Rode: Phys. Rev. B2, 1012 (1970).
228 H.M. Woodbury: Phys. Rev. B2, 5188 (1974).
229 W. Scott: J. Appl. Phys. 43, 1055 (1972).
230 B.R. Nag: Phys. Stat. Sol. (b) 66, 719 (1974).
231 D. Chattopadhyay, B.R. Nag: J. Appl. Phys. 45, 1463 (1974);
D. Chattopadhyay, B.R. Nag: Phys. Rev. B12, 5846 (1975).
232 L. Makowski, M. Glieksman: J. Phys. Chem. Sol. 34, 487 (1973).
233 R. Granger: Phys. Stat. Sol. 16, 599 (1966).
234 C.T. Elliott, I.L. Spain: J. Phys. C7, 727 (1974).
235 B.L. Gel'mont, V.I. Ivanov-Omskii, B.T. Kolomiets, V.K. Ogorodnikov, K.P. Smekalova: Sov. Phys. Semicond. 5, 228 (1971).
236 E.I. Georgitsê, V.I. Ivanov-Omskīi, B.T. Kolomiets, A.A. Mal'kova, K.P. Smekalova: Sov. Phys. Semicond. 6, 1122 (1973).
237 E.I. Georgitsê, V.I. Ivanov-Omskīi, B.T. Kolomiets, A.A. Mal'kova, K.P. Smekalova: Sov. Phys. Semicond. 5, 1541 (1972).
238 G. Nimtz, G. Bauer, R. Dornhaus, K.H. Müller: Phys. Rev. B10, 3302 (1974).
239 L. Genzel, T.P. Martin, C.H. Perry: Phys. Stat. Sol. (b) 62, 83 (1974).
240 C. Erginsoy: Phys. Rev. 79, 1013 (1950).
241 H.S.W. Massey, B.L. Moiseīwitsch: Phys. Rev. 78, 180 (1950).
242 T. Shimizu: J. Phys. Soc. Jap. 18, 1838 (1963).
243 A. Morita: J. Phys. Soc. Jap. 18, 1437 (1963).
244 T.C. Harman: in /81/, p. 982.
245 T.A. Aliev, F.M. Gashimzade, S.A. Aliev, T.G. Gadzhiev, E.M. Aliev, M.I. Aliev: Sov. Phys. Semicond. 5, 274 (1971).
246 V.I. Ivanov-Omskii, B.T. Kolomiets, V.K. Ogorodnikov, K. Smekalova: Sov. Phys. Semicond. 4, 214 (1970).
247 H. Wagini, B. Reiss: Phys. Stat. Sol. 15, 457 (1966).
248 L. Liu, E. Tosatti: Phys. Rev. B2, 1926 (1970).
249 L. Liu, D. Brust: Phys. Rev. 173, 777 (1968).
250 V.I. Ivanov-Omskii, B.T. Kolomiets, V.K. Ogorodnikov, K.P. Smekalova: Sov. Phys. Semicond. 1, 232 (1967).
251 R. Dornhaus, G. Nimtz, W. Schlabitz, P. Zaplinski: Solid State Commun. 15, 495 (1974).
252 I.M. Tsidilkovskii, W. Giriat, G.I. Kharus, E.A. Neifeld: Phys. Stat. Sol. (b) 64, 717 (1974).
253 N.N. Berchenko, M.V. Pashkovskii: Sov. Phys. Sol. State 15, 2479 (1974).
254 M. Tokumoto, C. Yamanouchi, K. Yoshihiro: J. Phys. Soc. Jap. 36, 914 (1974).
255 V.L. Gurevich, Y.A. Firsov: Sov. Phys. JETP 13, 137 (1961).
256 D.R. Cassiday, H.N. Spector: Phys. Rev. B9, 2618 (1974).
257 E.N. Adams, T.D. Holstein: J. Phys. Chem. Sol. 10, 254 (1959).
258 P.N. Argyres: Phys. Rev. 109, 1115 (1958).
259 R. Gerhardts, J. Hajdu: Z. Physik 245, 126 (1971).
260 H.I. Tolpygo: Phys. Stat. Sol. 42, 155 (1970).
261 H.I. Tolpygo: Phys. Stat. Sol. (b) 61, 311 (1974).
262 R.R. Gerhardts: Z. Physik B22, 327 (1975).
263 J. Hajdu: Lecture on Int. Conf. on the Appl. of High Magnetic Fields in Semicond. Physics, Würzburg (1974).
264 B.P. Pal, S.S. Sharma, S.K. Sharma: Phys. Stat. Sol. (a) 31, K 103 (1975).
265 L.S. Dubinskaya: Sov. Phys. JETP 29, 436 (1963).
266 R. Gerhardts, J. Hajdu: Solid State Commun. 9, 1607 (1971).
267 L.S. Dubinskaya: Sov. Phys. Sol. State 7, 2280 (1966).
268 B. Magnusson: Phys. Stat. Sol. (b) 56, 269 (1973).
269 P.N. Argyres: J. Phys. Chem. Sol. 4, 19 (1958).
270 S. Sharma, U.P. Phadke: Phys. Rev. Lett. 29, 272 (1972).
271 R.G. Agaeva, B.M. Askerov, F.M. Gashimzada: Sov. Phys. Semicond. 7, 1085 (1974).
272 L. Liu, M. Tan: Phys. Rev. B9, 632 (1974).
273 V.I. Ivanov-Omskii, N.N. Konstantinova, R.V. Parfen'ev, V.V. Sologub, I.G. Tagiev: Sov. Phys. Semicond. 7, 436 (1973).

274 K. Suizu, S. Narita: Phys. Lett. 43A, 353 (1973).
275 I.M. Tsidilkovskii, A.I. Ponomarev: in /81/, p. 1103.
276 R.I. Bashirov, R.M. Gadzhieva: Sov. Phys. Semicond. 4, 1659 (1971).
277 R. Dornhaus, G. Nimtz, W. Schlabitz: Verh. DPG 5, 428 (1975);
R. Dornhaus, G. Nimtz, W. Schlabitz, H. Burkhard: Solid State Commun. 17, 837 (1975).
278 M. Yamamoto: J. Phys. Soc. Jap. 21, 2428 (1966);
M. Yamamoto: J. Phys. Soc. Jap. 24, 73 (1968).
279 R.A. Stradling, G.A. Antcliffe: in Proc. Int. Conf. Phys. Semicond., Kyoto (1966);
R.A. Stradling, G.A. Antcliffe: J. Phys. Soc. Jap. 21, 374 (1966).
280 A.A. Galkin, E.N. Ukraintsev, V.D. Prozorovskii, V.I. Ivanov-Omskii: Phys. Stat. Sol. 38, K 101 (1970).
281 J. Furdyna: Phys. Rev. Lett. 16, 646 (1966).
282 G.A. Antcliffe, R.T. Bate, R.A. Reynolds: in /97/, p. 499.
283 R.V. Pomortsev, A.I. Ponomarev, G.I. Harus, I.M. Tsidilkovskii: in /122/, p. 720.
284 T.O. Poehler, J.R. Apel: Phys. Lett. 32A, 268 (1970).
285 J. Tuchendler, M. Grynberg, Y. Conder, H. Thomé, R. Le Toullec: Phys. Rev. B8, 3884 (1973).
286 B.D. McCombe, R.J. Wagner, G.A. Prinz: Solid State Commun. 8, 1687 (1970).
287 B.D. McCombe, R.J. Wagner: in /38/, p. 321.
288 E.I. Georgitsé, V.I. Ivanov-Omskii, B.T. Kolomiets, A.A. Mal'kova, K.P. Smekalova: Sov. Phys. Semicond. 6, 392 (1972).
289 Y. Guldner, C. Rigaux, M. Grynberg, A. Mycielski: Phys. Rev. B8, 3875 (1973).
290 R. Dornhaus, G. Nimtz, M. Schifferdecker: Verhandl. DPG, HL 219, 7/1976.
291 K.I. Amirkhanov, R.I. Bashirov: Sov. Phys. Sol. State 8, 1733 (1967).
292 D. Sherrington, W. Kohn: Phys. Rev. Lett. 21, 153 (1968).
293 J.G. Broerman: Phys. Rev. B5, 397 (1972).
294 M. Grynberg, R. Le Toullec, M. Balkanski: Phys. Rev. B9, 517 (1974).
295 S. Goettig: Phys. Stat. Sol. (b) 72, 91 (1975).
296 P. Kacman, W. Zawadzki: Phys. Stat. Sol. (b) 47, 629 (1971).
297 M. Takeshima: J. Appl. Phys. 44, 4717 (1973).
298 W. Zawadzki, J. Kowalski: Solid State Commun. 15, 303 (1974).
299 B. Ancker-Johnson, C.L. Dick: in /97/, p. 523.
300 V.I. Ivanov-Omskii, B.T. Kolomiets, K.P. Smekalova, V.A. Smirnov: Sov. Phys. Semicond. 2, 1004 (1969).
301 V.N. Kobyzev, A.S. Tager: JETP Lett. 14, 107 (1971).
302 J. Kolodziejczak, L. Sosnowski: Acta Physica Polonica 21, 399 (1962).
303 T.C. Harman, J.M. Honig, B.M. Tarmy: J. Phys. Chem. Sol. 24, 835 (1963).
304 W. Zawadzki, R. Kowalczyk, J. Kolodziejczak: Phys. Stat. Sol. 10, 513 (1965).
305 L.L. Korenblit, D.W. Mashovec, S.S. Shalyt: Sov. Phys. Sol. State 7, 673 (1965).
306 V.V. Sologub, V.I. Ivanov-Omskii, V.M. Muzhdaba, S.S. Shalyt: Sov. Phys. Sol. State 13, 1452 (1971).
307 V.I. Ivanov-Omskii, B.T. Kolomiets, A.A. Mal'kova: Sov. Phys. Sol. State 6, 1140 (1964).
308 V.I. Ivanov-Omskii, B.T. Kolomiets, V.M. Melnik, V.K. Ogorodnikov: Sov. Phys. Sol. State 11, 2076 (1969).
309 V.I. Ivanov-Omskii, B.T. Kolomiets, V.K. Ogorodnikov, Y.V. Rud, V.M. Tsmots: Phys. Stat. Sol. (a) 13, 61 (1972).
310 V.I. Ivanov-Omskii, B.T. Kolomiets, V.K. Ogorodnikov, K.P. Smekalova, V.M. Tsmots: Phys. Stat. Sol. (a) 14, 51 (1972).
311 M. Steinhaus: unpublished (1975);
W. Schlabitz, M. Steinhaus, G. Nimtz: Verh. DPG 5, 427 (1975).
312 J.A. Krumhansl, M. Brooks: Bull. Amer. Phys. Soc. II, 1/3 FS (1956).
313 B.T. Kolomiets, A.A. Mal'kova: Sov. Phys. Sol. State 5, 889 (1963).
314 S.A. Ignatowicz, A. Kisiel, M. Zimnal: Thin Solid Films 21, 231 (1971).
315 D.L. Carter, M.A. Kinch, D.D. Buss: in /97/, p. 273.
316 R.S. Kim, S. Narita: J. Phys. Soc. Jap. 31, 613 (1971).
317 J. Baars, F. Sorger: Solid State Commun. 10, 875 (1972).

318 H. Burkhard, P. Grosse, G. Nimtz: to be published.
319 G. Lucovsky, M.H. Brodsky, E. Burstein: Phys. Rev. B2, 3295 (1970).
320 A. Mooradian, T.C. Harman: in /97/, p. 297.
321 H.W. Verleur, A.S. Barker: Phys. Rev. 149, 715 (1966).
322 A.S. Barker, A.J. Sievers: Rev. Mod. Phys. 47, Suppl. 2 (1975).
323 F. Sorger: Dissertation, Universität Freiburg (1972).
324 J.S. Blakemore: Semiconductor Statistics (Pergamon Press, Oxford, London, N.Y., Paris 1962).
325 R.A. Reynolds, C.G. Roberts, R.A. Chapman, H.B. Bebb: in /99/, p. 217.
326 M.A. Kinch, M.J. Brau, A. Simmons: J. Appl. Phys. 44, 1649 (1973).
327 G. Nimtz, K.H. Müller: Phys. Stat. Sol. (a) 22, K 215 (1974).
328 F. Bartoli, R. Allen, L. Esterowitz, M. Kruer: J. Appl. Phys. 45, 2150 (1974).
329 M.A. Kinch, S.R. Borello: Infrared Physics 15, 111 (1975).
330 P.E. Petersen: J. Appl. Phys. 41, 3465 (1970).
331 J.C. Ayache, Y. Marfaing: Compt. Rend. Acad. Sci. B265, 568 (1967).
332 R.L. Williams, B.H. Breazeale, C.G. Roberts: in /99/, p. 237.
333 M. Rodot, C. Vérié, Y. Marfaing, J. Besson, H. Leberch: IEEE QE-2, 586 (1966).
334 P.W. Kruse: Appl. Optics 4, 687 (1965).
335 E.H. Putley: Phys. Stat. Sol. 6, 571 (1964).
336 E.L. Stelzer, J.L. Schmit, O.N. Tufte: IEEE Trans. ED-16, 880 (1969).
337 P.W. Kruse, M.D. Blue, J.H. Garfunkel, W.D. Saur: Infrared Physics 2, 53 (1962).
338 R.L. Williams: Infrared Physics 8, 337 (1968).
339 M. Smollet: Infrared Physics 8, 3 (1968).
340 B.E. Bartlett, D.E. Charlton, W.E. Dunn, P.C. Ellen, M.D. Jenner, M.H. Jervis: Infrared Physics 9, 35 (1969).
341 D. Long: Infrared Physics 7, 169 (1967).
342 M.P. Shchetinin, N.S. Baryshev, I.S. Aver'yanov, F.P. Volkova, A.P. Cherkasov: Sov. Phys. Semicond. 5, 1095 (1971).
343 M.P. Shchetinin, N.S. Baryshev, I.S. Aver'yanov, A.P. Cherkasov, F.P. Volkova: Sov. Phys. Semicond. 5, 2059 (1972).
344 I. Melngailis: J. of Luminescence 7, 501 (1973).
345 F.J. Bartoli, L. Esterowitz, M.R. Kruer, R.E. ALlen: Appl. Optics 14, 2499 (1975).
346 F.J. Bartoli, L. Esterowitz, M.R. Kruer, R.E. Allen: J. Appl. Phys. 46, 4519 (1975).
347 P.S. Kireev, V.P. Saperov: Sov. Phys. Semicond. 4, 1390 (1971).
348 J.E. Slawek, H.H. Randall: Infrared Physics 15, 339 (1975).
349 R. Allen, L. Esterowitz, M. Kruer, F. Bartoli: Infrared Physics 15, 265 (1975).
350 N.C. Aldrich, J.D. Beck: Appl. Optics 11, 2153 (1972).
351 H. Halpert, B.L. Musicant: Appl. Optics 11, 2157 (1972).
352 C. Vérié, R. Granger: Compt. Rend. Acad. Sci. 261, 3349 (1965).
353 C. Vérié, J. Ayas: Appl. Phys. Lett. 10, 241 (1967).
354 E.N. Figurovskii, P.S. Kireev, A.V. Vanynkov, Yu.Y. Erseev, A.P. Korovin: Sov. Phys. Semicond. 3, 1572 (1970).
355 C. Vérié, M. Sireix: IEEE QE-8, 180 (1972).
356 D.L. Spears, C. Freed: Appl. Phys. Lett. 23, 445 (1973).
357 J.M. Pawlikowski, P. Beda: Infrared Physics 15, 331 (1975).
358 L.A. Bovina, V.P. Meshcheryakova, L.K. Klyukin, V.I. Stafeev: Sov. Phys. Semicond. 7, 1532 (1974).
359 A.G. Foyt, T.C. Harman, J.P. Donnelly: Appl. Phys. Lett. 18, 321 (1971).
360 G. Fiorito, G. Gasparrini, F. Svelto: Appl. Phys. Lett. 23, 448 (1973).
361 G. Fiorito, G. Gasparrini, F. Svelto: Infrared Physics 15, 287 (1975).
362 J. Marine, C. Motte: Appl. Phys. Lett. 23, 450 (1973).
363 D.A. Soderman, W.H. Pinkston: Appl. Optics 11, 2162 (1972).
364 Y. Marfaing, J. Chevallier: IEEE Trans. ED-18, 465 (1971).
365 G. Cohen-Solal, Y. Marfaing: Solid State Electronics 11, 1131 (1968).
366 E. Tyssen, R. Dornhaus, G. Haider, P. Kokoschinegg, K.H. Müller, G. Nimtz, H. Happ: Solid State Commun. 17, 1459 (1975).

Received February 1976

367 M.H. Weiler: "Magnetooptical Properties" in *Semiconductors and Semimetals*, Vol. 16, ed. by R.K. Willardson, A.C. Beer (Academic, New York 1981) p. 119
368 P.W. Kruse, J.F. Ready: "Nonlinear Optical Effects", in *Semiconductors and Semimetals*, Vol. 16, ed. by R.K. Willardson, A.C. Beer (Academic, New York 1981) p. 194
369 H. Maier, J. Hesse: In *Crystals. Growth, Properties, and Applications*, Vol. 4, Organic Crystals. Germanates. Semiconductors, ed. by C.J.M. Rooijmans (Springer, Berlin, Heidelberg, New York 1980) p. 145
370 I.R. Cottam, G.A. Saunders: J. Phys. Chem. Solids 36, 187 (1975)
371 S.C. Lakkad: J. Appl. Phys. 42, 4277 (1971)
372 A.P. Rusakov: Sov. Phys. Solid State 13, 510 (1971)
373 J.G. Collins, G.K. White, J.A. Birch, T.F. Smith: J. Phys. C13, 1649 (1980)
374 O. Caporaletti, G.M. Graham: Appl. Phys. Lett. 39, 338 (1981)
375 T.F. Smith, G.K. White: J. Phys. C8, 2031 (1975)
376 J.A. Birch: J. Phys. C8, 2043 (1975)
377 S.A. Aliev, T.G. Gadzhiev, E.M. Aliev, D.G. Arasly, M.I. Aliev: Sov. Phys. Semicond. 5, 2044 (1972)
378 S.A. Aliev: Sov. Phys. Semicond. 7, 115 (1973)
379 C.R. Whitsett, D.A. Nelson: Phys. Rev. B5, 3125 (1972)
380 K. Kumazaki: phys. status solidi (a) 33, 615 (1976)
381 M. Lanir, C.C. Wang, A.H.B. Vanderwyck: Appl. Phys. Lett. 34, 50 (1979)
382 C.C. Wang, S.H. Shin, M. Chu, M. Lanir, A.H. Vanderwyck: J. Electrochem. Soc. 127, 175 (1980)
383 M. Chu, C.C. Wang: J. Appl. Phys. 51, 2255 (1980)
384 T.C. Harman: J. Electron Mater. 8, 191 (1979)
385 J.L. Schmit, J.E. Bowers: Appl. Phys. Lett. 35, 457 (1979)
386 A.V. Vanyukov, E.A. Mikolynk, I.I. Krotov, Yu.A. Figel'son, V.G. Yakovleva: Izv. Akad. Nauk SSSR, Neorg. Mater. 11, 1961 (1975)
387 Z. Nowak, J. Piotrowski, T. Piotrowski, Sadowski: Thin Solid Films 52, 405 (1978)
388a S.G. Yudin, I.S. Aver'yanov, Yu.G. Belaskov, I.M. Nesmelova: Izv. Akad. Nauk SSSR, Neorg. Mater. 11, 171 (1975)
388b J.P. Faurie, A. Million: J. Cryst. Growth 54, 582 (1981)
388c J.P. Faurie, A. Million: Appl. Phys. Lett. 41, 264 (1982)
388d J.P. Faurie, A. Million, J. Piagnet: Appl. Phys. Lett. 41, 713 (1982)
388e J.B. Mullin, S.J.C. Irvine: J. Phys. D14, L149 (1981)
388f S.J.C. Irvine, J.B. Mullin: J. Cryst. Growth 55, 107 (1981)
388g J.B. Mullin, S.J.C. Irvine: J. Vac. Sci. Technol. 21, 178 (1982)
389 J. Steininger: J. Cryst. Growth 37, 107 (1977)
390 J. Steininger: J. Electron Mater. 5, 299 (1976)
391 G. Fiorito, G. Gasparrini, D. Passoni: J. Electrochem. Soc. 125, 315 (1978)
392 G. Dittmar: Krist. Tech. 13, 6 (1978)
393 F. Schubert, G. Dittmar, B. Forbrig: Krist. Tech. 13, 10 (1978)
394 P.E. Vanier, F.H. Pollak, P.M. Raccak: Appl. Opt. 16, 2858 (1977)
395 V.A. Tyagai, O.V. Snitko, G.A. Fedorus, A.N. Krasiko, A.M. Evstigneev, K.V. Kolezhuk: Inorg. Mater. USSR 12, 1132 (1976)
396 V.G. Tiptsova-Yakovleva, A.M. Andreichuk, Yu.A. Figel'son: Inorg. Mater. USSR 11, 729 (1975)
397 G. Dittmar: Krist. Tech. 12, 631 (1977)
398 K.L. Bye: J. Mater. Sci. 14, 619 (1979)
399 V.T. Bublik, A.A. Zaitsev: phys. status solidi (a) 39, 345 (1977)
400 M. Lanir, A.H.B. Vanderwyck, C.C. Wang: J. Electron. Mater. 8, 175 (1979)
401 W. Scott, E.L. Stelzer, R.J. Hager: J. Appl. Phys. 47, 1408 (1976)
402 J. Stankicwicz, W. Giriat: Phys. Rev. B13, 665 (1976)
403a C. Verié, F. Raymond, F. Bailly, I. Jacquié, G. Weill, A. Kozacki, J. Rioux: In Ref. 403b, p. 241
403b B.L.H. Wilson (ed.): *Physics of Semiconductors:* Proceedings of the 14th In-

ternational Conference, Edinburgh, 1978 (Institute of Physics, Bristol 1979). Inst. Phys. Conf. Ser. 43 (1979)
404 S. Margalit, Y. Nemirovsky, I. Rotstein: J. Appl. Phys. 50, 6386 (1979)
405a B. Schlicht, A. Alpsancar, G. Nimtz, A.N.F. Schroeder: In Ref. 405b, p. 439
405b Proceedings of the 4th Int. Conf. on the Physics of Narrow-Gap Semiconductors, Linz, 1981. In: *Lecture Notes in Physics*, Vol. 152 (Springer Berlin, Heidelberg, New York)
405c C.A. Swarts, M.S. Daw, T.C. McGill: J. Vac. Sci. Technol. 21, 198 (1982)
405d C.L. Wang, S. Wu, D.S. Pan: to be published in J. Vac. Sci. Technol. (1983)
405e B. Schlicht, G. Nimtz: in Ref. 405b, p. 383
406 A.T. Hunter, D.L. Smith, T.C. McGill: Appl. Phys. Lett. 37, 200 (1980)
407 A.T. Hunter, T.C. McGill: J. Appl. Phys. 52, 5779 (1981)
408 C.E. Mallon, B.A. Green, R.E. Leadon, J.A. Naber: IEEE Trans. NS 22, 2283 (1975)
409 H. Ryssel, G. Lang, J. Biersack, K. Müller, W. Krüger: IEEE Trans. ED 27, 58 (1980)
410 E.S. Johnson, J.L. Schmit: J. Electron. Mater. 6, 25 (1977)
411 J.W. Lin: J. Solid State Chem. 15, 96 (1975)
412 P. Becla, J.M. Melanuik: Mater. Sci. IV/3, 103 (1978)
413 G. Bahir, R. Kalish: Appl. Phys. Lett. 39, 730 (1981)
414 H.R. Vydyanath: J. Electrochem. Soc. 128, 2609, 2619 (1981)
415 H.R. Vydyanath, J.C. Donovan, D.A. Nelson: J. Electrochem. Soc. 128, 2625 (1981)
416 H.R. Vydyanath, F.A. Kröger: J. Electron. Mater. 11, 111 (1982)
417 H.R. Vydyanath, R.C. Abbott: J. Appl. Phys. (1983), to be published
418 G.I. Kharus, N.G. Shelushima: Sov. Phys. Semicond. 12, 249 (1978)
419 V.K. Dugaev, P.P. Petrow: phys. status solidi (b) 96, 129 (1979)
420 G. Bastard: phys. status solidi (b) 80, 641 (1977)
421 G. Nimtz, B. Schlicht, R. Dornhaus: Appl. Phys. Lett. 34, 490 (1979)
422 J.L. Schmit, J.E. Bowers, M.W. Scott, J.B. Lindberg: In IRIS-DSGM Conference, June, 1979 (unpublished)
423 R.C. Farrow, G.R. Jones, G.M. Williams, P.W. Sullivan, W.J.O. Boyle, J.T.M. Wotherspoon: J. Phys. D12, L117 (1979)
424 H.M. Nitz, O. Ganschow, U. Kaiser, L. Wiedemann, A. Benninghoven: Surf. Sci. 104, 365 (1981)
425a J.L. Schmit, E.L. Stelzer: J. Electron. Mater. 7, 65 (1978)
425b C.L. Jones, M.J.T. Quelch, P. Capper, J.J. Gosney: J. Appl. Phys. 53, 9080 (1982)
426a W.F.H. Micklethwaite, R.F. Redden: Appl. Phys. Lett. 36, 379 (1980)
426b A.J. Syllaios, M.J. Williams: J. Vac. Sci. Technol. 21, 201 (1982)
427 P. Morgen, J. Silbermann, I. Lindau, W.E. Spicer, J.A. Wilson: J. Cryst. Growth 11, 161 (1982); J.A. Silbermann, P. Morgen, I. Lindau, W.E. Spicer: J. Vac. Sci. Technol. 21, 154 (1982)
428 J.L. Shay, W.E. Spicer: Phys. Rev. 161, 799 (1967)
429 N.J. Shevchik, J. Tejeda, M. Cardona, D.W. Langer: phys. status solidi (b) 59, 87 (1973)
430a B.A. Orlowski, H. Overhof, E.W. Spicer: In Ref. 430b, p. 170
430b F.G. Fumi (ed.): *Physics of Semiconductors:* Proceedings of the 13th International Conference, Rome 1976 (North-Holland, Amsterdam 1976)
430c J.A. Silbermann, P. Morgen, I. Lindau, W.E. Spicer: J. Vac. Sci. Technol. 21, 142 (1982)
430d J.A. Silbermann, P. Morgen, I. Lindau, W.E. Spicer, A.B. Chen, A. Sher, J.A. Wilson: J. Vac. Sci. Technol. 21, 467 (1982)
430e H. Ehrenreich, K.C. Hass: J. Vac. Sci. Technol. 21, 133 (1982)
430f A.B. Chen. A. Sher: J. Vac. Sci. Technol. 21, 138 (1982)
430g R.D. Graft: J. Vac. Sci. Technol. 21, 146 (1982)
430h K.C. Hass, H. Ehrenreich, B. Velicky: Phys. Rev. B27, 1088 (1983)
431 F. Casula, A. Kisiel: Nuovo Cimento 39B, 470 (1977)

432 M.H. Weiler, R.L. Aggarwal, B. Lax: Phys. Rev. B16, 3603 (1977)
433a Y. Nemirovsky, E. Finkman: J. Appl. Phys. 50, 8107 (1979)
433b G.L. Hansen, J.L. Schmit, T.N. Casselmann: J. Appl. Phys. 53, 7099 (1982)
433c G.L. Hansen, J.L. Schmit: J. Appl. Phys. 54, 1639 (1983)
434 W. Szuszkiewicz: phys. status solidi (b) 81, K119 (1977)
435 K. Sieranski, J. Szatkowski, E. Majchrowska: phys. status solidi (b) 77, K91 (1976)
436 K. Sieranski, J. Szatkowski, J.M. Pawlikowski, E. Majchrowska, E. Placzek-Popko: phys. status solidi (b) 81, K67 (1977)
437a A.L. Efros, B.L. Gelmont: In Ref. 437b, p. 435
437b J. Rauluszkiewicz, M. Gorska, E. Kaczmarek (ed.): *Physics of Narrow-Gap Semiconductors:* Proc. III Int. Conf., Warszawa 1977 (Elsevier, Amsterdam 1978)
438 B.L. Gelmont, A.L. Efros: Sov. Phys. Semicond. 11, 873 (1977)
439 B.L. Gelmont, V.I. Ivanov-Omskii, V.K. Ogorodnikov: phys. status solidi (b) 69, 717 (1975)
440 B.L. Gelmont: Sov. Phys. Semicond. 9, 1257 (1976)
441 A.A. Abrikosov: J. Low Temp. Phys. 18, 185 (1975)
442 J. Szymanski: In Ref. 437b, p. 441
443 I.M. Nesmelova, Z.K. Panasenko, N.I. Makarova: Sov. Phys. Semicond. 13, 830 (1979)
444 B.R. Nag, A.N. Chakravarti: phys. status solidi (b) 71, K45 (1975)
445 Y.F. Karalyanskas, A.Y. Shileika: Sov. Phys. Semicond. 13, 671 (1979)
446 J.M. Pawlikowski, E. Popko: Solid State Commun. 22, 231 (1977)
447 E. Popko, J.M. Pawlikowski: phys. status solidi (a) 46, K9 (1978)
448 V.A. Fedorin: Teor. Mat. Fiz. 37, 265 (1978)
449 K.I. Amirkhanov, Y.B. Magomedov, S.N. Emirov, R.M. Gadzhieva: Sov. Phys. Solid State 17, 1956 (1976)
450 M. Averous: phys. status solidi (b) 95, 9 (1979)
451 P.I. Baranskii, Y.N. Gavrilyuk, A.I. Elizarov, V.A. Kulik: Sov. Phys. Semicond. 11, 916 (1977)
452 P.I. Baranskii, A.I. Elizarov, V.A. Kulik, K.R. Kurbanov: Sov. Phys. Semicond. 13, 490 (1979)
453 M.P. Shchetinin, N.S. Baryshev: Sov. Phys. Semicond. 10, 867 (1976)
454 K. Takita, K. Onabe, S. Tanaka: phys. status solidi (b) 92, 297 (1979)
455a L. Liu, W. Leung: Phys. Rev. B12, 2336 (1975)
455b L. Liu, W. Leung: Phys. Rev. Lett. 33, 1145 (1974)
456 B.I. Halperin, T.M. Rice: Rev. Mod. Phys. 40, 755 (1968)
457 M. Balkanski: In *Elementary Excitations in Solids, Molecules and Atoms*, ed. by J.T. Devreese, A.B. Kunz, T.C. Collins (Plenum, London 1974) p. 229
458 A. Polian, R. Le Toullec, M. Balkanski: Phys. Rev. B13, 3558 (1976)
459 L. Liu, D. Brust: Phys. Rev. Lett. 20, 651 (1968)
460 L. Liu, E. Tosatti: Phys. Rev. Lett. 23, 772 (1969)
461 J.G. Broerman, L. Lui, K.N. Pathak: Phys. Rev. B4, 664 (1971)
462 Y.I. Tsipivka, H. Neumann, K. Unger: phys. status solidi (b) 101, 531 (1980)
463 D. Sherrington, W. Kohn: Rev. Mod. Phys. 40, 767 (1968)
464 W. Zawadski, W. Szymanska: phys. status solidi (b) 45, 415 (1971)
465 M. Hoclet, P. Plumelle, M. Vandevyver, R. Triboulet, Y. Marfaing: phys. status solidi (b) 92, 545 (1979)
466 R.S. Kim, S. Narita: phys. status solidi (b) 73, 741 (1976)
467 Y. Guldner, C. Rigaux, A. Mycielsky, Y. Couder: phys. status solidi (b) 81, 615 (1977); (b) 82, 149 (1977)
468 S. Uchida, S. Tanaka: J. Phys. Soc. Jpn. 40, 118 (1976)
469 K. Suzuki, N. Miura, S. Uchida, S. Tanaka: phys. status solidi (b) 76, 787 (1976)
470 K. Shimizu, S. Narita, Y. Nisida, V.I. Ivanov-Omskii: Solid State Commun. 32, 327 (1979)
471 K. Pastor, M. Grynberg, Y. Couder: In Ref. 437b, p. 97
472 M.H. Weiler, R.L. Aggarwal, B. Lax: In Ref. 437b, p. 137

473 R.L. Aggarwal: In *Semiconductors and Semimetals*, Vol. 9, ed. by R.K. Willardson, A.C. Beer (Academic, New York 1972) p. 151
474 M.H. Weiler, R.L. Aggarwal, B. Lax: Phys. Rev. B17, 3269 (1978)
475 M.H. Weiler: J. Magn. Magn. Mat. 11, 131 (1979)
476 C.R. Pidgeon, R.N. Brown: Phys. Rev. 146, 575 (1966)
477 J.M. Luttinger, W. Kohn: Phys. Rev. 97, 869 (1955)
478 R.R. Gerhardts: Private communication
479 J. Mycielski, G. Bastard, C. Rigaux: In Ref. 430b, p. 451
480 S.H. Groves, R.N. Brown, C.R. Pidgeon: Phys. Rev. 161, 779 (1967)
481 S. Uchida: 2nd Conf. Physics of Semimetals and Narrow-Gap Semiconductors, Nice 1975
482 C.R. Pidgeon, S.H. Groves: In Ref. 81
483 P. Byszewski, K. Szlenk, R.R. Galazka: In Ref. 437b, p. 191
484 L. Liu: In Ref. 437b, p. 151
485 G. Bastard: In Ref. 403b, p. 231
486 L. Liu, D. Brust: Phys. Rev. 157, 627 (1967)
487 B. Joos, K. Das, P. Wallace: In Ref. 437b, p. 167; to be published
488 L. Liu, C. Verie: Phys. Rev. Lett. 37, 453 (1976)
489 L. Liu, C. Verie: Solid State Commun. 20, 13 (1976)
490 G. Weill, A. Kozacki, C. Verie: In Ref. 403b, p. 173
491 U.G. Arapov, B.B. Ponikarov, I.M. Tsidilkovski: In Ref. 437b, p. 177
492 C. Fan, B.A. Lombos, J. Calas, M. Averous: In Ref. 437b, p. 183
493 M.A. Mekhtiev, T.G. Ismailov: phys. status solidi (b) 99, 507 (1980)
494 Z.W. Gortel, J. Szymanski, L. Swierkowski: In Ref. 403b, p. 261
495 C. Verie, F. Raymond, F. Bailly, I. Vacquie, G. Weill, A. Kozaki, J. Rioux: In Ref. 403b, p. 241
496 A. Kozacki, S. Otmezgnine, G. Weill, C. Verie: In Ref. 430b, p. 467
497 Y.G. Arapov, B.B. Ponikarov, I.M. Tsidilkovski, N.G. Shelushinina: Sov. Phys. Semicond. 13, 1126 (1980)
498 W. Giriat, E.A. Neifel'd, I.M. Tsidilkovski, N.A. Gorodilov: Sov. Phys. Semicond. 10, 37 (1976)
499 V.N. Brudnyi, A.V. Voitsekhovskii, Z.G. Grechnkh, M.A. Krivov, Y.V. Lilenko, L.N. Limarenko, M.V. Pashkovskii, A.S. Petrov, A.I. Potapov: Sov. Phys. Semicond. 11, 905 (1978)
500 L.A. Bovina, Y.N. Savchenko, V.I. Stafeev: Sov. Phys. Semicond. 9, 1362 (1976)
501 C. Fau, B.A. Lombos, J. Calas, M. Averous: In Ref. 437b, p. 183
502 R.V. Lutsiv, P.P. Petrov, N.A. Rud, M.V. Matviiv, I.M. Solskii: Isv. Akad. Nauk SSSR, Neorg. Mater. 12, 830 (1976)
503 S.A. Ignatowicz, A. Dajna, Z. Goliger, A. Mikulko, A. Piotrowska: Thin Solid Films 36, 399 (1976)
504 L.N. Korol, L.A. Bovina, V.I. Stafeyev: Radio Eng. and Electron. Phys. (USA) 22, 156 (1977)
505 A. Piotrowska: Electron Technol. 9, 83 (1976)
506 W. Szymanska, P. Boguslawski, W. Zawadzki: phys. status solidi (b) 65, 641 (1974)
507 W. Szymanska, T. Dietl: J. Phys. Chem. Solids 39, 1025 (1978)
508 P. Pfeffer, W. Zawadzki: phys. status solidi (b) 88, 247 (1978)
509 W. Walukiewicz: Phys. Rev. Lett. 33, 650 (1974); J. Phys. C9, 1945 (1976)
510 T. Dietl, J.J. Dubowski, W. Szymanska: In Ref. 403b, p. 245
511 J.J. Dubowski: phys. status solidi (b) 105, 399 (1981)
512 J. Kossut: phys. status solidi (b) 86, 593 (1978)
513 J.J. Dubowski: phys. status solidi (b) 85, 663 (1978)
514 W. Zawadzki, A. Mauger, S. Otmezguine, C. Verie: Phys. Rev. B15, 1035 (1977)
515 W. Zawadzki: In Ref. 437b, p. 281
516 P. Gossar: In Ref. 403b, p. 269
517 G. Weill, A. Kozacki, C. Verie: In Ref. 437b, p. 173
518 G.V. Shuster, A.I. Kozinskaya: Sov. Phys. Semicond. 9, 619 (1975)

519 S.A. Alekperov, C.O. Kadzhar, E.Y. Salaev: Sov. Phys. Semicond. 13, 827 (1979)
520 J.J. Dubowski, T. Dietl, W. Szymanska, R.R. Galazka: J. Phys. Chem. Solids 42, 351 (1981)
521 W. Szymanska: In Ref. 437b, p. 357
522 D.A. Nelson, W.M. Higgins, R.A. Lancaster, R. Roy, H.R. Vydyanath: Extended Abstracts of U.S. Workshop on the Physics and Chemistry of Mercury Cadmium Telluride, October 28-30, 1981, Minneapolis, Minnesota, p. 175
523 J.R. Meyer, F.L. Bartoli: J. Vac. Sci. Technol. 21, 237 (1982)
524 F.J. Bartoli, C.A. Hoffman, J.R. Meyer, R.E. Allen: In Ref. 405b, p. 126
525a F.J. Bartoli, R.E. Allen, L. Esterowitz, M.R. Kruer: Solid State Commun. 25, 963 (1978)
525b F.J. Bartoli, J.R. Meyer, C.A. Hoffmann, R.E. Allen: Phys. Rev. B, February (1983)
525c F.J. Bartoli, J.R. Meyer, R.E. Allen, C.A. Hoffmann: J. Vac. Sci. Technol. 21, 241 (1983)
526 V.K. Arora, M. Jaafarian: Phys. Rev. B13, 4457 (1976)
527 Chhi-Chong Wu, A. Chen: J. Appl. Phys. 51, 4855 (1980)
528 Chhi-Chong Wu, A. Chen: Phys. Rev. B21, 4757 (1980)
529 J. Kossut, J. Hajdu: Solid State Commun. 27, 1401 (1978)
530 N.N. Berchenko, A.S. Zakhar'yash, I.I. Izhnin, S.P. Kommissarchuk, M.V. Pashkovskii: Sov. Phys. Semicond. 13, 602 (1979)
531 L.N. Korol', V.S. Vershinin, L.A. Bovina, V.I. Stafeev: Sov. Phys. Semicond. 12, 275 (1978)
532 R. Dornhaus, G. Nimtz: Solid State Commun. 22, 41 (1977)
533 K. Heift, Y. Ono, J. Hajdu: In Ref. 405b, p. 372
534 R. Dornhaus: Ph.D. Thesis, University of Cologne, 1978 (unpublished)
535 G. Nimtz, B. Schlicht, E. Tyssen, R. Dornhaus, L.D. Haas: Solid State Commun. 32, 669 (1979)
536 G. Nimtz, B. Schlicht: Festkörperprobleme XX, 669 (1980)
537a G. Nimtz, B. Schlicht: In Ref. 537b, p. 313
537b Proc. 15th Int. Conf. Phys. Semicond., Kyoto 1980. J. Phys. Soc. Jpn. 49, Suppl. A (1980)
538 B. Schlicht, G. Nimtz: In Ref. 405b, p. 382
539 G. De Vos, F. Herlach: In Ref. 405b, p. 377
540 R.R. Gerhardts: Solid State Commun. 36, 397 (1980)
541 A. Raymond, R.L. Aulombard, L. Konczewicz, C. Bousquet, J.L. Robert, M. Royer: In Ref. 537b, p. 301
542 A. Raymond, J.L. Robert, R.L. Aulombard, C. Bousquet, O. Valassiades, M. Royer: In Ref. 405b, p. 386
543 M. Carvalho, C. Fau, M. Averous: In Ref. 405b, p. 430
544 C.K. Sarkar, D. Chattopadhyay: J. Phys. Chem. Solids 40, 565 (1979)
545 A.N. Chakravarty, A.K. Chowdhury, K.P. Ghatak, S. Ghosh, A. Dhar: Appl. Phys. 25, 105 (1981)
546 S.A. Aliev: Sov. Phys. Semicond. 9, 1438 (1976)
547 A.L. Efros: Sov. Phys. Solid State 7, 1206 (1965)
548 H. Keiter, J. Hajdu: phys. status solidi 38, 757 (1970)
549 E. Dudziak, P. Becla, J. Brzezinski, L. Jedral: Solid State Commun. 27, 987 (1978)
550 Y.N. Gavrilyuk, S.G. Gasan-zade, E.A. Sal'kov, G.A. Shepelskii: Sov. Phys. Semicond. 11, 923 (1977)
551 Y.N. Gavrilyuk, S.G. Gasan-zade, E.A. Sal'kov, G.A. Shepelskii: Sov. Phys. Semicond. 13, 64 (1979)
552 D.A. Kichigin, D.D. Khalameida: Sov. Phys. Semicond. 13, 1401 (1979)
553 H. Pascher, R. Grisar, G. Bauer, H. Burkhard: Solid State Commun. 38, 723 (1981)
554 J.R. Meyer, F.J. Bartoli, M.R. Kruer: Phys. Rev. B21, 1559 (1980)

555 V.G. Golubev, E.I. Georgitse, V.I. Ivanov-Omskii: phys. status solidi (b) 85, K39 (1978)
556 E. Shatkovskii, A. Tamasevicius, E. Matulionis: phys. status solidi (a) 51, K39 (1979)
557 V.I. Ivanov-Omskii, V.K. Ogorodnikov, T.T. Totieva: Sov. Phys. Semicond. 14, 412 (1980)
558 V.I. Ivanov-Omskii, R. Link, A.A. Malkova: Sov. Phys. Semicond. 12, 504 (1978)
559 E.E. Vdorkina, N.S. Baryshev, M.P. Shchetinin, A.P. Cherkasov, I.S. Averyanov: Sov. Phys. Semicond. 10, 109 (1976)
560 Y.N. Gavrilyuk, S.G. Gasan-zade, G.A. Shepelskii: Sov. Phys. Semicond. 13, 955 (1979)
561 S.G. Gasan-zade, E.A. Salkov, G.A. Shepelskii: In Ref. 437b, p. 251
562 E.B. Osipov: Sov. Phys. Solid State 21, 1191 (1979)
563 E.A. Mozhaev, V.I. Ivanov-Omskii, V.A. Maltseva, D.V. Mashovets, R.V. Parfenev: Sov. Phys. Semicond. 11, 1260 (1977)
564 B.L. Gelmont, V.I. Ivanov-Omskii, N.N. Konstantinova, D.V. Mashovets, R.V. Parfenev, I.N. Yassievich: In Ref. 430b, p. 471; in Ref. 437b, p. 337
565 Y.A. Bratashevskii, V.B. Tyutyunnik, K.S. Averyanov, I.M. Nesmelova: Sov. Phys. Semicond. 9, 114 (1975); Sov. J. Low Temp. Phys. 3, 586 (1977)
566 P.W. Kruse: Appl. Phys. Lett. 28, 90 (1976)
567 P.W. Kruse, J.F. Ready, M.A. Khan: Infrared Phys. 19, 497 (1979)
568 P. Norton, P.W. Kruse: Optics Commun. 22, 147 (1977)
569 W. Walukiewicz, J.H.M. Stoelinga, R.L. Aggarwal, B. Lax, P.W. Kruse: In Ref. 437b, p. 81
570 W. Walukiewicz, R.L. Aggarwal, J.H.M. Stoelinga: Phys. Rev. B19, 6350 (1979)
571 M.A. Khan, P.W. Kruse, J.F. Ready: Opt. Commun. 28, 374 (1979)
572 T.J. Bridges, E.G. Burkhardt, V.T. Nguyen: Opt. Commun. 30, 66 (1979)
573 R.K. Jain, D.G. Steel: Appl. Phys. Lett. 37, 1 (1980)
574 C.K.N. Patel, R.E. Slusher, P.A. Fleury: Phys. Rev. Lett. 17, 1011 (1966)
575 P.A. Wolff, G.A. Pearson: Phys. Rev. Lett. 17, 1015 (1966)
576 K.C. Rustagi: J. Phys. Chem. Solids 30, 2547 (1969)
577 S.D. Smith: In Ref. 405b, p. 113
578 C.R. Pidgeon, B.S. Wherrett, A.M. Johnston, J. Dempsey, A. Miller: Phys. Rev. Lett. 42, 1785 (1979)
579 C.R. Pidgeon, A.M. Johnston, J. Dempsey: In Ref. 405b, p. 126
580 S.A. Jamison, A.V. Nurmikko: Phys. Rev. B19, 5185 (1979)
581 S.A. Jamison, A.V. Nurmikko: Appl. Phys. Lett. 33, 182 (1978)
582 A.V. Nurmikko, S.A. Jamison: In Ref. 403b, p. 315
583a A.V. Nurmikko: Opt. Commun. 18, 522 (1976)
583b A.V. Nurmikko, B.D. Schwartz: J. Vac. Sci. Technol. 21, 229 (1982)
584 J.C. Matter, A.L. Smirl, M.O. Scully: Appl. Phys. Lett. 28, 507 (1976)
585 D. Chattopadhyay: Phys. Lett. 81A, 241 (1981)
586 L.A. Bovina, Y.N. Savchenko, V.I. Stafeev: Sov. Phys. Semicond. 10, 944 (1977)
587 V.S. Vershinin, V.I. Stafeev, L.A. Bovina: Sov. Phys. Semicond. 12, 529 (1978)
588 N.N. Kopylov: Sov. Phys. Semicond. 13,493 (1979)
589 I.M. Dykman, V.M. Rosenbaum, F.T. Vasko: phys. status solidi (b) 88, 385 (1978)
590 S.D. Beneslavskii, O. Ziep: phys. status solidi (b) 88, 221 (1978)
591 I.A. Lubashevskii, V.I. Ryzhii: Sov. Phys. Semicond. 13, 751 (1979)
592 V.K. Malyutenko, K.Y. Guga, Y.M. Malozovskii: phys. status solidi (a) 65 131 (1981)
593 A. Jedrzejezak, T. Dietl: phys. status solidi (b) 76, 737 (1976)
594 D.K. Amirkhanova, R.M. Gadzhieva: Sov. Phys. Semicond. 9, 1528 (1976)

595 K. Takita, G. Landwehr: In Ref. 437b, p. 343
596 L.A. Bovina, V.P. Ponomarenko, V.I. Stafeev: Sov. Phys. Semicond. 12, 1313 (1978)
597 K.I. Amirkhanov, D.K. Amirkhanova, P.P. Khokhlachev: Sov. Phys. Solid State 18, 1131 (1976)
598 M.M. Gadzhialiev, R.M. Gadzhieva: Sov. Phys. J. 10, 1364 (1979)
599 H. Kepa, W. Gebicki, T. Giebultowicz, B. Buras, K. Clausen: Solid State Commun. 34, 211 (1980)
600 H. Kepa, T. Giebultowicz, B. Buras, B. Lebach, K. Clausen: In Ref. 405b, p. 275
601 R. Bichard, Y. Guldner, P. Lavallard: In Ref. 437b, p. 245
602 B.L. Gelmont, V.I. Ivanov-Omskii, V.A. Maltseva, V.A. Smirnov: In Ref. 405b, p. 131
603 D.L. Polla, C.E. Jones: Solid State Commun. 36, 809 (1980)
604 D.L. Polla, C.E. Jones: In Ref. 405b, p. 434
605a D.L. Polla, C.E. Jones: J. Appl. Phys. 51, 6233 (1981)
605b C.E. Jones, V. Nair, J. Lindquist, D.L. Polla: J. Vac. Sci. Technol. 21, 187 (1982)
606 R.R. Gerhardts, R. Dornhaus, G. Nimtz: Solid State Electron. 21, 1467 (1978)
607 B.L. Gelmont: Phys. Lett. 66A, 323 (1978)
608 B.L. Gelmont: Sov. Phys. Semicond. 14, 1140 (1980)
609 T.N. Casselmann: J. Appl. Phys. 52, 848 (1981)
610 A.I. Vlasenko, Yu.N. Gavrilyuk, A.V. Lyubchenko, E.A. Sal'kov: Sov. Phys. Semicond. 13, 1274 (1980)
611 T.N. Casselmann, P.E. Peterson: Solid State Commun. 33, 615 (1980)
612 I.M. Baker, F.A. Cappocci, D.E. Charlton, J.T.M. Wotherspoon: Solid State Electron. 21, 1475 (1978)
613 G. Nimtz: Phys. Rep. 63, 265 (1980)
614 D.L. Polla, S.P. Tobin, M.B. Reine, A.K. Sood: J. Appl. Phys. 52, 5182 (1981)
615 D.A. Nelson, W.M. Higgins, R.A. Lancaster, R.P. Murosako, M. Reine, R. Roy, E. Smolinsky: Extended Abstracts of U.S. Workshop on the Physics and Chemistry of Mercury Cadmium Telluride, October 28-30, 1981, Minneapolis, Minnesota, p. 153
616 D.L. Polla, J.F. Shanley, J.A. Mroczkowski, M.B. Reine: Extended Abstracts of U.S. Workshop on the Physics and Chemistry of Mercury Cadmium Telluride, October 28-30, 1981, Minneapolis, Minnesota, p. 149
617 R. Dornhaus, G. Nimtz: Solid State Commun. 27, 575 (1978)
618 R. Dornhaus, G. Nimtz: Solid State Electron. 21, 1471 (1978)
619a P.A. Wolff, C. Verié, M.H. Weiler, S.T. Yuen, L.R. Ram-Mohan: In Ref. 405b, p. 135
619b P.A. Wolff, C. Verié, S.Y. Yuen: J. Vac. Sci. Technol. 21, 231 (1982)
620 A.R. Beattie, P.T. Landsberg: Proc. R. Soc. London A249, 1b (1959)
621 R. Dornhaus, K.H. Müller, G. Nimtz, M. Schifferdecker: Phys. Rev. Lett. 37, 710 (1976)
622 R.R. Gerhardts: Solid State Commun. 23, 137 (1977)
623 R.R. Gerhardts: In Ref. 437b, p. 103
624 M.A. Kinch, S.R. Borrello, A. Simmons: Infrared Phys. 17, 127 (1977)
625 M.A. Kinch, S.R. Borrello, B.H. Breaceale, A. Simmons: Infrared Phys. 17, 137 (1977)
626 T.N. Casselman: In Ref. 405b, p. 147
627 E. Igras, J. Piotrowski, T. Piotrowski: Infrared Phys. 19, 143 (1979)
628 J.M. Palmer: Infrared Phys. 19, 243 (1979)
629 M. Itoh, H. Takigawa, R. Ueda: IEEE Trans. ED-27, 150 (1980)
630 S. Borrello, M. Kinch, D. LaMont: Infrared Phys. 17, 121 (1977)
631 M.Y. Pines, D.M. Staesudd, P.B. Bratt: Infrared Phys. 19, 633 (1979)
632 S.A. Alekperov, Ch.O. Kadzhar, E.Yu. Salaev: Sov. Phys. Semicond. 13, 827 (1979)

633 W. Galns, T. Persak, J. Piotrowski: Infrared Phys. 19, 649 (1979)
634 J.M. Pawlikowski: Infrared Phys. 19, 179 (1979)
635 G. Nimtz, B. Schlicht, H. Lehmann, E. Tyssen: Appl. Phys. Lett. 35, 640 (1979)
636 B. Weber, S.M. Kulpa: Forth International Conference on Infrared and Millimeter Waves and Their Applications, Conference Digest: IEEE Cat. No. 7SCH1384-7 MTT, 148 (1979)
637 B. Schlicht, G. Nimtz: "A Sensitive HgCdTe Bolometer for the Detection of Millimeter Wave Radiation", in *Physics in High Magnetic Fields*, ed. by S. Chikazumi, M. Miura, Springer Series in Solid-State Sciences, Vol. 24 (Springer Berlin, Heidelberg, New York 1981), p. 257
638 J.M. Pawlikowski: Infrared Phys. 17, 1 (1979)
639 G. Fiorito, G. Gasparrini, F. Svelto: Infrared Phys. 17, 25 (1977)
640 J. Piotrowski: Sov. Phys. Semicond. 11, 643 (1977)
641 G. Fiorito, G. Gasparrini, F. Svelto: Appl. Phys. 17, 105 (1978)
642 A. Kolodny, I. Kidron: IEEE Trans. ED-27, 37 (1980)
643 J.Y. Wong: IEEE Trans. ED-27, 48 (1980)
644 S.P. Tobin, S. Iwasa, T.J. Tredwell: IEEE Trans. ED-27, 43 (1980)
645 J.M. Pawlikowski, J. Zylinski: Optica Applicata 8, 127 (1977)
646 M. Lanir, C.C. Wang, A.H.B. Vanderwyck: Appl. Phys. Lett. 34, 50 (1978)
647 C.C. Wang, M. Chu, S.H. Shin, W.E. Temant, J.T. Cheung, M. Lanir, A.H.B. Vanderwyck, G.M. Williams, L.O. Bubulac, R.J. Eisel: IEEE Trans. ED-27, 154 (1980)
648 G. Fiorito, G. Gasparrini, F. Svelto: Infrared Phys. 18, 59 (1978)
649 D.L. Spears: Infrared Phys. 17, 5 (1977)
650 D.L. Spears, R.H. Kingston: Appl. Phys. Lett. 34, 589 (1979)
651 M.A. Kinch, R.A. Chapman, A. Simmons, D.D. Buss, S.R. Borrello: Infrared Phys. 20, 1 (1980)
652 P. Felix, M. Moulin, B. Munnier, J. Portmann, J.P. Reboul: IEEE Trans ED-27, 175 (1980)
653 R.A. Chapman, M.A. Kinch, A. Simmons, S.R. Borrello, H.B. Morris, J.S. Wrobel, D.D. Buss: Appl. Phys. Lett. 32, 434 (1978)
654 R.A. Chapman, S.R. Borrello, A. Simmons, J.D. Beck, A.J. Lewis, M.A. Kinch, J. Hynecek, C.G. Roberts: IEEE Trans. ED-27, 134 (1980)
655 H. Takigawa, M. Doki, R. Ueda: IEEE Trans. ED-27, 146 (1980)
656 Y. Nemirovsky, I. Kidron: Solid State Electron. 22, 831 (1979)
657 Y. Nemirowsky, S. Margalit, I. Kidron: Appl. Phys. Lett. 36, 466 (1980)
658 M.A. Mekhtiev: Solid State Commun. 22, 433 (1977)
659a W.E. Spicer, J.A. Silberman, P. Morgen, I. Lindau, J. Vac. Sci. Technol. 21, 149 (1982)
659b P. Morgen, J.A. Silberman, I. Lindau, W.E. Spicer, J.A. Wilson: J. Vac. Sci. Technol. 21, 161 (1982)
660a U. Solzbach, H.J. Richter: Surf. Sci. 97, 191 (1980)
660b H.J. Richter, U. Solzbach, M. Seelmann, Eggebert, H. Brendecke, H. Maier, J. Ziegler, R. Krüger: *Insulating Films on Semiconductors*, Springer Series in Electron Physics, Vol. 7 (1981); (Proceedings of INFO '81)
660c M. Seelmann-Eggebert, G. Brandt, H.J. Richter: Extended Abstracts of U.S. Workshop on the Physics and Chemistry of Mercury-Cadmium-Telluride, February 1983, Dallas/Texas, p. 141
661 Y. Nemirovsky, E. Finkman: J. Electrochem. Soc. 126, 768 (1979)
662a B.K. Janousek, M.J. Daugherty, R.B. Schoolar: In *The Physics of MOS Insulators*, ed. by G. Lucovsky, S.T. Pantelides, F.L. Galeener (Pergamon, New York 1980) p. 217
662b B.K. Janousek, R.C. Carscallen: J. Vac. Sci. Technol. 21, 442 (1982)
663 S.P. Kowalczyk, J.T. Cheung: J. Vac. Sci. Technol. 18, 944 (1981)
664a T.S. Sun, S.P. Buchner, N.E. Byer: J. Vac. Sci. Technol. 17, 1067 (1980)

664b G.D. Davis, T.S. Sun, S.P. Buchner, N.E. Byer: J. Vac. Sci. Technol. 19, 472 (1981)
664c S.P. Buchner, G.D. Davis, N.E. Byer: J. Vac. Sci. Technol. 21, 446 (1982)
664d D.R. Rhiger, R.E. Kvaas: J. Vac. Sci. Technol. 21, 168 (1982)
664e D.R. Rhiger, R.E. Kvaas: J. Vac. Sci. Technol. 21, 448 (1982)
664f A. Lastras-Martinez, U. Lee, J. Zehnder, P.M. Raccah: J. Vac. Sci. Technol. 21, 157 (1982)
664g R.B. Schoolar, B.K. Janousek, R.L. Alt, R.C. Carscallen, M.J. Daughtery, A.A. Fote: J. Vac. Sci. Technol. 21, 164 (1982)
664h J.D. Beck, M.A. Kinch, E.J. Esposito, R.A. Chapman: J. Vac. Sci. Technol. 21, 172 (1982)
664i R.F.C. Farrow, P.N.D. Dennis, H.E. Bishop, N.R. Smart, J.T.M. Wotherspoon: Thin Solid Films 88, 87 (1982)
664j P.C. Catagnus, C.T. Baker, U.S. Pat. No. 3 977 018 (1976)
664k Y. Nemirovski, R. Goshen: Appl. Phys. Lett. 37, 813 (1980)
665 K. Takita, K. Masuda, H. Kudo, S. Seki: Appl. Phys. Lett. 37, 460 (1980)
666 F.A. Zaitov, A.V. Gorshkov, G.M. Shalyapina: Sov. Phys. Solid State 21, 1224 (1979)
667 W.L. Baun: Appl. Surf. Sci. 7, 46 (1981)
668 M. Michael, W.F. Leonard: Solid State Electron. 17, 71 (1974)
669a D.R. Choudhury, A.K. Choudhury, A.N. Chakravarti: phys. status solidi (a) 59, K69 (1980)
669b M. Daughtery, B.K. Janousek: Appl. Phys. Lett. 42, 290 (1976)
670 J.C. Thuillier, F. Bazenet: J. Vac. Sci. Technol. 16, 1417 (1979)
671 A.F. Tasch, R.A. Chapman, B.H. Breazeale: J. Appl. Phys. 41, 4202 (1970)
672 W.F. Leonard, M. Michael: J. Appl. Phys. 50, 1450 (1979)
673 M. Sirieix, M. Garcia, J. Farre, J. Simonne: Infrared Phys. 18, 233 (1978)
674 A.J. Steckl: Infrared Phys. 16, 65 (1976)
675 W.W. Anderson: Infrared Phys. 17, 147 (1977)
676 A.F. Tasch, D.D. Buss, R.T. Bate, B.H. Breazeale: Proc. 10th Int. Conf. on the Physics of Semiconductors, Cambridge, MA, 1970, ed. by S.P. Keller, J.C. Hensel, F. Stern (US Atomic Energy Comm., Washington 1970) p. 458
677 F.J. Ohkawa, Y. Uemura: J. Phys. Soc. Jpn. 37, 1325 (1974)
678 S. Narita, T. Kuroda: Nuovo Cimento B39, 834 (1977)
679 T. Kuroda, S. Narita: J. Phys. Soc. Jpn. 41, 709 (1976)
680 S. Narita, T. Kuroda, Y. Nisida: In Ref. 403b, p. 1235
681 Y. Takada, Y. Uemura: J. Phys. Soc. Jpn. 43, 139 (1977)
682 J.N. Schulman, T.C. McGill: Appl. Phys. Lett. 34, 663 (1979)
683 J.N. Schulman, T.C. McGill: J. Vac. Sci. Technol. 16, 1513 (1979)
684 J.N. Schulman, T.C. McGill: Solid State Commun. 34, 29 (1980)
685 T. Hagiwara, S. Tanaka: Jpn. J. Appl. Phys. Suppl. 2, Pt. 2, 433 (1974)
686 J. Niewodniczanska-Zawadzka, J. Rauluszkiewicz: In Ref. 430b, p. 443
687 J. Niewodniczanska-Zawadzka, J. Grembowicz: phys. status solidi (b) 91, K13 (1979)
688 J.M. Pawlikowski: Acta Phys. Pol. A49, 139 (1976)
689 J.M. Pawlikowski, P. Becla, K. Lubowski, K. Roszkiewicz: Acta Phys. Pol. A49, 563 (1976)
690 J.M. Pawlikowski: Acta Phys. Pol. A51, 95 (1977)
691 J.M. Pawlikowski: phys. status solidi (a) 40, 613 (1977)
692 J.M. Pawlikowski: phys. status solidi (a) 37, K183 (1976)
693 J.M. Pawlikowski, J. Zylinski: Opt. Appl. 7, 127 (1977)
694 K.W. Nill, J.N. Walpole, A.R. Calawa, T.C. Harman: In *Physics of Semimetals and Narrow-Gap Semiconductors*, Proc. Int. Conf., Dallas, 1970, ed. by D.L. Carter, R.T. Bate (Pergamon, New York 1971) p. 383
695 A.M. Cowley, S.M. Sze: J. Appl. Phys. 36, 3212 (1965)
696 J.M. Pawlikowski: Thin Solid Films 44, 241 (1977)
697 J. Szatkowski, K. Sieranski, J.M. Pawlikowski, E. Placzek-Popko, P. Becla, E. Dudziak: phys. status solidi (a) 42, 721 (1977)

698 J.M. Pawlikowski: phys. status solidi (a) 30, K147 (1975)
699 J. Szatkowski, K. Kaczmarski, B. Kochan, K. Sieranski, J.M. Pawlikowski: Thin Solid Films (to be published)
700 J.M. Pawlikowski, J.F. Kasprzak: Solid State Commun. 25, 645 (1978)
701 P. Becla, J.F. Kasprzak, J.M. Pawlikowski: Opt. Appl. 8, 21 (1978)
702 J.M. Pawlikowski: Infrared Phys. 19, 179 (1979)
703 L. Leibler, J. Mycielski, J.K, Furdyna: Phys. Rev. B11, 3037 (1975)
704 J.E. Bowers, J.L. Schmit, C.J. Speerschneider, R.B. Maciolek: IEEE Trans. ED-27, 24 (1980)
705 C.C. Wang, S.H. Chin, M. Chu, M. Lanir, A.H.B. Vanderwyck: J. Electrochem. Soc. 127, 175 (1980)
706 M. Chu, C.C. Wang: J. Appl. Phys. 51, 2255 (1980)
707 S.H. Shin, M. Chu, A.H.B. Vanderwyck, M. Lanir, C.C. Wang: J. Appl. Phys. 51, 3772 (1980)
708 M. Chu: J. Appl. Phys. 51, 5876 (1980)
709 R.H. Cornely, L. Suchow, T. Gabara, P. Diodato: IEEE Trans. ED-27, 29 (1980)
710 M. Averous, J. Calas, S. Charar, C. Fan, M. Royer: phys. status solidi (a) 58, 135 (1980)

List of Important Symbols

a, (b,c)	Lattice constant(s)
a_B	Bohr radius
$\underline{A}$	Vector potential (p.46,153)
b	$= \mu_n/\mu_p$, mobility ratio
B	Magnetic field; bulk modulus
$B_N^{\pm}$	Magnetic field for SdH effect maxima (p.224)
$C_{\alpha\beta}$, $C_{\alpha\beta\gamma}$	Elastic constants
c_e, c_i	Bowing parameters (p.182)
C_{AB}	Average ionic energy gap (p.181)
D	Deformation potential
D^*	Detectivity
D_{av}	Average valence band to d-band f sum (p.181)
e_{14}	Piezoelectric constant
e_T^{*2}	Lattice polarisability, (e_T^* transverse effective charge) (p.91)
E	Energy; electric field
E_F	Fermi energy
E_i	Interband gaps, i = 0,1,2
E_{ih}	Average homopolar energy gap (p.147)
E_p	$= (2m/\hbar^2)p^2$
E^j	Polar electron phonon coupling (p.209)
E_o	Fundamental gap
$\tilde{E}_o$	$= E_o + 2E_F$, effective band gap (p.99)
f_i	Phillips ionicity
f_o	Distribution function
g	Carrier generation rate (p.235)
g^*	Effective g factor of conduction band
g(E)	Density of states
g_ℓ^c	Longitudinal band-edge effective g factor (conduction band)
g_ℓ^v	Longitudinal band-edge effective g factor (valence band)
g_t^c	Transverse band-edge effective g factor (conduction band)
g_t^v	Transverse band-edge effective g factor (valence band)
g_ℓ^-	Far band contribution to the longitudinal band-edge g factor (conduction band)
g_ℓ^+	Far band contribution to the longitudinal band-edge g factor (valence band)
g_t^-	Far band contribution to the transverse band-edge g factor (conduction band)
g_t^+	Far band contribution to the transverse band-edge g factor (valence band)

G, G_o	Reciprocal lattice vectors
$H, \mathcal{H}$	Hamilton operator
j	Current density
$\underline{k}$, k_z	Wave vector of a carrier
k	Absorption coefficient
k_B	Boltzmann's constant
k_F	Free electron Fermi momentum
k_s	Thomas-Fermi screening wave number
K, K_m	Mass anisotropy factor, defined (3.18) (p.55)
K'	Mobility anisotropy (p.74)
K_τ	Relaxation time anisotropy (p.74)
$n^m_{L_k}$	Generalized Fermi integral
m, m_o	Free electron mass
m^*	Effective carrier mass
m^*_ℓ	Light hole effective mass
m^*_{hh}	Heavy hole effective mass
m^*_{so}	Split-off hole effective mass
m^*_o	Band-edge mass of conduction electron
m^{cyc}_t	Transverse cyclotron effective mass
m^c_d	Density of states effective mass (conduction band)
m^v_d	Density of states effective mass (valence band)
m^c_ℓ	Longitudinal band-edge effective mass (conduction band)
m^v_ℓ	Longitudinal band-edge effective mass (valence band)
m^c_t	Transverse band-edge effective mass (conduction band)
m^v_t	Transverse band-edge effective mass (valence band)
m^-_ℓ	Far band contribution to the longitudinal band-edge effective mass (conduction band)
m^+_ℓ	Far band contribution to the longitudinal band-edge effective mass (valence band)
m^-_t	Far band contribution to the transverse band-edge effective mass (conduction band)
m^+_t	Far band contribution to the transverse band-edge effective mass (valence band)
M_α	Ionic mass (p.172)
n	Electron, carrier concentration
n_i	Intrinsic carrier concentration
n_e	Excess carrier concentration
n_o	Thermal equilibrium carrier concentration
N	Magnetic quantum number
$\mathcal{N}^c$	Conduction band effective density of states
$\mathcal{N}^v$	Valence band effective density of states
p	Hole concentration, pressure
$\underline{p}$	Momentum operator
P	Momentum matrix element
P_ℓ	Direct longitudinal momentum matrix element
P_t	Direct transverse momentum matrix element
$\underline{r}$	Position vector
r_α	Covalent radius
R_D	Detector resistance
R_H	Hall coefficient

R_i	Primitive translation vectors of the lattice
S_a, S_e, S_o	Scattering rates (p.209)
S_j	Oscillator strength (p.209)
T	Temperature
T_c	Transition temperature, Curie temperature
T_b	Band inversion temperature
T_D	Dingle temperature
u_α	Mean square displacement
v	Carrier velocity
V	Potential or voltage; pseudopotential form factor
W_ℓ	Lattice thermal resistivity
x	Alloy composition
X	Reduced energies
z	E/k_BT reduced energy
Z	Valence number
α	Linear thermal expansion coefficient; fine structure constant thermoelectric power; Fröhlich coupling constant
β	$= k_BT/E_0$; warm electron coefficient; cluster parameter
$\gamma, \bar{\gamma}$	Luttinger parameter (p.229)
γ_0	Grüneisen parameter
δ, Δ	Spin-orbit splitting energies (Fig.18)
$\Delta_{\ell,t}^{c,v}$	Defined in (3.11) (p.51)
	Permittivity of free space
ε_p	Penn's dielectric constant (p.146)
ε_∞	Optical relative dielectric constant

ε_0	Static relative dielectric constant
ε_1	Real part of relative dielectric constant
ε_2	Imaginary part of relative dielectric constant
ζ	Chemical potential
η	$(= E_F/k_BT)$; quantum efficiency
η_i	Reduced intrinsic Fermi energy
Θ_D	Debye temperature (optical phonon)
$\varkappa$	Luttinger parameter (p.229)
λ	Wavelength
μ	Carrier mobility
μ_B	Bohr magneton
μ_n	Electron mobility
μ_p	Hole mobility
μ_H	Hall mobility
ν	Frequency
ν_{CR}	Cyclotron resonance frequency
ν_{CS}	Combined resonance frequency
ν_{LO}	Wavenumber longitudinal optical phonon
ν_S	Spin resonance frequency
Ξ	Optical deformation potential
Ξ_{cv}	Interband optical deformation potential
π_ℓ	Longitudinal piezoresistance coefficient (p.177)
ρ	Mass density; resistivity, magnetoresistivity
$\rho_\ell, \rho_\parallel$	Longitudinal magnetoresistivity ($\underline{j} \parallel \underline{B}$)
$\rho_t, \rho_\perp$	Transverse magnetoresistivity ($\underline{j} \perp \underline{B}$)

σ	Conductivity
$\underline{\sigma}$	Pauli spin operator (p.153)
τ	Momentum relaxation time; photoconductive decay time
τ_A	Auger lifetime
τ_{Ai}	Intrinsic Auger lifetime
τ_p	Plasma relaxation time
τ_R	Radiative lifetime
τ_{Ri}	Intrinsic radiative lifetime
τ_o	Carrier lifetimes
χ	Magnetic susceptibility
ω_C	Angular frequency of cyclotron resonance
ω_{LO}^i	Angular frequency of LO phonons, i = 1,2
ω_p; $\omega_{p\ell}$	Angular frequency of plasma resonance
$\tilde{\omega}_{TO}$	Effective renormalized TO-phonon frequency (p.99)
ω_{TO}	Angular frequency of TO phonons
Ω	Volume of unit cell (p.180)

Numerical Values of Important Quantities

h	=	$(4.135708 \pm 14)\cdot 10^{-15}$ eVsec
$\hbar$	=	$(1.05459193 \pm 123)\cdot 10^{-34}$ $Wsec^2$
	=	$6.58218292\cdot 10^{-16}$ eVsec
k_B	=	$8.61573\cdot 10^{-5}$ eV/K = 1.38046 10^{-23} Wsec/K
$\|e\|$	=	$(1.60219177 \pm 44)\cdot 10^{-19}$ Asec
m	=	$(9.10955854 \pm 600)\cdot 10^{-31}$ $Wsec^3/m^2$
$\|e\|/m$	=	$1.758834\cdot 10^{11}$ $m^2/Vsec^2$ = $1.758834\cdot 10^2$ GHz/T
N_{Av}	=	$6.0247\cdot 10^{23}$/mol
c	=	$(2.997925010 \pm 330)\cdot 10^8$ m/sec
$4\pi\varepsilon\hbar c/e^2$	=	α^{-1} = 137.0360221
ε	=	$8.859\cdot 10^{-12}$ Asec/Vm = $1/\mu_o c^2$
μ_o	=	$1.25602\cdot 10^{-6}$ Vsec/Am
$1/\varepsilon c = \mu_o c$	=	376.732 Ω
$a_B = 4\pi\varepsilon\hbar^2/me^2$	=	0.5291 Å
$\mu_B = e\hbar/2m$	=	$5.788\cdot 10^{-5}$ eV/T
$Ry = me^4/2(4\pi\varepsilon\hbar)^2$	=	13.607 eV

(Errors are in the last digits, after /51/)

Combined Subject Index